Lecture Notes in Geosystems Mathematics and Computing

More information about this series at http://www.springer.com/series/15481

Mathias Richter

Inverse Problems

Basics, Theory and Applications in Geophysics

Second Edition

 Birkhäuser

Mathias Richter
Faculty for Electrical Engineering
and Information Technology
Bundeswehr University Munich
Neubiberg, Germany

ISSN 2730-5996 ISSN 2512-3211 (electronic)
Lecture Notes in Geosystems Mathematics and Computing
ISBN 978-3-030-59316-2 ISBN 978-3-030-59317-9 (eBook)
https://doi.org/10.1007/978-3-030-59317-9

Mathematics Subject Classification: 65F22, 65M32, 65R32, 86A22

This book is published under the imprint Birkhäuser, www.birkhauser-science.com by the registered
company Springer Nature Switzerland AG
The registered company address is: Gewerbestrasse 11, 6330 Cham, Switzerland

Preface of the First Edition

The term "inverse problem" has no acknowledged mathematical definition, and its meaning relies on notions from physics. One assumes that there is a known mapping

$$T : \mathbb{U} \to \mathbb{W},$$

which models a physical law or a physical device. Here, $\mathbb{U}$ is a set of "causes" and $\mathbb{W}$ is a set of "effects." The computation of an effect $T(u)$ for a given cause u is called a **direct problem**. Finding a cause $u \in \mathbb{U}$ that entails a given effect $w \in \mathbb{W}$ is called an **inverse problem**. Solving an inverse problem thus means to ask for the solution of an equation $T(u) = w$.

Maybe a certain effect $w \in \mathbb{W}$ is desirable and one is looking for some $u \in \mathbb{U}$ to produce it. An inverse problem of this kind is called a **control problem**. In the following, it will be assumed that an effect is actually observed and that its cause has to be found. An inverse problem of this kind is called an **identification problem**. It arises when an interesting physical quantity is not directly amenable to measurements but can only be derived from observations of its effects. There are numerous examples of identification problems in science and engineering such as:

- **Inverse gravimetry**: Given its mass density distribution (cause), the gravitational force (effect) exercised by a body can be computed (direct problem). The inverse problem is to derive the mass density distribution from measured gravitational forces. An application is the detection of oil or gas reservoirs in geological prospecting.
- **Transmission tomography**: Given the density distribution of tissue (cause), the intensity loss (effect) of an X-ray traversing it can be computed (direct problem). Inversely, one tries to find the tissue's density from measured intensity losses of X-rays traversing it. Applications exist in diagnostic radiology and nondestructive testing.
- **Elastography**: From known material properties (cause), the displacement field (effect) of an elastic body under external forces can be computed (direct

problem). The inverse problem is to derive material properties from observed displacements. An application is the medical imaging of soft tissue.

- **Seismic tomography**: Knowing the spatial distribution of mechanical characteristics (cause) of the earth's crust and mantle, seismic energy propagation (effect) from controlled sound sources or earthquakes can be computed (direct problem). Inversely, one seeks to find mechanical characteristics from records of seismic data. A geophysical application is to map the earth's interior.

Very often, the solution u^* of an inverse problem $T(u) = w$ depends in an extremely sensitive manner on w, two very similar effects having very different causes. This would not be a principal problem, if w was known exactly. However, in practice, one never knows w perfectly well but can only construct some approximation $\tilde{w}$ of it from a limited number of generally inexact measurements. Then the true cause $u^* \in \mathbb{U}$ defined by $T(u) = w$ cannot even be found to good approximation, no matter the available computing power. This is so because w is not known, and, by the aforementioned sensitivity, solving $T(u) = \tilde{w}$ would produce a completely wrong answer $\tilde{u}$, far away from u^*. There is no way to miraculously overcome this difficulty, unless we are lucky and have some *additional information* of the kind "the true cause u^* has property P." If this is the case, then we might replace the problem of solving $T(u) = \tilde{w}$ by the *different problem* of finding a cause $\tilde{u}$ within the set of all candidates from $\mathbb{U}$ *also having property P*, such that some distance between $T(\tilde{u})$ and $\tilde{w}$ is minimized. If the solution $\tilde{u}$ of this replacement problem depends less sensitively on the effect $\tilde{w}$ than the solution u^* of the original problem depends on w but converges to u^* if $\tilde{w}$ converges to w, then one speaks of a **regularization** of the original inverse problem.

Scope

Many excellent books on inverse problems have been written, we only mention [Kir96, EHN96], and [Isa06]. These books rely on functional analysis as an adequate mathematical tool for a unified approach to analysis and regularized solution of inverse problems. Functional analysis, however, is not easily accessible to non-mathematicians. It is the first goal of the present book to provide an access to inverse problems without requiring more mathematical knowledge than is taught in undergraduate math courses for scientists and engineers. From abstract analysis, we will only need the concept of functions as vectors. Function spaces are introduced informally in the course of the text, when needed. Additionally, a more detailed, but still condensed, introduction is given in Appendix B. We will not deal with regularization theory for operators. Instead, inverse problems will first be discretized and described approximately by systems of algebraic equations and only then will be regularized by setting up a replacement problem, which will always be a minimization problem. A second goal is to elaborate on the single steps to be taken when solving an inverse problem: discretization, regularization, and practical solution of the regularized optimization problem. Rather than being as general as possible, we want to work out these steps for model problems from the fields of inverse gravimetry and seismic tomography. We will not delve into details of numerical algorithms, though, when high-quality software is readily available on which we can rely for computations. For the numerical examples in this book, the programming environment Matlab [Mat14][1] was used as well as C programs from [PTVF92].

Content

We start in Chap. 1 by presenting four typical examples of inverse problems, already illustrating the sensitivity issue. We then formalize inverse problems as equations in vector spaces and also formalize their sensitivity as "ill-posedness." In the chapter's last two sections we have a closer look on problems from inverse gravimetry and seismic tomography. We define specific model problems that we will tackle and solve in later chapters. Special attention will be paid to the question whether sought-after parameters (causes) can be uniquely identified from observed effects, at least in the hypothetical case of perfect observations.

All model problems introduced in Chap. 1 are posed in function spaces: effects and causes are functions. Chapter 2 is devoted to the discretization of such problems, that is to the question of how equations in function spaces can approximately be turned into equations in finite-dimensional spaces. Discretization is a prerequisite for solving inverse problems on a computer. A first section on spline approximation describes how general functions can be represented approximately by a finite set of parameters. We then focus on linear inverse problems. For these, the least squares method, investigated in Sect. 2.2, is a generally applicable discretization method. Two alternatives, the collocation method and the method of Backus–Gilbert, are presented in Sects. 2.3 and 2.4 for special, but important, classes of inverse problems, known as Fredholm equations of the first kind. Two of our model problems belong to this class. Even more specific are convolutional equations, which play a significant role not only in inverse gravimetry but also in other fields like communications engineering. Convolutional equations can be discretized in the Fourier domain, which leads to very efficient solution algorithms. This is discussed in Sect. 2.5. In the nonlinear case, discretizations are tailored to the specific problem to be solved. We present discretizations of two nonlinear model problems from inverse gravimetry and seismic tomography in Sect. 2.6.

The last two chapters investigate possibilities for a regularized solution of inverse problems in finite-dimensional spaces (i.e., after discretization). In Chap. 3 we treat the linear case, where the discretized problem takes the form of a linear system of equations $Ax = b$. More generally, one considers the linear least squares problem

---

[1] For the second edition, we used Release 2019b of Matlab.

of minimizing $\|b - Ax\|_2$, which is equivalent to solving $Ax = b$, if the latter *does* have a solution, and is still solvable, if the system $Ax = b$ no longer is. In the first two sections, the linear least squares problem is analyzed, with an emphasis on quantifying the sensitivity of its solution with respect to the problem data, consisting of the vector b and of the matrix A. A measure of sensitivity can be given, known as condition number in numerical analysis, which allows to estimate the impact of data perturbations on the solution of a linear least squares problem. If the impact is too large, a meaningful result cannot be obtained, since data perturbations can never be completely avoided. The least squares problem then is not solvable practically. Regularization tries to set up a new problem, the solution of which is close to the actually desired one *and* which can be computed reliably. The general ideas behind regularization are outlined in Sect. 3.3. The most popular regularization technique, Tikhonov regularization, is explained and analyzed in Sect. 3.4, including a discussion of how a regularized solution can be computed efficiently. Tikhonov regularization requires the choice of a so-called regularization parameter. The proper choice of this parameter is much debated. In Sect. 3.5 we present only a single method to make it, the discrepancy principle, which is intuitively appealing and often successful in practice. For alternative choices, we refer only to the literature. Tikhonov regularization can also be used to solve problems derived by the Backus–Gilbert method or for convolutional equations transformed to the Fourier domain. These topics are treated in Sects. 3.7 and 3.8. Very interesting alternatives to Tikhonov regularization are iterative regularization methods, which are attractive for their computational efficiency. Two of these methods, the Landweber iteration and the conjugate gradient method, are described in Sects. 3.9 and 3.10. It will be found that the Landweber iteration can be improved by two measures, which relates it back to Tikhonov regularization. One of these measures, a coordinate transform, can also be taken to improve the conjugate gradient method. Technical details about this coordinate transform are given in Sect. 3.6, which describes a transformation of Tikhonov regularization to some standard form. This is also of independent interest, since Tikhonov regularization in standard form is the easiest to analyze.

Regularization of nonlinear problems is studied in Chap. 4. In Sect. 4.1, nonlinear Tikhonov regularization is treated abstractly, whereas in Sect. 4.2 this method is applied to a model problem from nonlinear inverse gravimetry. Nonlinear Tikhonov regularization leads to an optimization problem, namely a nonlinear least squares problem. Section 4.3 discusses various possibilities for solving nonlinear least squares problem numerically. All of these methods require the computation of gradients, which can mean a prohibitive computational effort, if not done carefully. The "adjoint method," presented in Sect. 4.4, can sometimes drastically reduce the numerical effort to obtain gradients. The adjoint method is presented in the context of a nonlinear problem from seismic tomography, which is then solved in Sect. 4.5. A final section presents one example of a nonlinear iterative regularization method, the "inexact Newton-CG method." The usefulness of this method is illustrated for a problem of inverse gravimetry.

Appendix A lists required material from linear algebra and derives the singular value decomposition of a matrix. Appendix B gives a condensed introduction into functions spaces, i.e., into abstract vector spaces containing functions as elements. Appendix C gives an introduction into the subject of (multidimensional) Fourier transforms and their numerical implementation, including the case of non-equidistant sample points. Finally, Appendix D contains a technical proof outsourced from Chap. 3, which shows the regularizing property of the conjugate gradient method applied to linear least squares problems.

Acknowledgments

I am much indebted to my editor Dr. Clemens Heine from Birkhäuser-Springer publishers for his promotion of this book. I am greatly thankful to my friend and colleague Professor Stefan Schäffler for his critical advice when reading my manuscript and for having supported my work for well over twenty years.

Munich, Germany   Mathias Richter

Preface of the Second Edition

Obvious goals of any second edition are the correction of known errors and the improvement of the textual and graphical presentation of the material. This led to minor changes throughout the monograph. Beyond that, we reworked our nonlinear inverse model problems of inverse gravimetry and seismic tomography. After discretization, in both cases a nonlinear least squares problem is obtained, the *global* minimum point of which has to be found numerically. In the absence of convexity of the objective function, this is challenging. On the one hand, all efficient solvers for nonlinear least squares problems like the Gauß–Newton method, the Levenberg–Marquardt method, or the reflective Newton method of Coleman and Li are not fit to find a global minimum point. They iteratively improve an initial guess of a minimum point by proceeding into directions of decreasing function values—this way they can at best detect a local minimum point, where the iteration terminates. These solvers must be provided a "good" starting point, close enough to the global minimum point, which might be very hard to find. On the other hand, stochastic algorithms for finding global minima *do* exist, but we do not know of any such method that works efficiently when it has to face an optimization problem in *many* dimensions and with *many* local minima at the same time, as it is the case for seismic tomography and, to a lesser extent, for inverse gravimetry. While we have no all-purpose method for solving any nonlinear least squares problem, we offer solutions for our model problems.

For *inverse gravimetry*, **multiscale optimization**, as developed by Chavent and Kunisch (see, e.g., [Cha09]), is quite effective. It relies on **multiscale discretization**, which is presented in Sect. 2.6. In this section, we also give a reason for the success of multiscale optimization, when applied to inverse gravimetry.

We completely revised our treatment of *seismic tomography*. First of all, we now prefer to speak of **full-waveform seismic inversion (FWI)**. The latter name is commonly used in the literature for the problem we want to solve, whereas "seismic tomography" could be confounded with "travel-time tomography," which in fact means a different kind of inverse problem considered in seismology. In our previous treatment of FWI, we relied on the so-called travel-time transformation and on the method of characteristics to derive a rather well-behaved nonlinear least squares problem. However, this approach is limited to FWI in one space dimension. We still restrict ourselves to the one-dimensional case only but now base our inversion method on a standard finite difference discretization of the underlying initial/boundary-value problem. This can be generalized to FWI in two or three space dimensions. The new discretization is exposed in Sect. 2.7. It leads to an optimization problem more challenging to solve than the one from the first edition, due to the presence of multiple local minima of the objective function. Multiscale optimization is no remedy in this case, since the objective function is oscillatory even at large scales, as explained in the text. Instead, we resort to a reconstruction method that gets us the sought-after subsurface parameters layer by layer for increasing depths.

The structure of Chap. 4 was modified with a detailed discussion of our two model problems now pushed to the last two sections. We also added a new appendix on existence and uniqueness theorems for FWI.

I thank Dr. Rainer von Chossy, who carefully read parts of the first edition and pointed out several errors to me. I also wish to thank my editor Christopher Tominich from Birkhäuser-Springer publishers for his promotion of this book and the very pleasant cooperation.

Munich, Germany   Mathias Richter
February 2020

Contents

Chapter 1
Characterization of Inverse Problems

In Sect. 1.1 we present four typical examples of inverse problems from physics and engineering and give an idea of why these problems can be difficult to solve in practice. In all examples causes as well as effects are functions. Therefore, any useful notion of inverse problems as equations relating causes to effects must be general enough to include equations in function spaces, like differential or integral equations. In Sect. 1.2 we do indeed describe inverse problems as equations in vector spaces and define "ill-posedness," which is the primary concern when dealing with inverse problems. In Sects. 1.3 and 1.4, we formulate four model problems from inverse gravimetry and full-waveform seismic inversion, which will be taken up in later chapters to illustrate the methods presented in this book.

1.1 Examples of Inverse Problems

Example 1.1 (Determination of Growth Rates) The initial value problem

$$w'(t) = \frac{dw(t)}{dt} = u(t) \cdot w(t), \quad w(t_0) = w_0 > 0, \quad t_0 \le t \le t_1, \quad t_0 < t_1,$$

(1.1)

is a simple and well-known mathematical model for a growth process. Here, $w(t)$ might represent the population size of a colony of bacteria at time t. Then, w_0 represents the known initial population size and $u(t)$ represents a variable bacterial growth rate. For a given, continuous function $u : [t_0, t_1] \rightarrow \mathbb{R}$, (1.1) has a unique, continuously differentiable solution $w : [t_0, t_1] \rightarrow]0, \infty[$, to wit:

$$w(t) = w_0 \cdot e^{U(t)}, \quad U(t) = \int_{t_0}^{t} u(s) \, ds, \quad t_0 \le t \le t_1.$$

(1.2)

© The Author(s), under exclusive license to Springer Nature Switzerland AG 2020
M. Richter, *Inverse Problems*, Lecture Notes in Geosystems
Mathematics and Computing, https://doi.org/10.1007/978-3-030-59317-9_1

Thus, a cause u (the growth rate) entails an effect w (the population size). Formally, this can be described as a mapping $T : u \mapsto w$, parameterized by t_0, t_1, and w_0 (a mathematically correct definition of this mapping from one function space into another is deferred to Sect. 1.2). Inversely, for a given function w, one asks for a function u such that $T(u) = w$. This inverse problem also has a unique solution, explicitly given by

$$u(t) = \frac{w'(t)}{w(t)} = \frac{d}{dt} \ln(w(t)), \quad t_0 \leq t \leq t_1. \tag{1.3}$$

Function u is the input of the direct problem ("the data") and function w is its result. For the inverse problem, function w is the input and function u is the result. In practice, we never know the respective inputs exactly, since this would mean to exactly know infinitely many values $u(t)$ (for the direct problem) or $w(t)$ (for the inverse problem). Rather, we only have a finite number of measurements, subject to measurement errors. From these measurements, we can only approximate the true input function.

For the direct problem, such unavoidable errors do not have serious consequences. To see this, assume that $\tilde{u}$ is an approximation of the true input u with

$$\max\{|u(t) - \tilde{u}(t)|; \ t_0 \leq t \leq t_1\} \leq \varepsilon \quad \text{for some} \quad \varepsilon > 0.$$

Then, for the corresponding results $\tilde{w}$ and w we have

$$\max\{|w(t) - \tilde{w}(t)|; \ t_0 \leq t \leq t_1\} \leq \varepsilon C$$

with some constant C.[1] This means that deviations of results can be kept under control if one cares to keep deviations of inputs small enough. This quality of the mapping from inputs to outputs is called **stability** or **robustness** of the direct problem. The inverse problem behaves very differently. For example, take

$$\text{input} \quad \left\{ \begin{array}{rcl} w : [t_0, t_1] & \to & \mathbb{R} \\ t & \mapsto & e^{\sin(t)} \end{array} \right\} \quad \text{and result} \quad \left\{ \begin{array}{rcl} u : [t_0, t_1] & \to & \mathbb{R} \\ t & \mapsto & \cos(t) \end{array} \right\}$$

and consider the specific perturbed inputs

$$w_n : [t_0, t_1] \to \mathbb{R}, \quad t \mapsto w(t) \cdot \left(1 + \frac{1}{\sqrt{n}} \cos(nt) \right), \quad n \in \mathbb{N}, n \geq 2, \tag{1.4}$$

---

[1] From Theorem 12.V in [Wal98] we get

$$C = \frac{w_0}{\mu} e^{(\mu + \varepsilon)(t_1 - t_0)} (e^{\mu(t_1 - t_0)} - 1) \quad \text{where} \quad \mu := \max\{|u(t)|; \ t_0 \leq t \leq t_1\}.$$

with $\max\{|w_n(t) - w(t)|;\ t_0 \le t \le t_1\} \to 0$ for $n \to \infty$. Inputs w_n lead to results

$$u_n : [t_0, t_1] \to \mathbb{R}, \quad t \mapsto u(t) - \frac{\sqrt{n}\sin(nt)}{1 + \frac{1}{\sqrt{n}}\cos(nt)}, \quad n \in \mathbb{N}, n \ge 2,$$

with

$$\max\{|u_n(t) - u(t)|;\ t_0 \le t \le t_1\} \to \infty \quad \text{for} \quad n \to \infty.$$

The better $w_n(t)$ approximates $w(t)$, the larger becomes the deviation of $u_n(t)$ from $u(t)$! The reason for this is the differentiation operation in (1.3). Whereas integration as in (1.2) is a smoothing operation (which, for example, will level out sharp function peaks), the inverse differentiation operation necessarily must roughen all function details which are smoothed by integration. But this will eventually also blow up perturbations of the true input function. As a consequence, the explicit formula (1.3), albeit mathematically correct, is practically useless. Figure 1.1 illustrates this example for $n = 40$. The graphs of input w and result u are shown as red lines on the left and on the right, respectively. The wildly oscillating black lines are the graphs of w_n and u_n. ◊

The next example probably is the best known inverse problem.

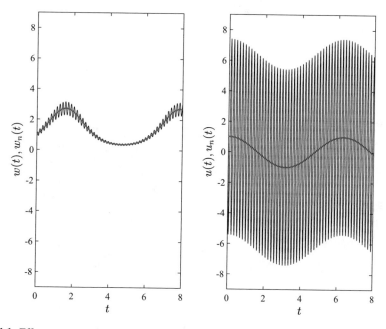

Fig. 1.1 Effects w, w_n and causes u, u_n in Example 1.1, $n = 40$

Example 1.2 (Computerized Tomography, CT)   A plane section of a body is characterized by a density function $f : \mathbb{R}^2 \to \mathbb{R}$, $x \mapsto f(x)$, where we assume that

$$f(x) = 0 \text{ for } x \notin D := \{x \in \mathbb{R}^2; \ \|x\|_2 < 1\} \subseteq \mathbb{R}^2,$$

which can always be achieved by proper scaling. The value $f(x)$ models the tissue's density at position x. Knowing $f(x)$ at every position $x \in D$ means to know the body's interior. If we knew f, then we also could compute the intensity loss of an X-ray sent through the body section along a straight line L: by Beer's law, the ray's intensity loss at position x is proportional to $f(x)$ and so the total loss corresponds to the line integral

$$\int_L f(x) \, ds,$$

see Fig. 1.2, left. According to this figure, f is assumed to be a piecewise constant function, equal function values being indicated by the same shading intensity with darker shades meaning denser tissue.

We will rewrite the line integral. For a given angle φ let us define vectors

$$\theta := \begin{pmatrix} \cos \varphi \\ \sin \varphi \end{pmatrix} \quad \text{and} \quad \theta^\perp := \begin{pmatrix} -\sin \varphi \\ \cos \varphi \end{pmatrix}.$$

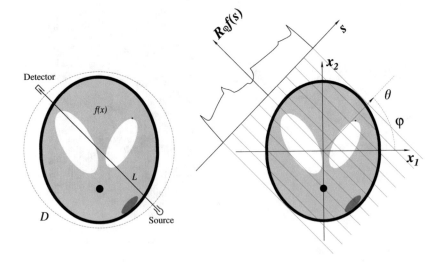

Fig. 1.2 Principle of CT

Since every straight line L through D can be parameterized in the form

$$L = \{s\theta + t\theta^\perp;\ t \in \mathbb{R}\} \quad \text{for} \quad \varphi \in [0, \pi[\quad \text{and} \quad s \in]-1, 1[,$$

the line integral becomes

$$Rf(\varphi, s) := R_\varphi f(s) := \int_{-\infty}^{\infty} f(s\theta + t\theta^\perp)\, dt. \tag{1.5}$$

Since $f(x) = 0$ for $x \notin D$, this is only seemingly an improper integral. Figure 1.2 in its right part illustrates $R_\varphi f$ for a fixed angle φ as a function of s. Note that local maxima appear where the total intensity loss reaches a local maximum (depending on the density of the traversed tissue). The function

$$Rf : [0, \pi[\times]-1, 1[\to \mathbb{R}, \quad (\varphi, s) \mapsto Rf(\varphi, s),$$

is called **Radon transform** of f. The role of the physical law modelled by a mapping T is now played by the transform R, function f is the "cause," and Rf is the "effect." The inverse problem consists in solving the equation $Rf = g$ for an unknown f when g is known. An explicit inversion formula was found in 1917 by the Austrian mathematician Johann Radon:

$$f(x) = \frac{1}{2\pi^2} \int_0^\pi \int_{-1}^1 \frac{\frac{d}{ds}g(\varphi, s)}{x \cdot \theta - s}\, ds\, d\varphi, \quad x \in D. \tag{1.6}$$

The appearance of derivatives in (1.6) is a hint that f depends sensitively on g, as it was the case in Example 1.1. ◇

The next example has the form of an **integral equation**. Consider a relation $T(u) = w$ between cause and effect $u, w : [a, b] \to \mathbb{R}$, which is given in the form

$$\int_a^b k(s, t, u(t))\, dt = w(s), \quad a \le s \le b, \tag{1.7}$$

with $k : [a, b]^2 \times \mathbb{R} \to \mathbb{R}$ a known, so-called **kernel function**. Equation (1.7) is called **Fredholm integral equation of the first kind**. A special case is the **linear Fredholm integral equation of the first kind** having the form

$$\int_a^b k(s, t)u(t)\, dt = w(s), \quad a \le s \le b, \tag{1.8}$$

where again $k : [a, b]^2 \to \mathbb{R}$ is given. If this kernel function has the special property

$$k(s, t) = 0 \quad \text{for} \quad t > s,$$

then (1.8) can be written in the form

$$\int_a^s k(s, t)u(t)\, dt = w(s), \quad a \le s \le b, \qquad (1.9)$$

and this equation is called **Volterra integral equation of the first kind**. Another special case are kernel functions with the property

$$k(s, t) = k(s - t).$$

In this case, a linear Fredholm integral equation is called **convolutional equation** and can be written as

$$\int_a^b k(s - t)u(t)\, dt = w(s), \quad a \le s \le b. \qquad (1.10)$$

One speaks of Fredholm or Volterra integral equations of the *second kind*, if the function u also appears outside the integral. An example is

$$u(s) + \lambda \int_a^s k(s, t)u(t)\, dt = w(s), \quad a \le s \le b, \quad \lambda \in \mathbb{R}.$$

Linear Fredholm equations of the first and second kind do have quite different properties. Technical details are given in [Eng97], Corollary 2.40. We only provide an informal explanation, as follows. If the kernel function k is "smooth" (continuous, for example), then the mapping $u \mapsto w$ defined by integral equations of the first kind also has a smoothing property. Thus, the solution of an integral equation, which means an inversion of integration, necessarily has to roughen the right-hand side, thereby also amplifying errors of approximations $\tilde{w}$ to w. In fact the computation of derivatives, which was found to be problematic in Example 1.1, can be interpreted as solving a Volterra integral equation:

$$u(t) = w'(t), \ w(t_0) = 0 \quad \Longleftrightarrow \quad w(t) = \int_{t_0}^t u(s)\, ds\ .$$

In contrast, integral equations of the second kind also contain an unsmoothed copy of u. The solution of such equations thus does not necessarily have to roughen a given right-hand side. Fredholm equations are also defined for functions having multiple arguments. In this case, the integration interval $[a, b]$ generalizes to a compact subset of $\mathbb{R}^s$, $s \in \mathbb{N}$, see the following example.

Example 1.3 (Inverse Gravimetry) Let $D \subset \mathbb{R}^3$ be a *bounded domain*, i.e., an open, connected set, and let $S := \overline{D}$ be its closure (consisting of all interior and boundary points of D). Let a body B occupy the compact region $S \subset \mathbb{R}^3$ in space and have a mass density given by a function $\rho : S \to \mathbb{R}$, $x \mapsto \rho(x) \geq 0$. The gravitational potential of B at any point $x \in \mathbb{R}^3 \setminus S$ is defined by the volume integral

$$V(x) = -G \int_S \frac{\rho(y)}{\|x - y\|_2} \, dy. \tag{1.11}$$

Here, $\|x - y\|_2$ is the Euclidean distance between x and y, G is the gravitational constant, and S and ρ are assumed to be regular enough to make (1.11) a properly defined (Lebesgue) integral. Equation (1.11) is a convolutional equation for functions of three independent variables. The gravitational force exercised by B on a unit mass located in $x \in \mathbb{R}^3 \setminus S$ is given by the negative gradient of the potential: $F(x) = -\nabla V(x)$.

It is unrealistic to assume that F can be measured everywhere outside B. We consider the following situation. Let $S \subset \Omega \subset \mathbb{R}^3$, where Ω is a *convex domain with sufficiently smooth boundary* (a ball or a half-space would do), let $\Gamma \subset \partial\Omega$ *contain an interior (with respect to $\partial\Omega$) point*, and assume that the magnitude of the gravitational force induced by B can be measured on Γ, as illustrated by Fig. 1.3. We thus assume that

$$g : \Gamma \to \mathbb{R}, \quad x \mapsto \|\nabla V(x)\|_2 \tag{1.12}$$

can be observed. Then, (1.11) and (1.12) define a mapping of the form

$$T : \rho \mapsto g \tag{1.13}$$

(a precise definition of this mapping from one function space into another will only be given in Sect. 1.3). The density function ρ is interpreted as being the cause of the observed effect g. The inverse problem of gravimetry consists in finding ρ such that $T(\rho) = g$, where g is given.

It can be shown that under appropriate conditions on S, Ω, Γ, and ρ, two potentials V_1 and V_2 necessarily coincide everywhere on $\mathbb{R}^3 \setminus S$, if $\|\nabla V_1(x)\|_2 = \|\nabla V_2(x)\|_2$ for all $x \in \Gamma$ (see Lemma 2.1.1 in [Isa90]). In this sense it is sufficient to observe $\|\nabla V(x)\|_2$ for $x \in \Gamma$. It is also known, however, that even complete knowledge of V on $\mathbb{R}^3 \setminus S$ is *not enough* to uniquely determine ρ. To achieve uniqueness, restrictions beyond Eq. (1.11) have to be imposed on ρ. For example, within the set

Fig. 1.3 Gravitational force
exercised by a body,
measured on a surface

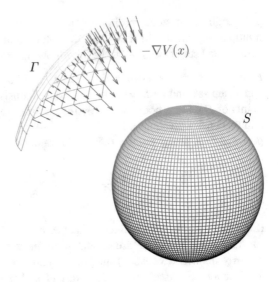

$$\mathcal{L} := \left\{ \rho : S \to [0, \infty); \ \rho \text{ solves } (1.11) \text{ and } \int_S \rho(x)^2 \, dx < \infty \right\}$$

(it is to be understood that all functions $\rho \in \mathcal{L}$ are assumed to be regular enough for all involved integrals to make sense), there exists a unique mass density distribution ρ_h, such that

$$\int_S \rho_h(x)^2 \, dx \leq \int_S \rho(x)^2 \, dx \quad \text{for all } \rho \in \mathcal{L}.$$

This function ρ_h is harmonic, i.e., the Laplace equation $\Delta \rho_h(x) = 0$ holds for all x in the interior of S, see, e.g., the survey article [MF08]. By the maximum principle for harmonic functions, this means that ρ_h takes its maximal values at the surface of S, which might be undesirable. For example, the mass density distribution of the earth's interior does not have maximal values at the surface. There may be situations, however, where we are only interested in knowing any ρ satisfying (1.11)—for example if the real goal is to construct $V(x)$ for all $x \notin S$ from observed data (1.12). This could in principle be done by finding ρ_h and then using (1.11) to compute V.

Another situation where uniqueness holds, and which is of evident interest in geological prospecting, is the one illustrated by Fig. 1.4, where a body of constant mass density is included in another body, also having constant mass density.[2] One

---

[2] Actually, Fig. 1.4 shows a two-dimensional plane section of a three-dimensional scenery. Thus the three-dimensional body B and its inclusion appear as two-dimensional regions. The corresponding section of the boundary part Γ, where gravitational forces are measured, is shown as a fat line.

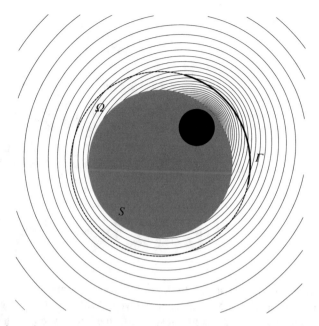

Fig. 1.4 Sections of equipotential surfaces of a body with inhomogeneous mass density

might think of the task to detect a subterranean ore deposit. In certain cases it is possible to determine the exact shape and location of the hidden body. This is the case if both mass densities are known and if the inclusion is convex "along one dimension." A set $S \subset \mathbb{R}^3$ is x_j-convex for some $j \in \{1, 2, 3\}$ and thus "convex along one dimension," if its intersection with any straight line parallel to the x_j-axis is an interval (convex sets are x_1-convex, x_2-convex, and x_3-convex at the same time). This uniqueness result goes back to the Russian mathematician Novikov and the year 1938. Technical details are described in [Isa90], see for example Theorem 3.1.4. We will consider the problem of finding a hidden body when formulating a model problem for inverse gravimetry in Sect. 1.3. Of course, the assumption of constant mass densities can only be approximately true in reality.   ◇

Example 1.4 (Full-Waveform Seismic Inversion) The goal of **full-waveform seismic inversion (FWI)**[3] is to obtain information about the earth's subsurface material parameters. To this end, seismic waves are generated at or near the surface and the pressure or the velocity field of the reflected waves is measured, usually on a part of the earth's surface or in the ocean. Figure 1.5 schematically illustrates a marine seismic experiment. An acoustic source, towed behind a survey ship, creates

---

Sections of equipotential surfaces, where the potential V is constant, are also shown. The gradient $\nabla V(x)$ at a point x is orthogonal to the equipotential surface through x.

[3] Also: waveform inversion, for short.

Fig. 1.5 Marine seismic
experiment

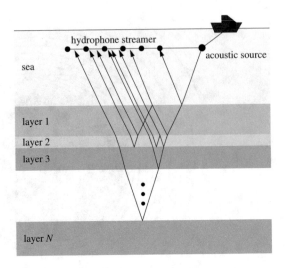

waves traveling into the subsurface. In the simplified situation shown in Fig. 1.5, the
medium is composed of plane homogeneous layers. Within each layer as well as in
sea water, waves propagate at a constant speed. At the interface between two layers,
a traveling wave is transmitted in part and reflected in part. Hydrophones record the
pressure field of upcoming reflected waves below the sea surface. Note that despite
the simplicity of the model, multiple reflections are considered.

Mathematically, the propagation of seismic waves in a domain $\Omega \subset \mathbb{R}^3$ is
modelled by the **elastic wave equation**, see, e.g., [Sch07] or [Sym09]. We will
exclusively use a simpler mathematical model, the **acoustic wave equation**, refer-
ring to [Sym09] for a discussion of the conditions under which this simplification
is adequate. Acoustic waves consist of small localized material displacements,
which propagate through the medium and lead to deviations from time-independent
equilibrium pressure. The displacement at location $x \in \Omega$ and at time $t \geq 0$ is
a vector $\mathbf{u}(x, t) \in \mathbb{R}^3$, defining a vector field $\mathbf{u} : \Omega \times [0, \infty[\to \mathbb{R}^3$. The **sound
pressure**, which is the deviation from equilibrium pressure, at time t and location
$x \in \Omega$ is a scalar $p(x, t) \in \mathbb{R}$, defining a scalar function $p : \Omega \times [0, \infty[\to \mathbb{R}$. Both
are related by **Hooke's law**

$$p = -\kappa \nabla \cdot \mathbf{u} + S, \tag{1.14}$$

where $\kappa = \kappa(x)$ is a material parameter, called **elasticity coefficient** or **bulk
modulus** and where

$$\nabla \cdot \mathbf{u} := \frac{\partial \mathbf{u}}{\partial x_1} + \frac{\partial \mathbf{u}}{\partial x_2} + \frac{\partial \mathbf{u}}{\partial x_3} = \text{div}(\mathbf{u})$$

is a convenient notation for the divergence of $\mathbf{u}$ *taken only with respect to x.* Further,
$S = S(x, t)$ is a time and space dependent "source term." This could be a source

injecting material into the medium and thus exciting the wave, like an air gun in the marine seismic experiment. A second relation between sound pressure and displacement is given by **Newton's law**

$$\nabla p = -\rho \frac{\partial^2 \mathbf{u}}{\partial t^2}, \tag{1.15}$$

where the gradient ∇p again is to be taken only with respect to the space variable x and where $\rho = \rho(x)$ is another material parameter, the **mass density**. In the acoustic model, knowing Ω, ρ, and κ means to completely know the medium. From (1.14) and (1.15) one can immediately derive the (scalar) acoustic wave equation for p, reading

$$\frac{1}{\rho c^2} \frac{\partial^2 p}{\partial t^2} - \nabla \cdot \left(\frac{1}{\rho} \nabla p \right) = \frac{1}{\kappa} \frac{\partial^2 S}{\partial t^2}, \tag{1.16}$$

where we have introduced the **wave velocity**

$$c = c(x) = \sqrt{\frac{\kappa(x)}{\rho(x)}}. \tag{1.17}$$

The value $c(x)$ is the acoustic wave's propagation speed at $x \in \Omega$. The differential equation (1.16) is complemented with initial conditions

$$p(x, 0) = 0, \quad \frac{\partial p}{\partial t}(x, 0) = 0, \quad x \in \Omega, \tag{1.18}$$

and boundary conditions

$$p(x, t) = 0, \quad x \in \partial\Omega, \quad t > 0. \tag{1.19}$$

Under appropriate assumptions concerning ρ, κ, Ω, and S, one can prove that a unique solution p of (1.16), (1.18), and (1.19) exists. This means that there is a mapping $F : (\rho, \kappa) \mapsto p$. Hydrophones measure the sound pressure $p(x, t)$ on a subset $M \subset \Omega \times [0, \infty[$, so we have another mapping

$$T : (\rho, \kappa) \mapsto p_{|M}$$

with $p_{|M}$ the restriction of p to the subset M. A precise definition of a variant of this mapping will be given in Sect. 1.4. The inverse problem takes the usual form of finding the cause (ρ, κ) responsible for an observed effect $p_{|M}$. This is an example of a parameter identification problem for a partial differential equation.

Similar to the situation in inverse gravimetry, it is known that T in general is not one-to-one, i.e., ρ and κ cannot be determined from $p_{|M}$. In Sect. 1.4 we will consider a much simplified situation, which is to determine κ alone, assuming ρ to be known. $\Diamond$

1.2   Ill-Posed Problems

We present Hadamard's definition of ill-posedness, formulating it specifically for identification problems, i.e., in the context of equation solving. "Solving an equation" has to be understood in a broad sense, including equations in function spaces, like differential or integral equations. Function spaces and mappings from one function space into another are studied in functional analysis. In the following, we do not need deep results from functional analysis. We only make use of some of its formalism to precisely state the essential difficulties coming up with inverse problems. A condensed introduction into function spaces is given in Appendix B.

Definition and Practical Significance of Ill-Posedness

The following definition refers to general normed spaces $(X, \| \bullet \|_X)$, as formally defined in Appendix B. For example, $X = \mathbb{R}^n$ is the Euclidean space, which can be equipped with the norm $\| \bullet \|_X = \| \bullet \|_2$ defined by $\|x\|_2 = \sqrt{x_1^2 + \ldots + x_n^2}$ for every $x \in \mathbb{R}^n$. As another example, $X = C[a, b] = \{f : [a, b] \to \mathbb{R}; \ f \text{ continuous}\}$ is the space of all continuous mappings from $[a, b] \subset \mathbb{R}$ into $\mathbb{R}$, which can be equipped with the maximum norm $\| \bullet \|_X = \| \bullet \|_{C[a,b]}$ defined by $\|f\|_{C[a,b]} = \max\{|f(t)|; \ t \in [a, b]\}$ for every $f \in C[a, b]$.

Definition 1.5 (Ill-Posed Inverse Problem)   Let $(X, \| \bullet \|_X)$ and let $(Y, \| \bullet \|_Y)$ be normed linear spaces and let

$$T : \mathbb{U} \subseteq X \to \mathbb{W} \subseteq Y$$

be a given function. Consider the inverse problem of solving the equation

$$T(u) = w, \quad u \in \mathbb{U}, \quad w \in \mathbb{W}, \tag{1.20}$$

for u, when w is given. This problem is called **well-posed (in the sense of Hadamard)**, or equivalently is called **properly posed**, if

(1)   for every $w \in \mathbb{W}$ a solution $u \in \mathbb{U}$ does exist (**condition of existence**),
(2)   the solution is unique (**condition of uniqueness**), and
(3)   the inverse function $T^{-1} : \mathbb{W} \to \mathbb{U}$ is continuous (**condition of stability**).

Otherwise, problem (1.20) is called **ill-posed**, or equivalently is called **improperly posed**.

Properties (1) and (2) are formally equivalent to the existence of an inverse function $T^{-1} : W \to U$. This means that for every effect $w \in W$ a unique cause $u = T^{-1}(w)$ can be identified. Continuity of T^{-1} means

$$\lim_{n \to \infty} \|w_n - w\|_Y = 0 \quad \Longrightarrow \quad \lim_{n \to \infty} \|T^{-1}(w_n) - T^{-1}(w)\|_X = 0. \qquad (1.21)$$

The significance of this stability condition lies in the following: the solution $u^* = T^{-1}(w)$ of $T(u) = w$ can be approximated arbitrarily well by the solution $\tilde{u} = T^{-1}(\tilde{w})$ of $T(u) = \tilde{w}$, if $\tilde{w} \in W$ approximates w arbitrarily well. If the stability condition is violated, we cannot hope that $\tilde{u}$ tells us anything about u^*, no matter how much effort we undertake to get a good approximation $\tilde{w}$ of w.

Example 1.6 Computing the inverse Fourier transform of a known square integrable function is a well-posed problem. Plancherel's Theorem (C.6) shows that the condition of stability is met.   ◇

Example 1.7 (Determination of Growth Rates, Part 2) In Example 1.1 the vector space $X = U = C[t_0, t_1]$ of continuous functions had been equipped with the norm $\| \bullet \|_X := \| \bullet \|_{C[t_0,t_1]}$, defined by $\|v\|_{C[t_0,t_1]} := \max\{|v(t)|; \ t \in [t_0, t_1]\}$. The space $Y = C^1[t_0, t_1]$ of continuously differentiable functions had been equipped with the same norm $\| \bullet \|_Y = \| \bullet \|_{C[t_0,t_1]}$. We also had used $W := \{w \in Y; \ w(t) > 0 \text{ for } t_0 \le t \le t_1\}$. Direct and inverse problem are determined by the mapping

$$T : U \to W, \quad u \mapsto w, \quad w(t) = w_0 e^{U(t)}, \quad U(t) = \int_{t_0}^{t} u(s) \, ds.$$

Formula (1.3) shows that a unique solution of the inverse problem exists for every $w \in W$. But the inverse operator[4] T^{-1} is not continuous, as was already observed in Example 1.1: for the sequence $(w_n)_{n \in \mathbb{N}}$ defined in (1.4) and for $w(t) = \exp(\sin(t))$ we had seen

$$\lim_{n \to \infty} \|w_n - w\|_Y = 0, \quad \text{but} \quad \lim_{n \to \infty} \|T^{-1}(w_n) - T^{-1}(w)\|_X = \infty.$$

Consequently, the determination of growth rates is an ill-posed problem, since the condition of stability is violated.   ◇

In practice, already the existence condition can be problematic, because one usually only disposes of an approximation $\tilde{w}$ of the true effect w and there is no guarantee that $\tilde{w} \in T(U)$. In the presence of modelling errors, it would even be possible that $w \notin T(U)$. This will happen if T is a simplified mathematical description of an existing physical law, such that not even the exact effect w can be described in the form $T(u) = w$. However, we will not consider this case. Rather, we will

---

[4]The term "operator" is used synonymously to "function," see Appendix B.

always assume $w \in T(\mathbb{U})$, tacitly adding an eventual modelling error to the data error $\tilde{w} - w$. Since we investigate identification problems, we cannot concede the uniqueness condition. If it was violated, then it would not be possible to *identify* a specific cause u responsible for the effect w. In such a case we would either have to observe more or other kinds of effects or to somehow restrict the set of possible causes in order to restore uniqueness. The latter was done in Example 1.3 for inverse gravimetry and will also be done for waveform inversion. Taking the uniqueness condition for granted, one could approximately solve an inverse problem by first finding a "best" approximation $\hat{w} \in T(\mathbb{U})$ of $\tilde{w}$, as defined, for example, by the projection theorem (see Theorem B.12), if $T(\mathbb{U})$ is closed and convex and if Y is a Hilbert space, and by then finding the unique element $\hat{u} \in \mathbb{U}$ with $T(\hat{u}) = \hat{w}$. Whether this approach produces a meaningful approximation $\hat{u}$ of the sought-after solution depends on the stability condition. Violation of the stability condition is the remaining problem and in fact is a major concern when dealing with identification problems.

Amendments to Definition 1.5

Definition 1.5 is academic and depends on a meaningful choice of $\mathbb{U}$, $\mathbb{W}$, $\| \bullet \|_X$, and $\| \bullet \|_Y$ to be useful. These choices must reflect the characteristics of the actual problem to be solved.

First of all, stability means continuity of the operator T^{-1}. Continuity, as defined in (1.21), depends on the chosen norms $\| \bullet \|_X$ and $\| \bullet \|_Y$. It is well possible that an operator is continuous with respect to one norm, but discontinuous with respect to another. If $\| \bullet \|_X$ and $| \bullet |_X$ are two norms defined on a linear space X and if there is a constant $C > 0$ with

$$\|x\|_X \le C|x|_X \quad \text{for all } x \in X,$$

then $| \bullet |_X$ is called **stronger** than $\| \bullet \|_X$ and $\| \bullet \|_X$ is called **weaker** than $| \bullet |_X$, since convergence of a sequence $(x_n)_{n\in\mathbb{N}}$ with respect to $| \bullet |_X$ entails convergence with respect to $\| \bullet \|_X$, but not vice versa. Consequently, if one replaces the norm $\| \bullet \|_Y$ on Y by a stronger norm $| \bullet |_Y$, then there are less sequences $(w_n)_{n\in\mathbb{N}}$ converging to w and thus there are less sequences for which the implication in (1.21) must hold. The condition of stability imposed on T^{-1} is thus weakened and a discontinuous operator T^{-1} may artificially be turned into a continuous one. Likewise, if the norm $\| \bullet \|_X$ on X is replaced by a weaker norm, then the implication in (1.21) is easier to fulfill. Again, this weakens the stability condition imposed on T^{-1}.

Example 1.8 (Determination of Growth Rates, Part 3) On the linear space $Y = C^1[t_0, t_1]$, we can replace the norm $\| \bullet \|_{C[t_0,t_1]}$ by the stronger one $\| \bullet \|_{C^1[t_0,t_1]}$, defined by $\|f\|_{C^1[t_0,t_1]} = \|f\|_{C[t_0,t_1]} + \|f'\|_{C[t_0,t_1]}$. The sequence $(w_n)_{n\in\mathbb{N}}$ from

(1.4) is no longer convergent with respect to $\| \bullet \|_{C^1[t_0,t_1]}$ and can no longer furnish an example for the discontinuity of T^{-1}. Quite to the contrary, T^{-1} *is continuous* with respect to the norms $\| \bullet \|_{C^1[t_0,t_1]}$ on Y and $\| \bullet \|_{C[t_0,t_1]}$ on X, compare Proposition B.9. We have turned an ill-posed inverse problem into a well-posed one! $\diamond$

Enforcing stability by changing norms is a mathematical trick, which is not helpful in practice. In Example 1.1 we may observe $w(t)$, but we *cannot* observe $w'(t)$, the determination of which is the essence of the inverse problem. If we could somehow observe w' directly, the inverse problem would be trivial. Its difficulty only comes up since w' cannot be observed and thus we cannot hope to find some approximation $\tilde{w}$ being close to w with respect to the norm $\| \bullet \|_{C^1[t_0,t_1]}$. Instead, (noisy) observations might get us a function $\tilde{w}$ which approximates well the function values of w, but at the same time $\tilde{w}'$ might approximate badly the values of w'.

A *second* warning concerns the dependence of Definition 1.5 on the choice of $\mathbb{U}$ and $\mathbb{W}$. These sets must be chosen appropriately for the application under consideration. For example, arguing that one is only interested in solving the problem $T(u) = w_0$ for a single, specific effect w_0, one could be tempted to define $\mathbb{W} = \{w_0\}$, a set containing only this single element. Continuity of T^{-1} then would become trivially true. But such an artificial choice of $\mathbb{W}$ does not reflect the fact that w_0 never is exactly known in practice. $\mathbb{W}$ will have to contain some neighborhood of w_0, if the mathematical model of the inverse problem is to be meaningful. If one was in fact only interested in the solution for a specific right-hand side $w_0 \in \mathbb{W}$, then the definition of well-posedness should be changed. One would still require existence and uniqueness of an element $u_0 \in \mathbb{U}$ such that $T(u_0) = w_0$, but the stability condition would have to be modified. One should demand the existence of $r > 0$ such that for any sequence $(u_n)_{n \in \mathbb{N}} \subseteq \mathbb{U}$ with $\|u_0 - u_n\|_X < r$ for all $n \in \mathbb{N}$, the following implication holds:

$$\|T(u_n) - T(u_0)\|_Y \overset{n \to \infty}{\longrightarrow} 0 \quad \implies \quad \|u_n - u_0\|_X \overset{n \to \infty}{\longrightarrow} 0.$$

In this case, the inverse problem is called **locally well-posed in** w_0.

A *third* objection to Definition 1.5 again concerns the stability condition. Continuity of the inverse T^{-1} just tells us that we can compute u to arbitrary precision, if w is known to arbitrary precision. In practice we only know w to some *finite* precision, i.e., we have at hand some $\tilde{w}$ and hopefully know some finite ε such that $\|\tilde{w} - w\|_Y \leq \varepsilon$. Then we would like to estimate how close $T^{-1}(\tilde{w})$ is to $T^{-1}(w)$, i.e., we would like to give a bound for $\|T^{-1}(\tilde{w}) - T^{-1}(w)\|_X$. Such bounds will be given in Sect. 3.2, where we introduce so-called **condition numbers** as a quantitative measure of ill-posedness for linear least squares problems in finite dimensions. The same could be done for general operators in vector spaces by means of "linearization," but this would require us to introduce operator derivatives, so-called Fréchet derivatives.

1.3 Model Problems for Inverse Gravimetry

Starting from Example 1.3, two model problems for inverse gravimetry will be
formulated, which were taken from [Sam11]. The situation is illustrated in Fig. 1.6
(note the orientation of the x_3-axis!). Here, for known constants a and $0 < h < H$,

$$S = \{x \in \mathbb{R}^3; \ -a \leq x_1 \leq a, \ -a \leq x_2 \leq a, \ h \leq x_3 \leq H\}$$

is (the location of) a given body with known constant mass density c_S. S contains
an inclusion D with *known*, constant density c_D and of (partially) unknown shape

$$D = \{x \in \mathbb{R}^3; \ -a \leq x_1 \leq a, \ -a \leq x_2 \leq a, \ h \leq u(x_1, x_2) \leq x_3 \leq H\}.$$

The shape of D is determined by a continuous function

$$u : [-a, a]^2 \to \mathbb{R}, \quad (x_1, x_2) \mapsto u(x_1, x_2),$$

which has to be found. One might think of a planar approximation of an outer
segment of the earth with D being the earth's mantle and $S \setminus D$ being the
earth's crust, the task being to determine their interface—the so-called **Mohorovičić
discontinuity**.

According to Fig. 1.6, we assume that D is "convex in x_3," meaning that the
intersection of D and every straight line parallel to the x_3-axis is an interval. In this
situation, Theorem 3.1.4 from [Isa90] tells us that it is indeed possible to determine

Fig. 1.6 A body S
containing an inclusion D
convex in x_3

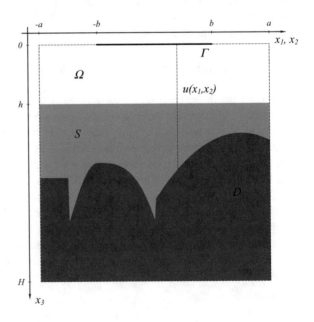

the unknown function u from knowledge of

$$\|\nabla V(x)\|_2, \quad x \in \Gamma = \{x \in \mathbb{R}^3; \; -b \le x_1 \le b, \; -b \le x_2 \le b, \; x_3 = 0\},$$

where $0 < b \le a$ is a known constant and where, omitting constants,

$$V(x) = \int_{-a}^{a} \int_{-a}^{a} \int_{h}^{u(t_1,t_2)} \frac{1}{\sqrt{(x_1 - t_1)^2 + (x_2 - t_2)^2 + (x_3 - t_3)^2}} \, dt_3 \, dt_2 \, dt_1$$

$$(1.22)$$

is the potential of $S \setminus D$. We assume that the first two components of $\nabla V(x)$ are of negligible size, so that $\|\nabla V(x)\|_2$, $x \in \Gamma$, is approximately determined by the vertical component $V_{x_3} = \partial V / \partial x_3$ of the gravitational force, measured at $x = (x_1, x_2, 0) \in \Gamma$. From the fundamental theorem of calculus[5] we get, differentiating under the integral sign:

$$V_{x_3}(x_1, x_2, 0) = -\int_{-a}^{a} \int_{-a}^{a} \frac{1}{\sqrt{(x_1 - t_1)^2 + (x_2 - t_2)^2 + u(t_1, t_2)^2}} \, dt_2 \, dt_1$$

$$+ \underbrace{\int_{-a}^{a} \int_{-a}^{a} \frac{1}{\sqrt{(x_1 - t_1)^2 + (x_2 - t_2)^2 + h^2}} \, dt_2 \, dt_1}_{=: \, V_C(x_1, x_2)} \,. \quad (1.23)$$

The values $V_{x_3}(x_1, x_2, 0)$ represent exact (satellite) measurements of the gravitational force, where the coordinate system is chosen such that the satellite's height is 0. Actually, one would observe the vertical component F_{x_3} of the joint gravitational force induced by $S \setminus D$ and D *together*. But since gravitational forces induced by two bodies simply add up, and since the constant mass densities c_D and c_S are known, one can easily find $V_{x_3}(x_1, x_2, 0)$, $(x_1, x_2) \in [-b, b]^2$, from the values $F_{x_3}(x_1, x_2, 0)$. Thus the left hand side of (1.23) can be considered a known function of x_1 and x_2 (although, in reality, we only dispose of a discrete set of noisy measurements). Likewise, the term V_C in (1.23) can be computed analytically and therefore also is a known function. We are led to our first model problem:

Problem 1.9 (Nonlinear Inverse Gravimetry) Solve the nonlinear Fredholm equation of the first kind

$$w(x_1, x_2) = \int_{-a}^{a} \int_{-a}^{a} k(x_1, x_2, t_1, t_2, u(t_1, t_2)) \, dt_2 \, dt_1 \quad (1.24)$$

(continued)

---

[5] $f(b) - f(a) = \int_{a}^{b} f'(x) \, dx.$

Problem 1.9 (continued)

for the unknown, continuous function $u : [-a, a]^2 \to [h, H]$. Here,

$$w(x_1, x_2) = V_C(x_1, x_2) - V_{x_3}(x_1, x_2, 0), \quad (x_1, x_2) \in [-b, b]^2, \quad (1.25)$$

is a known function (since V_{x_3} ("observation") as well as V_C defined in (1.23) are known). Further, the kernel function is defined by

$$k(x_1, x_2, t_1, t_2, u) := \frac{1}{\sqrt{(x_1 - t_1)^2 + (x_2 - t_2)^2 + u^2}} \quad (1.26)$$

for $x_1, x_2, t_1, t_2 \in \mathbb{R}$ and $0 < h \le u \le H$.

Remark Problem 1.9 is an inverse problem, which can formally be described using the notation of Definition 1.5. Namely, set $X = C([-a, a]^2)$ with norm $\| \bullet \|_X = \| \bullet \|_{C([-a,a]^2)}$ and set $Y = C([-b, b]^2)$ with norm $\| \bullet \|_Y = \| \bullet \|_{C([-b,b]^2)}$. Set

$$\mathbb{U} := \{u \in C([-a, a]^2); \ 0 < h \le u(x_1, x_2) \le H \text{ for } (x_1, x_2) \in [-a, a]^2\}, \quad (1.27)$$

set $\mathbb{W} = Y = C([-b, b]^2)$ and define the mapping T by

$$T : \mathbb{U} \to \mathbb{W}, \quad u \mapsto w, \quad w(x_1, x_2) = \int_{-a}^{a} \int_{-a}^{a} k(x_1, x_2, t_1, t_2, u(t_1, t_2)) \, dt_2 \, dt_1, \quad (1.28)$$

where the kernel function k is given by (1.26).

Next we will consider a special case, assuming that the values $u(x_1, x_2)$ do not deviate much from a constant average value $u_0 > h > 0$, i.e.,

$$u(x_1, x_2) = u_0 + \Delta u(x_1, x_2), \quad (x_1, x_2) \in [-a, a]^2,$$

$\Delta u(x_1, x_2)$ being "small." Evidently, this is *not* true in the situation of Fig. 1.6, but may be a reasonable assumption in case u describes the Mohorovičić discontinuity. Then we might *linearize* (1.24) by replacing the integrand k using the first order approximation:

$$k(x_1, x_2, t_1, t_2, u_0 + \delta u) \overset{\bullet}{=} k(x_1, x_2, t_1, t_2, u_0) + \frac{\partial k(x_1, x_2, t_1, t_2, u_0)}{\partial u} \cdot \delta u \ .$$

The unknown function Δu will then be computed by solving the equation

$$w(x_1, x_2) = \int_{-a}^{a} \int_{-a}^{a} \frac{1}{\sqrt{(x_1 - t_1)^2 + (x_2 - t_2)^2 + u_0^2}} \, dt_2 \, dt_1 \qquad (1.29)$$

$$- \int_{-a}^{a} \int_{-a}^{a} \frac{u_0}{\left((x_1 - t_1)^2 + (x_2 - t_2)^2 + u_0^2\right)^{3/2}} \Delta u(t_1, t_2) \, dt_2 \, dt_1,$$

which *approximates* (1.24). The first integral on the right-hand side of (1.29) can be computed analytically. We are thus lead to

Problem 1.10 (Linear Inverse Gravimetry) Solve the *linear* Fredholm equation of the first kind (the convolution equation)

$$f(x_1, x_2) = u_0 \int_{-a}^{a} \int_{-a}^{a} k(x_1 - t_1, x_2 - t_2) \Delta u(t_1, t_2) \, dt_2 \, dt_1, \qquad (1.30)$$

for an unknown, continuous function $\Delta u : [-a, a]^2 \to \mathbb{R}$, where

$$k(r, s) := \left(r^2 + s^2 + u_0^2\right)^{-3/2}, \quad r, s \in \mathbb{R}, \quad u_0 > h > 0, \qquad (1.31)$$

and where $f \in C([-b, b]^2)$ is known.

From (1.29) one actually gets

$$f(x_1, x_2) = \int_{-a}^{a} \int_{-a}^{a} \frac{1}{\sqrt{(x_1 - t_1)^2 + (x_2 - t_2)^2 + u_0^2}} \, dt_2 \, dt_1 - w(x_1, x_2)$$

$$(1.32)$$

for $(x_1, x_2) \in [-b, b]^2$ and for w from (1.24).

1.4   Model Problems for Full-Waveform Seismic Inversion

Waveform inversion "in the acoustic approximation" and in the setting of marine experiments was introduced in Example 1.4. The ill-posedness of this problem in general is proven in [KR14]. In the following, we will consider *land* experiments. In this case, geophones are used to record the vertical component of particle velocity on the earth's surface (whereas hydrophones access sound pressure in the ocean). We introduce a drastic simplification by considering only *plane* acoustic waves propagating in a *horizontally layered* medium, which means that we actually only consider one-dimensional wave propagation. The results presented below were obtained by Alain Bamberger, Guy Chavent, and Patrick Lailly, see [BCL77] and [BCL79]. The following mathematical model is used:

$$\rho(z)\frac{\partial^2 u(z,t)}{\partial t^2} - \frac{\partial}{\partial z}\left(\kappa(z)\frac{\partial u(z,t)}{\partial z}\right) = 0, \qquad z > 0, \ t > 0, \qquad (1.33)$$

$$u(z,t) = 0, \quad \frac{\partial u(z,t)}{\partial t} = 0, \qquad z > 0, \ t = 0, \qquad (1.34)$$

$$-\kappa(z)\frac{\partial u(z,t)}{\partial z} = g(t), \qquad z = 0, \ t > 0. \qquad (1.35)$$

Here, $t \geq 0$ means time and $z \geq 0$ means the depth of an infinitely deep wave propagation medium. Equation (1.33) can be derived from (1.14) and (1.15) with $S = 0$ (no source in the interior of the propagation medium) and where $u = \mathbf{u}$ now is a scalar displacement field. The source term was renamed and moved into the (Neumann) boundary condition (1.35), which implies that waves are excited at the surface $z = 0$ of the medium only. From (1.14) one can see that $g(t)$ means the pressure applied at time t on the surface. Equation (1.34) defines initial conditions. The mass density ρ and the elasticity coefficient κ were already introduced in Example 1.4. Under appropriate regularity conditions on ρ, κ, and g, the system of equations (1.33)–(1.35) has a unique solution u, which is differentiable with respect to t. So we may set

$$Y_d(t) := \frac{\partial u}{\partial t}(0,t), \quad t \geq 0, \qquad (1.36)$$

which is the velocity of particle displacement at the surface $z = 0$ and thus models the seismogram obtained, i.e., the observed effect of the seismic land experiment. In reality, of course, we will only dispose of a discrete set of noisy observations.

We start with a negative result: It is not possible in general to infer both functions ρ and κ from knowledge of Y_d. To see this, we will transform the system of equations (1.33)–(1.35) into an equivalent system of equations. Let us define

$$\psi : [0,\infty[\rightarrow [0,\infty[, \quad z \mapsto \psi(z) := \int_0^z \frac{ds}{c(s)} =: x, \qquad (1.37)$$

with $c = c(z) = \sqrt{\kappa(z)/\rho(z)}$ the wave velocity, as introduced in (1.17). By definition, $\psi(z)$ means the time a wave moving at speed c needs to travel from the surface of the medium down to depth z. Therefore, ψ is called **travel-time transformation**. Since c may be assumed to be positive and continuous almost everywhere, ψ is monotonously increasing, differentiable almost everywhere and has the property $\psi(0) = 0$. It will be used to re-parameterize the solution u of (1.33), (1.34), and (1.35), setting

$$y(x,t) := u(z,t) \quad \text{with} \quad x = \psi(z).$$

One finds that the system (1.33), (1.34), and (1.35) transforms into

$$\sigma(x)\frac{\partial^2 y(x,t)}{\partial t^2} - \frac{\partial}{\partial x}\left(\sigma(x)\frac{\partial y(x,t)}{\partial x}\right) = 0, \qquad x > 0, \ t > 0, \qquad (1.38)$$

$$y(x,0) = 0, \quad \frac{\partial y(x,0)}{\partial t} = 0, \qquad\qquad\qquad x > 0, \qquad (1.39)$$

$$-\sigma(0)\frac{\partial y(0,t)}{\partial x} = g(t), \qquad\qquad\qquad\qquad t > 0, \qquad (1.40)$$

with only a single parameter function remaining, the so-called **acoustical impedance**

$$\sigma(x) = \sqrt{\rho(z)\kappa(z)}, \quad x = \psi(z) . \qquad (1.41)$$

Because of $\psi(0) = 0$, one has $Y_d(t) = \partial u(0,t)/\partial t = \partial y(0,t)/\partial t$, meaning that the solutions u of (1.33), (1.34), and (1.35) and y of (1.38), (1.39), and (1.40) produce the same seismogram Y_d. Therefore, any two pairs of functions (ρ, κ) and $(\tilde{\rho}, \tilde{\kappa})$ with $\rho \cdot \kappa = \tilde{\rho} \cdot \tilde{\kappa}$ will produce the same seismogram Y_d.

To establish uniqueness, we will introduce a further idealization:

$$\rho \equiv \rho_0 \quad : \quad \text{density is a } known \text{ constant.}$$

Under this additional assumption, κ and σ determine each other mutually.[6] Thus, for constant density, if σ is identifiable from Y_d, as will be stated below, then so is κ.

Now assume that values $Y_d(t)$ are observed only for $0 < t < T_0$, where T_0 is a fixed point in time. According to (1.38), (1.39), and (1.40), waves are excited

---

[6]To be more precise, it is clear from (1.37) and from (1.41), that κ (as a function of z) determines σ (as a function of x), if $\rho \equiv \rho_0$ is known. Conversely, assume $\rho \equiv \rho_0$ and assume σ to be a known function of x. From (1.41), we get

$$\kappa(z) = \frac{\sigma^2(x)}{\rho_0}. \qquad (1.42)$$

Further, we have

$$\psi^{-1}(0) = 0 \quad \text{and} \quad \left(\psi^{-1}\right)'(x) = \frac{1}{\psi'(z)} \overset{(1.37)}{=} \sqrt{\frac{\kappa(z)}{\rho_0}} \overset{(1.41)}{=} \frac{\sigma(x)}{\rho_0}.$$

From the latter two equations we get the explicit formula

$$z = \psi^{-1}(x) = \int_0^x \sigma(t)/\rho_0 \, dt. \qquad (1.43)$$

Formulae (1.42) and (1.43) show that σ determines κ.

exclusively at the medium's surface $x = 0$ and any upcoming wave reaching the surface must be a reflection of a wave originally sent down from the surface. By Eq. (1.38) all waves travel at a constant speed $\sqrt{\sigma/\sigma} = 1$ and consequently, no waves reflected from "depths" greater than $x = T_0/2$ can reach the surface and have influence on the seismogram during the time interval $[0, T_0]$. We may as well "cut off the medium" at some $x = X_0$, if $X_0 \geq T_0/2$, and switch to the following model:

$$\sigma(x)\frac{\partial^2 y(x, t)}{\partial t^2} - \frac{\partial}{\partial x}\left(\sigma(x)\frac{\partial y(x, t)}{\partial x}\right) = 0, \qquad 0 < x < X_0, \; 0 < t < T_0,$$
(1.44)

$$y(x, 0) = 0, \quad \frac{\partial y(x, 0)}{\partial t} = 0, \qquad\qquad\qquad\qquad 0 < x < X_0,$$
(1.45)

$$-\sigma(0)\frac{\partial y(0, t)}{\partial x} = g(t), \quad \frac{\partial y(X_0, t)}{\partial x} = 0 \qquad\qquad 0 < t < T_0.$$
(1.46)

This system of equations needs only to be solved in a bounded domain. The new boundary condition introduced at $x = X_0$ changes the solution $y(x, t)$ as compared to the solution of (1.38), (1.39), and (1.40), but this does not have any influence on the observed seismogram $Y_d(t)$ for $0 < t < T_0$. The "artificial" Neumann condition at $x = X_0$ could be replaced by the Dirichlet condition $y(X_0, t) = 0$ as well.

System (1.44), (1.45), and (1.46) is investigated in [BCL77]. To formulate the results of Bamberger, Chavent, and Lailly, we informally introduce two function spaces. One is $L_2(a, b)$, $-\infty \leq a < b \leq \infty$, the space of **square integrable functions** defined on the interval $]a, b[\subset \mathbb{R}$. One may think of its elements as being functions $f :]a, b[\rightarrow \mathbb{R}$ which are regular enough such that

$$\|f\|_{L_2(a,b)} := \left(\int_a^b |f(t)|^2 \, dt\right)^{1/2}$$
(1.47)

exists, although this notion of $L_2(a, b)$ is not entirely correct (see Appendix B for a formal definition). If $-\infty < a < b < \infty$, then every $f \in C[a, b]$ belongs to $L_2(a, b)$. A function $f : [a, b] \rightarrow \mathbb{R}$ also belongs to $L_2(a, b)$, if it is continuous but at exceptional points $t_1, \ldots, t_n$, and if the one-sided limits

$$\lim_{t\downarrow t_i} f(t) = \lim_{t\rightarrow t_i+} f(t) \quad \text{and} \quad \lim_{t\uparrow t_i} f(t) = \lim_{t\rightarrow t_i-} f(t), \quad i = 1, \ldots, n,$$

exist. Here $t \uparrow t_i$ (or, equivalently, $t \rightarrow t_i-$) means $t \rightarrow t_i$ and $t \leq t_i$. Step functions are examples of non-continuous elements of $L_2(a, b)$.

The other space needed below is the **Sobolev space** $H^1(a, b)$. Although it is again not entirely correct, one may think of its elements as being functions $f :]a, b[\to \mathbb{R}$, which are regular enough such that they can be differentiated everywhere but at exceptional points ("almost everywhere") and such that

$$\|f\|_{H^1(a,b)} := \left(\int_a^b |f(t)|^2 \, dt + \int_a^b |f'(t)|^2 \, dt \right)^{1/2} \tag{1.48}$$

exists (as a finite number). For example, the function $f :]-1, 1[\to \mathbb{R}, t \mapsto |t|$, which can be differentiated everywhere except at $t = 0$, belongs to $H^1(-1, 1)$. A formal definition of Sobolev spaces is given in Appendix B. If $-\infty < a < b < \infty$, then it can be shown that

$$H^1(a, b) \subset C[a, b]. \tag{1.49}$$

The spaces $L_2(a, b)$ and $H^1(a, b)$ contain many more elements than those from $C[a, b]$ and $C^1[a, b]$, respectively, which makes it possible to find a solution of a (differential or integral) equation in $H^1(a, b)$, say, if it cannot be found in $C^1[a, b]$. Moreover, these spaces have an important property that $C[a, b]$ and $C^1[a, b]$ fail to have: they are Hilbert spaces—see Appendix B.

Returning to waveform inversion, let us now define the following set of feasible acoustic impedances:

$$\mathscr{S} := \{\sigma \in H^1(0, X_0); \ 0 < \sigma_- \leq \sigma(x) \leq \sigma_+ < \infty, \ \|\sigma'\|_{L_2(0,X_0)} \leq M\}, \tag{1.50}$$

where

$$X_0 \geq T_0/2, \tag{1.51}$$

and where σ_-, σ_+, and M are fixed constants. Note that for $\sigma \in H^1(0, X_0)$, the restriction $\sigma_- \leq \sigma(x) \leq \sigma_+$ makes sense because of (1.49). On the other hand, $\sigma \in H^1(0, X_0)$ does *not* imply differentiability of σ and therefore (1.44) has to be understood in some "generalized" or "weak" sense. In Appendix E, we explain what is meant by a weak solution of (1.44), (1.45), and (1.46) and state the following results from [BCL77].

For $\sigma \in \mathscr{S}$ and $g \in L_2(0, T_0)$:

(1) A unique solution y of (1.44), (1.45), and (1.46)—to be understood in a generalized sense—exists. The solution is differentiable in a weak sense with respect to t at $x = 0$, such that $Y_d \in L_2(0, T_0)$, where $Y_d(t) = \partial y(0, t)/\partial t$.

(continued)

(2) The mapping

$$T : \mathscr{S} \to L_2(0, T_0), \quad \sigma \mapsto Y_d,$$

is injective, meaning that σ can be uniquely identified from Y_d.

Since we have assumed $\rho \equiv \rho_0$ to be a constant, this result translates back to an identifiability result for κ. We want to formulate a corresponding inverse problem starting directly from our original model (1.33), (1.34), and (1.35), since this model can be extended to the cases of 2D and 3D waveform inversion, see [Sym09]. First we limit the space–time domain to $[0, Z_0] \times [0, T_0]$ and therefore have to introduce an additional, artificial boundary condition at $z = Z_0$. Z_0 and X_0 are related by

$$X_0 = \int_0^{Z_0} \sqrt{\frac{\rho_0}{\kappa(s)}} \, ds.$$

However, for unknown κ, this relation does not tell us explicitly how to choose Z_0 such that $X_0 \geq T_0/2$ is achieved. If Z_0 is chosen too small[7] with respect to T_0, then from a certain time $t = t'$ on, actual seismograms would exhibit effects of wave reflections from depths below Z_0, whereas simulated seismograms (based on solutions of (1.33), (1.34), and (1.35) in bounded domains) will contain artifacts. To mitigate this effect, one commonly uses a so-called absorbing boundary condition at $z = Z_0$. The purpose of an absorbing boundary condition is to at least suppress artificial wave reflections from the boundary at $z = Z_0$ back into the domain of interest. Without going into details, we use the so-called **Reynold's absorbing boundary condition**, see, e.g., [BG09]. It reads

$$\kappa(z) \frac{\partial u(z, t)}{\partial z} = -\sqrt{\rho_0 \kappa(z)} \, \frac{\partial u(z, t)}{\partial t}, \quad z = Z_0, \ t \in]0, T_0[.$$

Altogether, this leads to the model

$$\rho_0 \frac{\partial^2 u(z, t)}{\partial t^2} - \frac{\partial}{\partial z} \left(\kappa(z) \frac{\partial u(z, t)}{\partial z} \right) = 0, \qquad 0 < z < Z_0, \ 0 < t < T_0, \quad (1.52)$$

$$u(z, t) = 0, \quad \frac{\partial u(z, t)}{\partial t} = 0, \qquad\qquad\qquad 0 < z < Z_0, \ t = 0, \quad (1.53)$$

---

[7]This can be avoided when bounds on κ are known, which can be assumed in practice.

$$-\kappa(z)\frac{\partial u(z,t)}{\partial z} = g(t), \qquad\qquad z = 0,\ 0 < t < T_0, \qquad (1.54)$$

$$\kappa(z)\frac{\partial u(z,t)}{\partial z} = -\sqrt{\rho_0\kappa(z)}\,\frac{\partial u(z,t)}{\partial t}, \qquad z = Z_0,\ 0 < t < T_0. \qquad (1.55)$$

We arrive at the following model problem for waveform inversion.

> **Problem 1.11 (Nonlinear Waveform Inversion)** Let $T_0 > 0$ be a given point in time and let $Z_0 > 0$ be "large enough." Let
>
> $$\mathscr{K} := \{\kappa \in H^1(0, Z_0);\ 0 < \kappa_- \leq \kappa(z) \leq \kappa_+ < \infty,\ \|\kappa'\|_{L_2(0,Z_0)} \leq M\}, \qquad (1.56)$$
>
> where κ_-, κ_+, and M are constants. Let $g \in L_2(0, T_0)$ and define the mapping
>
> $$T : \mathscr{K} \to L_2(0, T_0), \quad \kappa \mapsto Y_d, \quad Y_d(t) := \frac{\partial u(0, t)}{\partial t}, \qquad (1.57)$$
>
> where u is the weak solution of (1.52), (1.53), (1.54), and (1.55). Given Y_d, find $\kappa \in \mathscr{K}$ such that $T(\kappa) = Y_d$.

Remarks

- Problem 1.11 is an inverse problem, which can be described using the notation of Definition 1.5. Namely, set $X = H^1(0, Z_0)$ with norm $\|\bullet\|_X = \|\bullet\|_{H^1(0,Z_0)}$ given by

$$\|f\|_{H^1(0,Z_0)}^2 = \int_0^{Z_0} |f(t)|^2\, dt + \int_0^{Z_0} |f'(t)|^2\, dt$$

and set $Y = L_2(0, T_0)$ with norm $\|\bullet\|_Y = \|\bullet\|_{L_2(0,T_0)}$ given by

$$\|f\|_{L_2(0,T_0)}^2 = \int_0^{T_0} |f(t)|^2\, dt.$$

Set $\mathbb{U} = \mathscr{K} \subset X$ and $\mathbb{W} = L_2(0, T_0) = Y$ and define $T : \mathbb{U} \to \mathbb{W}$ by (1.57).
- When considering Problem 1.11 in the sections to follow, we will always assume

$$g \in H^1(0, T_0). \qquad (1.58)$$

This is in accordance with assumption (1.61) about the linearized problem, see below. It will also assure that $g \in C[0, T_0] \subset H^1(0, T_0)$, so it makes sense to sample the function g.

* In the literature (e.g., in [BG09]), we found Problem 1.11 in a slightly modified form. Namely, it was assumed that one does not observe $\partial u(0, t)/\partial t$, but rather directly the function values $u(0, t)$ for $0 \le t \le T_0$. We accordingly adapt T to become an operator

$$T : \mathcal{K} \to H^1(0, T_0) \subset C[0, T_0], \quad \kappa \mapsto u(0, \bullet), \tag{1.59}$$

where u is the weak solution of (1.52), (1.53), (1.54), and (1.55). Since the function $\partial u(0, \bullet)/\partial t$ is uniquely determined by $u(0, \bullet)$, it is clear that κ also is uniquely determined by $u(0, \bullet)$.

We finally consider a linearized version of the above model problem. To derive it, we restart from the original equations (1.33), (1.34), and (1.35), i.e., we return to the unbounded space/time domain. We make the following assumptions:

$$g \in H^1(0, \infty), \quad g(0) = 0, \tag{1.60}$$

and we extend g to become a function $g \in H^1(\mathbb{R})$ by setting $g(t) = 0$ for $t < 0$. Further, we assume that

$$\kappa = \kappa_0(1 + f), \quad f \in H^1(0, \infty), \quad f(0) = 0, \quad |f| \text{ "small,"} \tag{1.61}$$

where κ_0 is a *known constant*. The following calculations are certainly valid if f and g are assumed to be continuously differentiable. But they also hold in a weakened sense if f and g are only functions in H^1. We do not go into the corresponding mathematical details, but refer the reader to Chap. 5 of [Eva98] for a very readable exposition of calculus in Sobolev spaces.

Let u_0 be the unique solution of (1.33), (1.34), and (1.35) corresponding to the constant value $\kappa = \kappa_0$ and let u be the (generalized) solution corresponding to $\kappa = \kappa_0(1 + f)$. For $w := u - u_0$, one derives the following partial differential equation:

$$\frac{\partial^2 w(z, t)}{\partial t^2} - c_0^2 \frac{\partial^2 w(z, t)}{\partial z^2} = c_0^2 \frac{\partial}{\partial z}\left(f(z)\frac{\partial u(z, t)}{\partial z}\right), \tag{1.62}$$

where

$$c_0 := \sqrt{\frac{\kappa_0}{\rho_0}}.$$

The **Born approximation** consists in replacing u on the right-hand side of (1.62) by u_0, which is known, since κ_0 is known. Setting

$$h(z, t) := c_0^2 \frac{\partial}{\partial z} \left(f(z) \frac{\partial u_0(z, t)}{\partial z} \right), \tag{1.63}$$

an approximation v of w is determined as the unique solution of the partial differential equation

$$\frac{\partial^2 v(z, t)}{\partial t^2} - c_0^2 \frac{\partial^2 v(z, t)}{\partial z^2} = h(z, t), \quad z > 0, \quad t > 0, \tag{1.64}$$

with initial values

$$v(z, 0) = 0, \quad \frac{\partial v(z, 0)}{\partial t} = 0, \quad z > 0, \tag{1.65}$$

and boundary values

$$\frac{\partial v(0, t)}{\partial z} = 0, \quad t > 0. \tag{1.66}$$

The solution v of this system depends *linearly* on f. This means that if v_1 is the solution of (1.64), (1.65), and (1.66) for $f = f_1$ and v_2 is the solution for $f = f_2$, then $\alpha v_1 + \beta v_2$ is the solution for $f = \alpha f_1 + \beta f_2$. The inverse problem consists in determining f from the observation of

$$V_d(t) := \frac{\partial v(0, t)}{\partial t} = \frac{\partial u(0, t)}{\partial t} - \frac{\partial u_0(0, t)}{\partial t}, \quad t > 0. \tag{1.67}$$

We show that this problem takes the form of solving an integral equation. Integral equations, especially the so-called **Lippmann–Schwinger** integral equation, also play a prominent role in inverse scattering theory, see, e.g., [Kir11]. First, the solution u_0 of (1.33), (1.34), and (1.35) for constant $\kappa = \kappa_0$ is explicitly given by

$$u_0(z, t) = \frac{1}{\sigma_0} G(t - z/c_0) \quad \text{with} \quad G(z) := \int_0^z g(s) \, ds, \tag{1.68}$$

where $\sigma_0 := \sqrt{\rho_0 \cdot \kappa_0}$ (recall that g was continued by $g(t) := 0$ for $t \leq 0$ to become a function defined on the real line $\mathbb{R}$). From this, one immediately gets a formula for h in (1.63):

$$h(z, t) = -\frac{1}{\rho_0} \frac{\partial}{\partial z} \left(f(z) g(t - z/c_0) \right) = -\frac{1}{\rho_0} f'(z) g(t - z/c_0) + \frac{1}{\sigma_0} f(z) g'(t - z/c_0). \tag{1.69}$$

Second, the system of equations (1.64), (1.65), and (1.66) has the form (E.11) considered in Appendix E. Thus, explicit formulae for v can be taken directly from (E.12) and (E.13). Especially for $z \leq c_0 t$ we get, according to (E.13)

$$v(z, t) = \int_0^{t-z/c_0} v(z, t; s)\, ds + \int_{t-z/c_0}^t v(z, t; s)\, ds, \tag{1.70}$$

where

$$v(z, t; s) = \begin{cases} \frac{1}{2c_0} \int_{z-c_0(t-s)}^{z+c_0(t-s)} h(y, s)\, dy, & z \geq c_0(t-s) \geq 0 \\ \frac{1}{2c_0} \int_0^{z+c_0(t-s)} h(y, s)\, dy + \frac{1}{2c_0} \int_0^{c_0(t-s)-z} h(y, s)\, dy, & 0 \leq z \leq c_0(t-s) \end{cases} \tag{1.71}$$

Third, an explicit formula for $\partial v(0, t)/\partial t$ can be derived from (1.70) and (1.71):

$$\frac{\partial v(0, t)}{\partial t} = \int_0^t h(c_0(t - s), s)\, ds \overset{(1.69)}{=} -\frac{1}{2\rho_0} \int_0^t g(2s - t) f'(c_0(t - s))\, ds,$$

where we used partial integration and made use of $f(0) = 0$ and $g(t) = 0$ for $t \leq 0$.
Substituting $\tau = c_0(t - s)$, from this last equation we get

$$V_d(t) = -\frac{1}{2\sigma_0} \int_0^{c_0 t} f'(\tau) g(t - 2\tau/c_0)\, d\tau = -\frac{1}{4\rho_0} \int_0^{2t} g(t - s) f'(c_0 s/2)\, ds. \tag{1.72}$$

In practice, the left hand side of (1.72) will only be observed for a finite time interval $t \in [0, T_0]$, $T_0 > 0$. Since $g(t) = 0$ for $t \leq 0$, we can rewrite (1.72) in the form

$$V_d(t) = -\frac{1}{4\rho_0} \int_0^{T_0} g(t - s) f'(c_0 s/2)\, ds, \quad 0 \leq t \leq T_0. \tag{1.73}$$

This integral equation must be solved for $f'(c_0 s/2)$, $0 \leq s \leq T_0$. This will get us the function f' on the interval $[0, c_0 T_0/2]$, from where f can be determined on the same interval, since $f(0) = 0$. The Born approximation thus leads to

Problem 1.12 (Linear Waveform Inversion) Let $T_0 > 0$, assume that (1.60) holds for g and that $\text{supp}(g) \subset [0, T_0]$. Let $\rho_0 > 0$, $\kappa_0 > 0$, $\sigma_0 := \sqrt{\rho_0 \cdot \kappa_0}$, and $c_0 := \sqrt{\kappa_0/\rho_0}$ be known constants. Determine $f \in H^1(0, c_0 T_0/2)$ from $f(0) = 0$ and from solving the equation

$$V_d(t) = -\frac{1}{4\rho_0} \int_0^{T_0} g(t - s) f'(c_0 s/2)\, ds \tag{1.74}$$

for f', where $V_d \in H^1(0, T_0)$ is a known function, as defined in (1.67).

Equation (1.74) is a linear Fredholm equation of the first kind and even a convolutional equation for the function $s \mapsto f'(c_0 s/2)$ (from which f can be derived).

Solving Problem 1.12 essentially means to solve a convolutional equation. Since Problem 1.10 also belongs to this class of equations, it is justified to highlight the treatment of integral equations and notably of convolutional equations below.

Chapter 2
Discretization of Inverse Problems

Parameter identification problems can be formulated as equations

$$T(u) = w, \quad u \in \mathbb{U} \subseteq X, \quad w \in \mathbb{W} \subseteq Y.$$

In many interesting cases, X and Y are infinite-dimensional spaces of functions—this was so for all model problems presented in Chap. 1. Although we generally suppose that a unique solution $u^* \in \mathbb{U}$ exists, explicit formulae for its computation are only rarely available, so that one has to be satisfied with the construction of an approximate solution by numerical methods. In practice, not even the equation $T(u) = w$ itself is perfectly known, if w is a function. Rather, an approximation of w has to be constructed on the basis of a finite number of (inexact) measurements (observations). All numerical solution methods for inverse problems are based on **discretization**, by which we mean an approximate description and solution of the inverse problem $T(u) = w$ in spaces of *finite dimension*. To achieve this, we choose spaces $X_n \subseteq X$ and $Y_m \subseteq Y$ of finite dimension, approximate w by an element $w_m \in Y_m$, approximate T by an operator $T_{n,m} : X_n \rightarrow Y_m$, and find $u_n \in X_n$ such that $T_{n,m}(u_n)$ approximates w_m. Then, u_n will be considered an approximation of the exact solution u^*.

There are many ways to implement these general ideas, and choosing a good discretization is not trivial. It not only decides how well the solution u^* can be approximated, but also shapes the resulting finite-dimensional problem and determines which practical solution methods are applicable and efficient. Necessarily, choosing a good discretization depends on the problem to be solved. We let ourselves be guided by the four model problems posed in Sects. 1.3 and 1.4 and only highlight some aspects of discretization. The more sophisticated approach of multiscale approximation will only be dealt with by way of example in one special case. Adaptive approximation and approximation of functions defined on balls and spheres—topics of obvious interest in geomathematics—are not discussed at all. The reader will find information on these topics in Chap. 3 of [Cha09] and in

© The Author(s), under exclusive license to Springer Nature Switzerland AG 2020
M. Richter, *Inverse Problems*, Lecture Notes in Geosystems
Mathematics and Computing, https://doi.org/10.1007/978-3-030-59317-9_2

[FNS10]. In Sect. 2.1, we present spaces of piecewise constant and of piecewise (bi-)linear functions as *candidates* for the choice of X_n and Y_m. These are very easy to handle and at the same time can approximate well a large class of functions. In Sect. 2.2, we discuss the least squares method to find an approximant u_n of u^* in the special case where T is a *linear* mapping and where $\mathbb{U} = X$. Special attention will be paid to an analysis of the error $u^* - u_n$. In Sect. 2.3, the collocation method is presented, which, in the context of Fredholm integral equations, can be interpreted as a special case of the least squares method. Section 2.4 again focuses on linear mappings T implicitly defined by Fredholm integral equations and introduces the method of Backus and Gilbert as an approach to approximately invert such operators. Fourier transform methods are efficient and therefore attractive in the important case of linear problems defined by convolutional Fredholm equations. These methods are considered and analyzed in Sect. 2.5. Finally, in Sects. 2.6 and 2.7, specific discretizations for the nonlinear model problems of gravimetry and waveform inversion from Sects. 1.3 and 1.4 are derived. All model problems will be reformulated in a discretized version.

2.1   Approximation of Functions

In this section, we discuss the approximation of univariate and bivariate functions (i.e., of functions having one or two arguments) by **spline functions** of low order. Splines are only *one* possible choice of candidates for the approximation of functions (alternatives include polynomials, Fourier sums, and wavelets). This choice often is very successful in practice, since splines are quite easy to handle on a computer and at the same time can approximate well a large class of functions. Splines can be generalized to any space dimension $s \in \mathbb{N}$.

Approximation in One Space Dimension

Definition 2.1 (Univariate Splines of Orders 1 and 2) Let $a < b$ and let $a = t_1 < t_2 < \ldots < t_m = b$ be a partitioning of the interval $[a, b]$. A function $s : [a, b] \to \mathbb{R}$ having the properties

$$s(t) = c_i \in \mathbb{R}, \quad t_i \leq t < t_{i+1}, \quad i = 1, \ldots, m - 1,$$

and $s(t_m) = s(t_{m-1})$ is called a **(univariate) spline function of order 1**. If s is continuous and has the properties

$$s(t) = a_i t + b_i, \quad a_i, b_i \in \mathbb{R}, \quad t_i \leq t \leq t_{i+1}, \quad i = 1, \ldots, m - 1,$$

then it is called a **(univariate) spline function of order 2**. The numbers t_i are called **knots**. The set of all spline functions of order k defined by the knots $t_1, \ldots, t_m$ is denoted by $\mathscr{S}_k(t_1, \ldots, t_m)$.

Remark Splines of order 1 are step functions, and splines of order 2 (also called "linear splines") are polygonal lines. More generally, linear splines could be allowed to have discontinuities at the knots. Also, one could define univariate splines composed of higher order polynomial pieces.

The set $\mathscr{S}_k(t_1, \ldots, t_m)$ is a vector space, since $\alpha_1 s_1 + \alpha_2 s_2 \in \mathscr{S}_k(t_1, \ldots, t_m)$, if $s_1, s_2 \in \mathscr{S}_k(t_1, \ldots, t_m)$ and $\alpha_1, \alpha_2 \in \mathbb{R}$. The dimension of this space is

$$\dim \mathscr{S}_k(t_1, \ldots, t_m) = m + k - 2. \tag{2.1}$$

A basis of $\mathscr{S}_1(t_1, \ldots, t_m)$ is given by the functions

$$N_{j,1}(t) := \begin{cases} 1, & t_j \leq t < t_{j+1} \\ 0, & \text{else} \end{cases}, \quad j = 1, \ldots, m-1 \tag{2.2}$$

(with additional agreement: $N_{m-1,1}(t_m) := 1$). A basis of $\mathscr{S}_2(t_1, \ldots, t_m)$ is given by the "hat functions"

$$N_{j,2}(t) := \begin{cases} \dfrac{t - t_{j-1}}{t_j - t_{j-1}}, & t \in [t_{j-1}, t_j] \quad (\text{if } j \geq 2) \\[2ex] \dfrac{t_{j+1} - t}{t_{j+1} - t_j}, & t \in [t_j, t_{j+1}] \quad (\text{if } j \leq m-1) \\[2ex] 0, & \text{else} \end{cases}, \quad j = 1, \ldots, m, \tag{2.3}$$

having the property $N_{j,2}(t_j) = 1$ for $j = 1, \ldots, m$. These basis functions are called **B-splines** of order 1 or 2, respectively. Any spline function $s \in \mathscr{S}_k(t_1, \ldots, t_m)$ can uniquely be written as a linear combination of B-splines:

$$s(t) = \sum_{j=1}^{m+k-2} \alpha_j N_{j,k}(t), \quad a \leq t \leq b. \tag{2.4}$$

A convenient way to describe approximation by splines from $\mathscr{S}_k(t_1, \ldots, t_m)$ is the introduction of approximation operators. For example, one may set

$$I_1 : L_2(a, b) \to \mathscr{S}_1(t_1, \ldots, t_m), \quad f \mapsto I_1(f) := \sum_{j=1}^{m-1} \alpha_j N_{j,1}, \tag{2.5}$$

with coefficients α_j defined by

$$\alpha_j := \left(\frac{1}{t_{j+1} - t_j} \int_{t_j}^{t_{j+1}} f(t)\, dt \right), \quad j = 1, \ldots, m-1.$$

I_1 thus is an operator, mapping any square integrable function f to a spline $I_1(f)$ of order 1, which is considered an approximant of f.[1] The approximant $I_1(f)$ has the same local mean values as f. Another example would be the **interpolation operator**

$$I_2 : C[a,b] \to \mathscr{S}_2(t_1, \ldots, t_m), \quad f \mapsto I_2(f) = \sum_{j=1}^{m} f(t_j) N_{j,2}, \tag{2.6}$$

which maps a continuous function f to its linear spline interpolant. The name interpolation operator is chosen because $s = I_2(f)$ evidently has the property

$$s(t_i) = f(t_i), \quad i = 1, \ldots, m. \tag{2.7}$$

Figure 2.1 illustrates the approximation schemes (2.5) and (2.6). The **approximation error** $f - I_k(f)$ in general can become arbitrarily large. Bounds can only be given under additional (smoothness) conditions on f. For example, if $f \in C^2[a,b]$, one can show that

$$\|f - I_2(f)\|_{C[a,b]} \le \frac{1}{.8} h^2 \|f''\|_{C[a,b]} \quad \text{with} \quad h := \max_{i=1,\ldots,m-1} \{(t_{j+1} - t_j)\}, \tag{2.8}$$

see [dB90], p. 37. The bound given in (2.8) is no longer useful if f is not a C^2-function. A different bound, given in the following theorem, applies to a larger class of functions f than does (2.8), since $C[a,b] \supset H^1(a,b) \supset C^2[a,b]$. Here, the second inclusion is evident, and the first one—already stated in (1.49)—is a consequence of Sobolev's embedding theorem, see, for example, Theorem A.5 in [LT03].

Theorem 2.2 (Approximation Errors) *Let $k \in \{1, 2\}$, let $a = t_1 < \ldots < t_m = b$ and let $h := \max_{i=1,\ldots,m-1}\{t_{i+1} - t_i\}$. Let $I_k : C[a,b] \to \mathscr{S}_k(t_1, \ldots, t_m)$ be defined by (2.5) for $k = 1$ and by (2.6) for $k = 2$. Then, for $f \in H^1(a,b)$, the bound*

$$\|f - I_k(f)\|_{L_2(a,b)} \le kh \, \|f\|_{H^1(a,b)} \tag{2.9}$$

on the approximation error holds.

---

[1]Of course, this is not the only possibility. Requiring $s(t_j) = f(t_j)$ to hold for $j = 1, \ldots, m-1$, would define a different spline approximant $s \in \mathscr{S}_1(t_1, \ldots, t_m)$ of $f \in C[a,b]$.

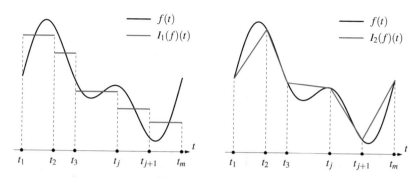

Fig. 2.1 Piecewise constant and linear approximation

We do not give a proof of this result, which is not trivial, but well known in approximation theory. From Theorem 2.2, it can be seen that $I_k(f) \in \mathscr{S}_k(t_1, \ldots, t_m)$ converges to f with respect to the norm $\| \bullet \|_{L_2(a,b)}$, if $h := \max\{t_{i+1} - t_i\}$ tends to 0.

Approximation in Two Space Dimensions

The definition of a two-dimensional analogon of spline functions requires partitionings of two-dimensional sets $D \subset \mathbb{R}^2$. For the sake of simplicity, we restrict ourselves to polygonal regions D.

Definition 2.3 (Triangulation, Rectangular Partitioning) Let $D := \overline{\Omega}$ be the closure of a bounded polygonal domain $\Omega \subset \mathbb{R}^2$. A **triangulation (rectangular partitioning)** of D is a set $\mathscr{T} = \{T_1, \ldots, T_m\}$ consisting of closed plane triangles (closed plane rectangles), which meet the following three conditions:

(1) $D = \bigcup_{i=1}^{m} T_i$.

(2) If $T_i \cap T_j$ contains only a single point, then this is a vertex of both T_i and T_j.

(3) If $T_i \cap T_j$, $i \neq j$, contains more than a single point, then $T_i \cap T_j$ is a common edge of T_i and T_j.

Examples of triangulations and rectangular partitionings are shown in Fig. 2.2. Talking of a partitioning is not fully justified, since the elements T_i can never be mutually disjoint. A triangulation of a polygonal region D does always exist, and a rectangular partitioning does exist for rectangles. Triangulations can be generalized to three (and four, ...) space dimensions using simplices. In the following, the word "partitioning" is used to designate either a triangulation or a rectangular partitioning.

We will use the notation $\mathscr{T}_h$ to designate any finite partitioning $\{T_1, \ldots, T_m\}$, when the maximal diameter of $T_i \in \mathscr{T}_h$ is equal to h. A family $\{\mathscr{T}_h\}$ of partitionings

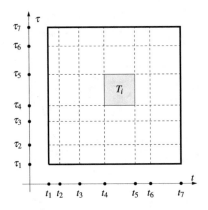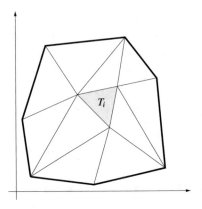

Fig. 2.2 Partitioning of a rectangle into rectangles and of a polygon into triangles

(consisting of partitionings for different values h) is called **quasi-uniform**, if there exists a constant $\kappa > 0$ such that for every $h > 0$ every $T \in \mathscr{T}_h$ contains a ball of radius ρ_T with

$$\rho_T \geq h_T/\kappa, \quad h_T := \text{diam}(T), \tag{2.10}$$

and it is called **uniform**, if there exists a constant $\kappa > 0$ such that

$$\rho_T \geq H/\kappa, \quad H := \max_{T \in \mathscr{T}_h} \{\, \text{diam}(T) \,\}. \tag{2.11}$$

Quasi-uniform partitionings cannot contain arbitrarily narrow elements. In a uniform family of partitionings, all elements $T \in \mathscr{T}_h$ shrink at the same rate, if $h \to 0$. Any uniform family is quasi-uniform, but not vice versa.

We will designate by $\mathscr{S}_1(\mathscr{T}_h)$ the set of all piecewise constant functions for a given partitioning $\mathscr{T}_h$. This means that $s \in \mathscr{S}_1(\mathscr{T}_h)$ shall take a constant value c_i in the interior of each $T_i \in \mathscr{T}_h$. We leave it open what values s shall take on edges.

If $\mathscr{T}_h$ is a *rectangular* partitioning of $D = [a, b] \times [c, d]$, then it is defined by $a = t_1 < \ldots < t_{m_1} = b$ and $c = \tau_1 < \ldots < \tau_{m_2} = d$, as shown in Fig. 2.2. We defined linear B-splines $N_{i,2} \in \mathscr{S}_2(t_1, \ldots, t_{m_1})$ and $N_{j,2} \in \mathscr{S}_2(\tau_1, \ldots, \tau_{m_2})$ in the last paragraph. From these, one can build **bilinear B-splines**

$$N_{i,j,2} : [a, b] \times [c, d] \to \mathbb{R}, \quad (t, \tau) \mapsto N_{i,j,2}(t, \tau) := N_{i,2}(t)N_{j,2}(\tau).$$

Because of the appearance of terms $t \cdot \tau$, these functions are not piecewise linear. Figure 2.3 shows an example of a bilinear B-spline for the special case where $t_{i+1} - t_i = \tau_{j+1} - \tau_j = h$ for all i and j. The fat line shows the boundary of the support of another one. Let us define the space

$$\mathscr{S}_2^B(\mathscr{T}_h) := \text{span}\{N_{i,j,2}; \ i = 1, \ldots, m_1, \ j = 1, \ldots, m_2\}, \tag{2.12}$$

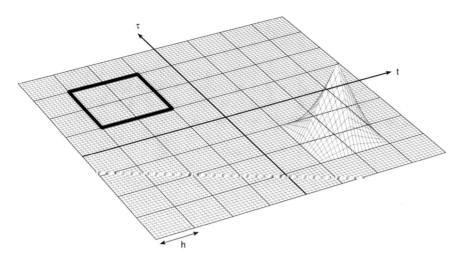

Fig. 2.3 Bilinear B-splines on an equidistant grid

of **bilinear spline functions** with basis $\{N_{i,j,2};\ i = 1, \ldots, m_1,\ j = 1, \ldots, m_2\}$. All elements $s \in \mathscr{S}_2^B(\mathscr{T}_h)$ belong to $C(D)$, i.e., they are continuous functions. An approximation operator can be defined by

$$I_2^B : C(D) \to \mathscr{S}_2^B(\mathscr{T}_h), \quad f \mapsto \sum_{i=1}^{m_1} \sum_{j=1}^{m_2} f(t_i, \tau_j) N_{i,j,2}. \tag{2.13}$$

I_2^B is in fact an interpolation operator, since $I_2^B(f)(t_i, \tau_j) = f(t_i, \tau_j)$ for all i and j.

As in the univariate case, the approximation error $f - I_2^B(f)$ cannot be bounded unless some additional regularity of f is assumed, which will be expressed by differentiability conditions. For a domain $\Omega \subset \mathbb{R}^2$ and for $k \in \mathbb{N}_0$, let (informally and not fully correctly) $H^k(\Omega)$ be the Sobolev space of functions $f : \Omega \to \mathbb{R}$, for which

$$\|f\|_{H^k(\Omega)} := \left(\sum_{|\alpha| \le k} \int_\Omega |D^\alpha f(x)|^2 \, dx \right)^{1/2} \tag{2.14}$$

exists. The sum extends over all partial derivatives of f up to order k, see Appendix B for the multi-index notation used here. All partial derivatives up to order k have to exist in a sense which gives a meaning to the above integrals and makes them have finite values. As in one dimension, this does not imply a pointwise existence of partial derivatives (as it is required for functions $f \in C^k(\Omega)$). The function

$$f : \Omega \to \mathbb{R} \cup \{\pm\infty\}, \quad (x, y) \mapsto \ln\left(\ln\left(\frac{2}{x^2 + y^2}\right)\right), \tag{2.15}$$

where $\Omega = \{(x, y) \in \mathbb{R}^2;\ x^2 + y^2 < 1\}$, is an example of a $H^1(\Omega)$-function. This function has a pole at $(x, y) = (0, 0)$ and cannot with full right be considered as "defined pointwise." See Appendix B for a formal definition of $H^k(\Omega)$. If Ω is a bounded polygonal domain, then it can be shown that

$$H^2(\Omega) \subset C(D), \quad D := \overline{\Omega}, \tag{2.16}$$

implying that $f \in H^2(\Omega)$ must have point values defined for every $x_0 \in D$. In two space dimensions, this is *not true* for $k = 1$, as shown by (2.15). Note that $H^0(\Omega) = L_2(\Omega)$.

Let us return now to the operator I_2^B and the space $\mathscr{S}_2^B(\mathscr{T}_h)$. The latter is contained in $H^1(\Omega)$, as proven, e.g., in [Bra07], p. 62. The following error estimate is proven in [Bra07], p. 81.

Theorem 2.4 (Approximation Error for Bilinear Spline Interpolation) *Let $\Omega \subset \mathbb{R}^2$ be an open, bounded rectangle with closure D. Let $f \in H^2(\Omega)$ and let $\{\mathscr{T}_h\}$ be a family of quasi-uniform rectangular partitionings of D. Then there exists a constant $c = c(\kappa)$ (depending only on κ from (2.10)) such that*

$$\|f - I_2^B(f)\|_{L_2(\Omega)} \le c \cdot h^2 \|f\|_{H^2(\Omega)}. \tag{2.17}$$

We finally consider a bounded polygonal domain $\Omega \subset \mathbb{R}^2$ with closure D and a *triangular* partitioning $\mathscr{T}_h$ of D. Let us define

$$\mathscr{S}_2(\mathscr{T}_h) := \{s \in C(D);\ s_{|T} \text{ is linear for all } T \in \mathscr{T}_h\}, \tag{2.18}$$

which is a space of piecewise linear bivariate functions. It can be shown that $\mathscr{S}_2(\mathscr{T}_h) \subset H^1(\Omega)$, see [Bra07], p. 62. Similarly to (2.13), one can define an interpolation operator $I_2 : C(D) \to \mathscr{S}_2(\mathscr{T}_h)$, and an error estimate of the form (2.17) also holds for I_2. But one cannot define an interpolation operator on $H^1(\Omega)$, since $H^1(\Omega) \not\subset C(D)$, so point values of $H^1(\Omega)$-functions are not defined. However, approximation operators $H^1(\Omega) \to \mathscr{S}_2(\mathscr{T}_h)$ do exist. The following theorem is proven in [Bra07], p. 83 ff.

Theorem 2.5 (Clément's Operator) *Let $\Omega \in \mathbb{R}^2$ be a bounded polygonal domain with closure D. Let $\{\mathscr{T}_h\}$ be a family of quasi-uniform triangulations of D. Then there exists a mapping $I_h : H^1(\Omega) \to \mathscr{S}_2(\mathscr{T}_h)$ and a constant $c = c(\kappa)$ such that*

$$\|u - I_h u\|_{H^m(\Omega)} \le ch^{1-m} \|u\|_{H^1(\Omega)}, \quad u \in H^1(\Omega), \quad m = 0, 1. \tag{2.19}$$

Theorem 2.5 tells us that any function $u \in H^1(\Omega)$ can be approximated arbitrarily well by a function $s \in \mathscr{S}_2(\mathscr{T}_h)$ with respect to the norm $\| \bullet \|_{L_2(\Omega)}$, if h is chosen small enough.

2.2 Discretization of Linear Problems by Least Squares Methods

In this section—which is based on Chap. 3 of [Kir11]—we present a discretization method which is applicable to a large class of *linear* inverse problems of the form $Tu = w$.[2] We make the following:

Assumption 2.6 *Let* $(X, \langle \bullet | \bullet \rangle_X)$ *and* $(Y, \langle \bullet | \bullet \rangle_Y)$ *be Hilbert spaces over the field of real numbers. Let* $T : X \to Y$ *be linear, continuous, and injective.*

Refer to Appendix B for the definitions of Hilbert spaces and of linear and continuous operators. The most important restriction here is the required linearity of the operator T. Linearity implies that T is defined on a linear space, which we have directly taken as X, not as a subspace. Considering subsets $\mathbb{U} \subset X$ as a constraint for a solution u^* of the inverse problem will lead to nonlinearity and is only considered in the context of nonlinear problems (see Sects. 2.6 and 2.7 and Chap. 4). It was already argued in Sect. 1.2 that injectivity of the map T is the one quality of well-posed problems that cannot be conceded when the goal is to identify parameters. Bijectivity of T is *not* assumed, however, nor do we assume continuity of the inverse of $T : X \to T(X)$. Requiring that X and Y are real spaces is not essential; the following could be generalized to complex vector spaces.

Description of the Method

Let $w \in T(X)$ and let u^* be the unique solution of $Tu = w$. Choose some d_n-dimensional subspace $X_n \subset X$:

$$X_n = \mathrm{span}\{\varphi_1, \ldots, \varphi_{d_n}\} \subset X, \quad \varphi_1, \ldots, \varphi_{d_n} \in X, \text{ linearly independent.} \tag{2.20}$$

The **least squares method** determines an approximant $u_n \in X_n$ of u^* by requiring that

$$\|Tu_n - w\|_Y \leq \|Tv - w\|_Y \quad \text{for all} \quad v \in X_n, \tag{2.21}$$

---

[2]For *linear* operators, it is common usage to write Tu instead of $T(u)$.

where $\| \bullet \|_Y$ is the norm induced by $\langle \bullet | \bullet \rangle_Y$. Solving this problem conceptually can be split into two steps.

- First, find the best approximation w_n of w in the d_n-dimensional subspace $Y_n := T(X_n)$ of Y with respect to the norm $\| \bullet \|_Y$. According to Theorem B.13, w_n is determined by the system of n equations

$$\langle w_n | T \varphi_i \rangle_Y = \langle w | T \varphi_i \rangle_Y \quad \text{for} \quad i = 1, \ldots, d_n. \tag{2.22}$$

- Second, find the unique $u_n \in X_n$ with $T u_n = w_n$. This works, since by Assumption 2.6 the map $T : X_n \rightarrow Y_n$ is bijective.

Both steps can be assembled into a single one by making the ansatz

$$u_n = \sum_{j=1}^{d_n} x_j \varphi_j, \quad x_j \in \mathbb{R}. \tag{2.23}$$

The parameters x_j need to be determined. Inserting into (2.22) leads to

$$\sum_{j=1}^{d_n} x_j \langle T \varphi_j | T \varphi_i \rangle_Y = \langle w | T \varphi_i \rangle_Y, \quad i = 1, \ldots, d_n. \tag{2.24}$$

This system can be written in matrix form:

$$Ax = b, \quad A_{i,j} = \langle T \varphi_j | T \varphi_i \rangle_Y, \quad b_i = \langle w | T \varphi_i \rangle_Y, \tag{2.25}$$

where $A \in \mathbb{R}^{d_n, d_n}$ is symmetric and positive definite and where $x, b \in \mathbb{R}^{d_n}$. By the properties of A, a unique solution of this system exists for any right-hand side. It defines a unique $u_n \in X_n$, the **least squares solution** of $T(u) = w$.

The same idea works if one only knows an approximation $\tilde{w} \approx w$. This approximation can be any vector $\tilde{w} \in Y$, even a non-attainable one (which means that $\tilde{w} \notin T(X)$ is permissible), because any $\tilde{w} \in Y$ can be projected on $T(X_n)$. The only thing to change is to replace w in (2.24) by $\tilde{w}$. This leads to a system $Ax = \tilde{b}$ with inaccuracies shifted from $\tilde{w}$ to $\tilde{b}$. Another possible source of inaccuracy is the numerical evaluation of the scalar products $\langle w | T \varphi_i \rangle_Y$ (or rather $\langle \tilde{w} | T \varphi_i \rangle_Y$), for example, if scalar products are defined by integrals, as it is the case for Sobolev spaces. Not distinguishing between measurement errors, approximation errors, and evaluation errors for scalar products, we will simply assume that we are given approximate discrete values

$$b_i^\delta \approx b_i = \langle w | T \varphi_i \rangle_Y, \quad i = 1, \ldots, d_n,$$

with the total data error bounded by

$$\|b^\delta - b\|_2 \le \delta, \tag{2.26}$$

whence the notation b^δ. The number $\delta \ge 0$ is a numerical bound for the overall inaccuracy in the discrete data. A more sophisticated, but also much more complicated, approach would be to use stochastic error models, but we will not go into this. Let us define

$$x^\delta := A^{-1}b^\delta, \quad u_n^\delta := \sum_{j=1}^{d_n} x_j^\delta \varphi_j \tag{2.27}$$

for the solution of the least squares problem computed for inaccurate data. In practice, x^δ will not be computed by inverting A, but by solving the linear system of equations $Ax = b^\delta$. To summarize:

Least Squares Method for a Linear Problem $Tu = w$.

- Choose a d_n-dimensional subspace $X_n \subset X$ with basis $\{\varphi_1, \dots, \varphi_{d_n}\}$.
- Set up the matrix A with components $A_{i,j} = \langle T\varphi_j | T\varphi_i \rangle_Y$ as in (2.24). Set up the vector b^δ with components $b_i^\delta \approx \langle w | T\varphi_i \rangle_Y$ from the available measured values of w.
- Compute the solution x^δ of $Ax = b^\delta$ and get an approximant

$$u_n^\delta = \sum_{j=1}^{d_n} x_j^\delta \varphi_j$$

of u^* as in (2.27).

Application to Model Problem 1.12: Linear Waveform Inversion

To recall, for $T_0 > 0$ and $Z_0 = c_0 T_0 / 2$, the task is to find $f \in H^1(0, Z_0)$ from observing

$$V_d(t) = -\frac{1}{4\rho_0} \int_0^{T_0} g(t-s) f'(c_0 s/2) ds = -\frac{1}{2\sigma_0} \int_0^{Z_0} g(t-2s/c_0) f'(s) ds, \tag{2.28}$$

for $0 \le t \le T_0$, where σ_0 and c_0 are positive constants, where $g \in H^1(\mathbb{R})$ is a known function with $g(t) = 0$ for $t \le 0$ and where $f(0) = 0$. Since the determination of f from f' is straightforward, we simplify our task and only ask for f'. This means we seek the solution u^* of $Tu = w$, where

$$T : X \to Y, \quad u \mapsto w, \quad w(t) = -\frac{1}{2\sigma_0} \int_0^{Z_0} g(t - 2s/c_0)u(s)\, ds, \qquad (2.29)$$

with $X = L_2(0, Z_0)$ and $Y = L_2(0, T_0)$, which are both Hilbert spaces, when equipped with the scalar products $\langle \bullet | \bullet \rangle_X = \langle \bullet | \bullet \rangle_{L_2(0, Z_0)}$ and $\langle \bullet | \bullet \rangle_Y = \langle \bullet | \bullet \rangle_{L_2(0, T_0)}$, respectively. Since $g \in H^1(\mathbb{R})$ is continuous, so is $w = Tu$ for any $u \in X$. For a discretization, we choose a parameter $n \in \mathbb{N}$ and define

$$h := Z_0/n \quad \text{and} \quad \tau_j := jh, \ j = 0, \ldots, n.$$

As an $(n+1)$-dimensional subspace of X, we use the spline space

$$X_{n+1} := \mathscr{S}_2(\tau_0, \ldots, \tau_n)$$

from Definition 2.1. A basis of X_{n+1} is given by the linear B-splines $\varphi_j = N_{j,2}$, $j = 0, \ldots, n$, see (2.3). $T\varphi_j$ is defined by the integral values

$$T\varphi_j(t) = -\frac{1}{2\sigma_0} \int_0^{Z_0} g(t - 2s/c_0)N_{j,2}(s)\, ds$$

for $0 \le t \le T_0$. Evaluation of the scalar products

$$\langle T\varphi_i | T\varphi_j \rangle_Y = \int_0^{T_0} T\varphi_i(t) \cdot T\varphi_j(t)\, dt, \quad b_i = \langle w | T\varphi_i \rangle_Y = \int_0^{T_0} w(t)T\varphi_i(t)\, dt$$

means integration again.

Example 2.7 As a kernel function, we use the **Ricker pulse**

$$g(t) = aG(f_0 t - 1) \qquad (2.30)$$

where a is the amplitude and f_0 is the "center frequency" of the pulse. The function G is proportional to the second derivative of the Gaussian function, namely:

$$G(\theta) = (1 - 2\pi^2\theta^2)e^{-\pi^2\theta^2}, \quad \theta \in \mathbb{R}.$$

Note that $g(t) = 0$ is only approximately true for $t \le 0$. We let $c_0 = 1$, $\sigma_0 = \frac{1}{2}$, $T_0 = 1$, $a = 1$, $f_0 = 5$, and take w such that

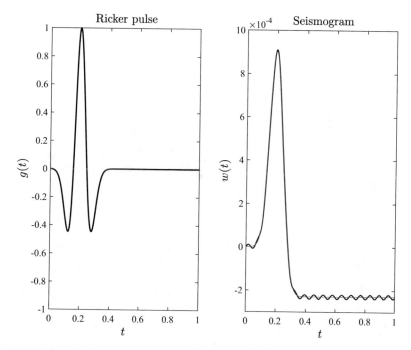

Fig. 2.4 Ricker pulse and exact and noisy seismogram

$$u^* : \left[0, \frac{1}{2}\right] \to \mathbb{R}, \quad t \mapsto 2t(1 - 2t) \tag{2.31}$$

is the exact solution of $Tu = w$. The functions $T\varphi_j$ and $w = Tu^*$ were evaluated on a fine grid with mesh size 10^{-4} by numerical integration using the trapezoidal rule. Based on these approximate values, the scalar products $\langle T\varphi_i | T\varphi_j \rangle$ and $\langle w | T\varphi_j \rangle$ were evaluated by the same numerical integration method using the same grid. Figure 2.4 shows the Ricker pulse function (to the left) and a noisy function w^δ (to the right, in black), which was constructed from the exact effect w (to the right, in red) by adding the function $10^{-5} \sin(100t)$. From w^δ values, $b_i^\delta = \langle w^\delta | T\varphi_i \rangle_Y$ were computed (by numerical integration as above) and an approximation u_n^δ of u^* was constructed as in (2.27). Figure 2.5 shows reconstructions u_n^δ (in black) we got for various values of n, as compared to the exact solution u^* (in red). Figure 2.6 shows one more reconstruction (left) and the L_2-error $\|u^* - u_n^\delta\|_X$ for various values of n (right). Visibly, the reconstruction at first gets better with n growing and then gets worse, the result for $n = 25$ already being totally useless. This behavior of discretized solutions for inverse problems is typical and will be explained below. Unluckily, we have no practical way to determine the optimal value of n (without knowing the solution of the inverse problem). $\diamond$

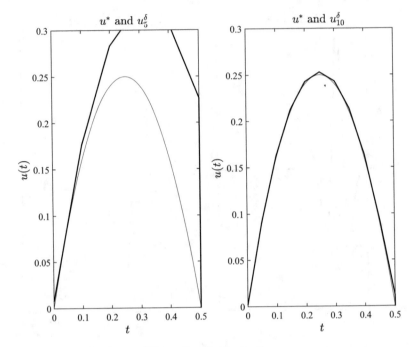

Fig. 2.5 Two reconstructions at different discretization levels

Analysis of the Method

The following theorem gives an error estimate for least squares reconstructions.

Theorem 2.8 (Error Estimates for Least Squares Method) *Let Assumption 2.6 hold. Let X_n be a d_n-dimensional subspace of X with basis $\{\varphi_1, \ldots, \varphi_{d_n}\}$ and let R_n be the linear "reconstruction operator"*

$$R_n : Y \to X_n, \quad y \mapsto u_n = \sum_{j=1}^{d_n} x_j \varphi_j, \tag{2.32}$$

where $x = (x_1, \ldots, x_{d_n})^T$ is the solution of the linear system

$$Ax = b, \quad A_{i,j} = \langle T\varphi_j | T\varphi_i \rangle_Y, \quad b_i = \langle y | T\varphi_i \rangle_Y, \tag{2.33}$$

as in (2.25). Let $w \in T(X)$ and let $u^* \in X$ be the corresponding unique solution of $Tu = w$. Let $w^\delta \in Y$ with $\|w - w^\delta\|_Y \leq \delta$ for some $\delta > 0$, then

$$\|u^* - R_n w^\delta\|_X \leq \|R_n\|\delta + \|R_n T u^* - u^*\|_X. \tag{2.34}$$

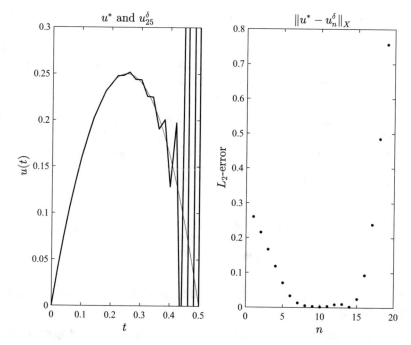

Fig. 2.6 Reconstruction errors as a function of n

Further, let b^δ be a perturbation of b, defined by $b_i = \langle w | T\varphi_i \rangle_Y$, such that $\|b^\delta - b\|_2 \leq \delta$. Let u_n^δ be defined by b^δ according to (2.27). Then, the following error estimates hold

$$\|u^* - u_n^\delta\|_X \leq \frac{a_n}{\sigma_n}\delta + \|R_n T u^* - u^*\|_X, \tag{2.35}$$

$$\|u^* - u_n^\delta\|_X \leq b_n\|R_n\|\delta + \|R_n T u^* - u^*\|_X, \tag{2.36}$$

where

$$a_n := \max\left\{ \left\|\sum_{j=1}^{d_n} \rho_j \varphi_j\right\|_X ; \ \sum_{j=1}^{d_n} |\rho_j|^2 = 1 \right\}, \tag{2.37}$$

$$b_n := \max\left\{ \sqrt{\sum_{j=1}^{d_n} |\rho_j|^2}; \ \left\|\sum_{j=1}^{d_n} \rho_j \cdot T(\varphi_j)\right\|_Y = 1 \right\}, \tag{2.38}$$

and where σ_n is the smallest singular value of the matrix A defined in (2.25).

Remark $R_n w^\delta$ is computed in the same way as $R_n w$, but with $y := w^\delta$ instead of $y := w$ in (2.33). The first error estimate (2.34) concerns a situation where the true effect w is pertubed to become w^δ. For the second error estimates (2.35) and (2.36), nothing is assumed about w^δ, errors are directly attributed to the discrete data. This is the reason why an additional factor b_n appears in (2.36).

Proof Using the identity $u^* - R_n w^\delta = u^* - R_n T u^* + R_n w - R_n w^\delta$, (2.34) easily follows from the triangle inequality. From the triangle inequality, one also gets $\|u^* - u_n^\delta\|_X \leq \|u_n^\delta - R_n w\|_X + \|R_n w - u^*\|_X$. Since $T u^* = w$, this means that only $\|u_n^\delta - R_n w\|_X$ has to be estimated in (2.35) and (2.36). Using $R_n w = \sum_{j=1}^{d_n} x_j \varphi_j$, one can write $u_n^\delta - R_n w = \sum_{j=1}^{d_n} (x_j^\delta - x_j) \varphi_j$ and estimate

$$\|u_n^\delta - R_n w\|_X \leq a_n \|x^\delta - x\|_2 = a_n \|A^{-1}(b^\delta - b)\|_2$$

$$\leq a_n \|A^{-1}\|_2 \|b^\delta - b\|_2 \leq \frac{a_n}{\sigma_n} \delta,$$

with $\|A^{-1}\|_2 \leq 1/\sigma_n$ by Theorem A.2. This shows (2.35). For the proof of (2.36), refer to [Kir11], p. 72.   □

Note that R_n is defined on Y, not only on $T(X)$, so we can reconstruct approximative causes for noisy effects $\tilde{w} \notin T(X)$. The estimates of Theorem 2.8 show that the total reconstruction error can be bounded by the sum of two terms. The first term on the right-hand sides of (2.35) and (2.36) tells us how much the discrete data error $b^\delta - b$ is amplified by the reconstruction and thus gives a measure of the robustness of the reconstruction. The factors a_n and b_n depend on the basis of X_n and can be made equal to 1 if $\{\varphi_1, \ldots, \varphi_{d_n}\}$ or $\{T\varphi_1, \ldots, T\varphi_{n_d}\}$ are chosen to be an orthonormal basis of X_n or $T(X_n)$, respectively. The second term $\|R_n T u^* - u^*\|_X$ is an estimate of the discretization error alone, disregarding data errors. It tells us how well R_n approximates the inverse of $T : X \to T(X)$. Natterer has investigated under which conditions $\|R_n T u^* - u^*\|_X$ can be made arbitrarily small, see [Nat77]. We cite one of his results from [Kir11], Theorem 3.10, omitting the proof.

Theorem 2.9 (Convergence of Least Squares Method) *Let Assumption 2.6 hold. Let $(X_n)_{n \in \mathbb{N}}$ be a sequence of d_n-dimensional subspaces of X and let $(R_n)_{n \in \mathbb{N}}$ be a corresponding sequence of reconstruction operators as in Theorem 2.8. Define*

$$\gamma_n := max\{\|z_n\|_X; \ z_n \in X_n, \ \|T z_n\|_Y = 1\}. \tag{2.39}$$

Beyond Assumption 2.6, assume that for every $x \in X$ and for every $n \in \mathbb{N}$ there exists an element $\tilde{x}_n \in X_n$ with

$$\|x - \tilde{x}_n\|_X \to 0 \quad for \quad n \to \infty \tag{2.40}$$

and also assume that there is a constant $c > 0$ such that

$$\min_{z_n \in X_n} \left\{ \|x - z_n\|_X + \gamma_n \|T(x - z_n)\|_Y \right\} \le c\|x\|_X \quad \text{for all } x \in X . \tag{2.41}$$

Under these two conditions, the reconstruction is convergent, i.e.,

$$\|R_n T u - u\|_X \to 0 \quad \text{for all } u \in X \text{ and for } n \to \infty, \tag{2.42}$$

and the operators R_n are bounded: $\|R_n\| \le \gamma_n$.

Some comments are in order.

- Equation (2.42) is called "pointwise convergence" of R_n to $T^{-1} : T(X) \to X$. As a prerequisite for this convergence to hold, we have, naturally, condition (2.40), *but we also need* the technical condition (2.41). There are in fact examples showing that (2.40) alone *is not sufficient for convergence.*
- Since every $z_n \in X_n$ can be written as $z_n = T^{-1}(y_n)$ for some $y_n \in T(X_n)$, we have

$$\gamma_n = \max\{\|T^{-1}(y_n)\|_X; \ y_n \in T(X_n), \ \|y_n\|_Y = 1\}. \tag{2.43}$$

Therefore, for $y_n, z_n \in T(X_n)$, we can estimate

$$\|T^{-1}(y_n) - T^{-1}(z_n)\|_X \le \gamma_n \|y_n - z_n\|_Y$$

and interpret γ_n as a "measure of stability" for the inverse of $T_{|X_n}$.
- The operators R_n evidently have the "projection property" $R_n T \tilde{x}_n = \tilde{x}_n$ for all $\tilde{x}_n \in X_n$. For any linear mapping $R_n : Y \to X_n$ with this quality, we have by definition of the operator norm:

$$\|R_n\| = \sup \left\{ \frac{\|R_n y\|_X}{\|y\|_Y}; \ y \in Y, \ y \ne 0 \right\} \ge \sup \left\{ \frac{\|R_n T x\|_X}{\|T x\|_Y}; \ x \in X, \ x \ne 0 \right\}$$

$$\ge \sup \left\{ \frac{\|R_n T v\|_X}{\|T v\|_Y}; \ v \in X_n, \ v \ne 0 \right\} = \sup \left\{ \frac{\|v\|_X}{\|T v\|_Y}; \ v \in X_n, \ v \ne 0 \right\}$$

$$= \gamma_n. \tag{2.44}$$

This means that among all convergent reconstruction processes having the (desirable) projection property, the least squares method is the most robust one, i.e., the one which leads to the least amplification of data errors.
- If $T^{-1} : T(X) \to X$ is not continuous (the usual case for inverse problems) then

$$\gamma_n \longrightarrow \infty \quad \text{for} \quad n \to \infty. \tag{2.45}$$

This can be seen as follows. If we had $\gamma_n \le C$ for all $n \in \mathbb{N}$, then $\|v\|_X \le C\|Tv\|_Y$ for all $v \in X_n$ and all n. Because of (2.40), we then would also have

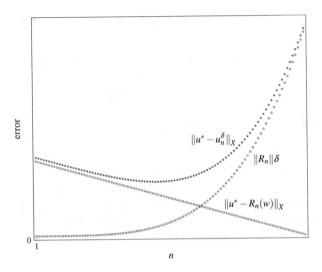

Fig. 2.7 Total reconstruction error; $T(u^*) = w$; δ, R_n and u_n^δ as in Theorem 2.8

$\|x\|_X \le C\|Tx\|_Y$ for all $x \in X$ and therefore $\|T^{-1}(y)\|_X \le C\|y\|_Y$ for all $y \in T(X)$ in contradiction to the unboundedness of $T^{-1}: T(X) \to X$.

In (the usual) case $\gamma_n \to \infty$ for $n \to \infty$, (2.44) and (2.36) tell us that the total error $\|u^* - u_n^\delta\|_X$ in general cannot become arbitrarily small for a finite value $\delta > 0$ and rather may blow up if n is chosen too big. This is illustrated in Fig. 2.7 and explains the error behavior observed in Example 2.7.

It is not always easy to apply Theorem 2.9 in practical situations. Condition (2.40) evidently is fulfilled if $X = H^1(\Omega)$ for some bounded (polygonal) domain $\Omega \subset \mathbb{R}$ or $\Omega \subset \mathbb{R}^2$, if $n \sim \frac{1}{h}$, and if X_n is chosen as a space of spline functions $\mathcal{S}_2(\mathcal{T}_h)$ as in Definition 2.1 for the one-dimensional case or in (2.18) for the two-dimensional case, compare Theorems 2.2 and 2.5. The difficulty comes from condition (2.41), which not only depends on X_n, but also on T itself. Natterer investigates the condition (2.41) in [Nat77], but his results cannot be directly applied to the operator T from Example 2.7.

2.3 Discretization of Fredholm Equations by Collocation Methods

The least squares method of Sect. 2.2 does not yet tell us how to fully discretize a linear inverse problem, since it does not specify how to compute the scalar products $\langle w | T\varphi_i \rangle_Y$ required to set up the matrix equation $Ax = b$ in (2.25). In the present section—based again on Chap. 3 of [Kir11]—we will present a fully discrete method, assuming that function samples of w are available. Although collocation methods exist for other inverse problems too, we will only discuss them in the context of linear Fredholm integral equations of the first kind. Precisely, we assume the following:

Assumption 2.10 *Let* $\mathbb{U} = X \subseteq L_2(a, b)$ *be a real Hilbert space and let* $k \in C([a, b]^2)$ *be such that*

$$T : X \to C[a, b], \quad u \mapsto w, \ w(t) = Tu(t) = \int_a^b k(t, s)u(s) \, ds \qquad (2.46)$$

is linear, continuous, and injective. Let $a \leq t_1 < t_2 < \cdots < t_m \leq b$ *and assume that samples* $w(t_i)$, $i = 1, \ldots, m$ *are given.*

Under Assumption 2.10, $Y := T(X)$ is a Hilbert space, when equipped with the scalar product

$$\langle y|z \rangle_Y := \langle T^{-1}(y)|T^{-1}(z) \rangle_{L_2(a,b)}. \qquad (2.47)$$

Assumption 2.10 and the collocation method could be generalized to the case of multi-dimensional Fredholm equations.

Description of the Method

The numbers t_i are called **collocation points**. We set up the **collocation equations**

$$Tu(t_i) = w(t_i), \quad i = 1, \ldots, m, \qquad (2.48)$$

as a substitute for equation $Tu = w$, which we actually want to solve. Choosing an n-dimensional subspace

$$X_n := \langle \varphi_1, \ldots, \varphi_n \rangle \subset X, \quad \varphi_1, \ldots, \varphi_n \text{ linearly independent,}$$

as in Sect. 2.2, one can rate an approximant of u^* (the solution of $Tu = w$) as

$$u_n = \sum_{j=1}^n x_j \varphi_j \in X_n, \quad x_j \in \mathbb{R}. \qquad (2.49)$$

Requiring (2.48) to hold, one obtains the linear system of equations

$$Tu_n(t_i) = \sum_{j=1}^n x_j \cdot T(\varphi_j)(t_i) = w(t_i), \quad i = 1 \ldots, m, \qquad (2.50)$$

which can be written in matrix form with $A \in \mathbb{R}^{m,n}$ and $b \in \mathbb{R}^m$:

$$Ax = b, \quad A_{ij} = T\varphi_j(t_i), \quad b_i = w(t_i). \tag{2.51}$$

There is no guarantee that a unique solution of (2.51) exists. In case $n > m$, (2.51) most likely is underdetermined and will have an infinite number of solutions (the data set $\{w(t_1), \ldots, w(t_m)\}$ is not sufficient to determine a unique u_n). In case $n < m$, (2.51) most likely is overdetermined and has no solution at all. We therefore will replace (2.51) by the minimization problem

$$\text{minimize } \|b - Ax\|_2, \quad x \in \mathbb{R}^n, \tag{2.52}$$

which always has a solution. If (2.51) does have a unique solution, then (2.52) will have the same unique solution. A detailed discussion of minimization problems of the form (2.52) is to be followed in Chap. 3.

No use was made so far of the special form (2.46) of T. Note that for any function $u \in L_2(a, b)$

$$Tu(t_i) = \int\limits_a^b k(t_i, s)u(s)\, ds = \langle k_i | u \rangle_{L_2(a,b)}, \quad k_i(s) := k(t_i, s), \ i = 1, \ldots, m.$$
$$\tag{2.53}$$

Since $w = Tu^*$ for the exact solution u^* of the inverse problem, the collocation equations (2.48) may equivalently be written in the form

$$\langle k_i | u^* \rangle_{L_2(a,b)} = \langle k_i | u \rangle_{L_2(a,b)} \quad \Longleftrightarrow \quad \langle k_i | u^* - u \rangle_{L_2(a,b)} = 0. \tag{2.54}$$

This situation is depicted in Fig. 2.8: The solid line symbolizes the linear subspace $\langle k_1, \ldots, k_m \rangle$ of $L_2(a, b)$, which is spanned by $k_1, \ldots, k_m$ and which, like every linear subspace, contains the zero element of $L_2(a, b)$. The dashed line symbolizes the affine subspace L of $L_2(a, b)$, which is the solution set of the collocation equations (2.48) and which, by virtue of (2.54), is orthogonal to $\langle k_1, \ldots, k_m \rangle$. As stated by Theorem B.13, there exists a unique **minimum-norm solution** $\hat{u}$ of the collocation equations, i.e., there exists a unique square integrable function $\hat{u} \in L_2(a, b)$ characterized by

$$\|\hat{u}\|_{L_2(a,b)} = \min\{\|z\|_{L_2(a,b)}; \ z \in L_2(a, b) \text{ satisfies } (2.48)\}. \tag{2.55}$$

This function $\hat{u}$ can be interpreted as the projection of u^* into the linear space spanned by $k_1, \ldots, k_m$. Any other function $\tilde{u} \in L$ (i.e., any other L_2-solution of the collocation equations) has larger norm (i.e., greater distance from 0). If $k_1, \ldots, k_m$ are linearly independent, we can find $\hat{u}$ by making the ansatz

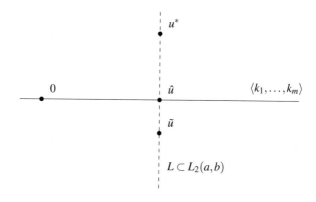

Fig. 2.8 Minimum-norm solution of collocation equations

$$\hat{u} = \sum_{j=1}^{m} x_j k_j \qquad (2.56)$$

and solving the linear system (2.51) with $n = m$ and $\varphi_j = k_j$, $j = 1, \ldots, m$. Finally, using (2.47), we can write the collocation equations (2.54) in the form

$$\langle Tk_i | Tu \rangle_Y = \langle Tk_i | w \rangle_Y, \quad i = 1, \ldots, m. \qquad (2.57)$$

This directly compares to (2.22) if $n = m$, $X_n = \langle k_1, \ldots, k_m \rangle$, and $k_1, \ldots, k_m$ are linearly independent. In this special case, the collocation method applied to linear Fredholm integral equations coincides with the least squares method.

Application to Model Problem 1.12: Linear Waveform Inversion

We take up again the problem of linear waveform inversion in the simplified form presented in Sect. 2.2, where we had

$$T : X \to Y, \quad u \mapsto w, \quad w(t) = -\frac{1}{2\sigma_0} \int_0^{Z_0} g(t - 2s/c_0)u(s)\, ds \qquad (2.29)$$

with $X = L_2(0, Z_0)$ and $Y = L_2(0, T_0)$, and where $T_0 > 0$ and $Z_0 = c_0 T_0/2$ are fixed parameters. Since $g \in H^1(\mathbb{R})$ is continuous, an effect w always is a continuous function, which can be sampled. Let us therefore assume that $m + 1$ samples

$$w(t_i) = -\frac{1}{2\sigma_0} \int_0^{Z_0} g(t_i - 2s/c_0)u(s)\, ds, \quad 0 \le t_0 < \ldots < t_m \le T_0,$$

are given, $m \in \mathbb{N}$. As observed above, we could choose the functions defined by $k_i(s) = g(t_i - 2s/c_0)$ as a basis to construct an approximate solution of $T(u) = w$ and would end up with a least squares solution. But we can also choose a parameter $n \in \mathbb{N}$, define $h = Z_0/n$ and $\tau_j := jh$, $j = 0, \ldots, n$, and work with the $(n+1)$-dimensional subspace

$$X_{n+1} := \mathscr{S}_2(\tau_0, \ldots, \tau_n) = \langle N_{0,2}, \ldots, N_{n,2} \rangle$$

of X, which is spanned by the linear B-splines $N_{j,2}$, $j = 0, \ldots, n$. The collocation equations take the form of system (2.51) with matrix $A \in \mathbb{R}^{m+1,n+1}$ having components

$$A_{i,j} = -\frac{1}{2\sigma_0} \int_0^{Z_0} g(t_i - 2s/c_0) N_{j,2}(s)\, ds, \quad i = 0, \ldots, m, \quad j = 0, \ldots, n.$$
(2.58)

The system $Ax = b$ corresponding to (2.51) and (2.58) is much simpler to set up than the corresponding system for the least squares solution, since only a single integration has to be carried out per matrix component. But we can no longer guarantee that a solution of the system exists—we have to substitute it by (2.52).

Example 2.11 Like in Example 2.7, we use the Ricker pulse defined in (2.30) as kernel function. Our choice of parameters is the same as in Example 2.7: $T_0 = 1$, $a = 1$, $f_0 = 5$, $\sigma_0 = \frac{1}{2}$, and $c_0 = 1$. Also, we choose w such that

$$u^* : \left[0, \frac{1}{2}\right] \to \mathbb{R}, \quad t \mapsto 2t(1 - 2t)$$
(2.31)

is the exact solution of $T(u) = w$. We fix the value $m = 20$ and use

$$t_i := i\frac{T_0}{m}, \quad i = 0, \ldots, m$$

as collocation points. Further, we let $n \in \mathbb{N}$ and define

$$h := \frac{1}{2n}, \quad \tau_i := hi, \quad i = 0, \ldots, n,$$

to define the spline space $X_{n+1} := \mathscr{S}_2(\tau_0, \ldots, \tau_n)$. With these values, the components of $A \in \mathbb{R}^{m+1,n+1}$ in (2.58) were computed by exact (and not by numerical) integration. A perturbed version w^δ of w was constructed just as in Example 2.7. It defines samples $b_i^\delta = w^\delta(t_i)$, $i = 0 \ldots, m$. A solution $x^\delta \in \mathbb{R}^{n+1}$ of the minimization problem (2.52) with b replaced by b^δ was determined, which defines the spline function $u_n^\delta(t) = \sum_{j=0}^n x_j^\delta N_{j,2}(t)$ approximating u^*. In Fig. 2.9, the best reconstruction result is shown (in black, as compared to the exact solution, shown in red), which was obtained for $n = 9$ and which is of comparable quality

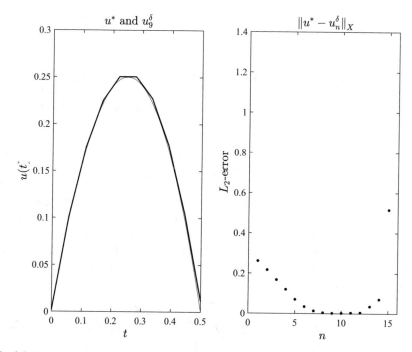

Fig. 2.9 Best collocation reconstruction and reconstruction error as a function of n

to the best reconstruction achieved by the least squares method. We also plot the reconstruction error versus the discretization level parameterized by n. The behavior is quite similar to the one observed for the least squares method. Again, we have no practical way to determine the optimal parameter n. But choosing n too small or too large will produce an unacceptable result.   ◊

Analysis of the Method

For an analysis, especially concerning error estimates analogous to Theorem 2.8, see Theorems 3.21 and 3.24 in [Kir11]. The analysis confirms a similar qualitative behavior of the error as for the least squares method. This was observed in Example 2.11—see the error plot in Fig. 2.9.

2.4   The Backus–Gilbert Method and the Approximative Inverse

This method will be presented in the context of Fredholm integral equations of the first kind, following Sect. 3.6 of [Kir11]. Assumption 2.10 is made again. Using the functions $k_i \in C[a, b]$ introduced in (2.53), let us define the linear operator

$$K : L_2(a, b) \to \mathbb{R}^m, \quad u \mapsto (y_1, \ldots, y_m), \quad y_i := \int_a^b k_i(s)u(s)ds, \ i = 1, \ldots, m,$$

(2.59)

mapping a cause u to a finite number of samples of its effect w. The collocation equations (2.48) can be written compactly in form $Ku = y$, where $y = (w(t_1), \ldots, w(t_m))$. If all L_2-functions are admitted as candidates for a solution of these equations, an additional requirement is needed to make a solution unique. This could be a minimum-norm requirement, as in (2.55). A different idea is to look for a linear operator of the form

$$S : \mathbb{R}^m \to L_2(a, b), \quad y \mapsto Sy, \quad Sy(t) = \sum_{j=1}^m y_j \psi_j(t),$$

(2.60)

which "approximately inverts" K such that—in a sense to be made precise

$$SKx \approx x \quad \text{for all } x \in L_2(a, b).$$

(2.61)

Here, $\psi_1, \ldots, \psi_m$ are fixed functions to be defined pointwise (and not as functions in $L_2(a, b)$) such that (2.60) is meaningful. The (yet imprecise) "inversion requirement" (2.61) replaces the collocation equations. From the definitions of S and K, we get

$$x_m(t) := SKx(t) = \sum_{j=1}^m \psi_j(t) \int_a^b k_j(s)x(s)\, ds = \int_a^b \underbrace{\left(\sum_{j=1}^m k_j(s)\psi_j(t) \right)}_{=:\, \varphi(s, t)} x(s)\, ds.$$

(2.62)

Depending on how the magnitude of the difference $SKx - x$ is measured, different answers will be given as for how to choose well the functions ψ_j. One idea is to require a pointwise approximation of x by SKx, i.e., to require

$$SKx(t) = x_m(t) \approx x(t).$$

(2.63)

Ideally, this should hold for all $t \in [a, b]$ and "for all" x. However, it certainly makes no sense to require (2.63) for all $x \in L_2(a, b)$, since point values are not defined for

L_2-functions. A precise formulation of what (2.63) actually shall mean will only be given below. For the moment, one might think of (2.63) as demanded for continuous functions x only. If (2.63) was to hold, then according to (2.62) a point value $x(t)$ should be produced from averaging the values $x(s)$ for $a \leq s \leq b$—which is why φ is called an **averaging kernel**. It is clear that $SKx(t) \approx x(t)$ can be expected to hold in good approximation at some point t, if the averaging kernel is normalized such that $\int_a^b \varphi(s, t)\, ds = 1$, if it is sharply peaked at $s = t$ and if it is close to zero for $s \neq t$. The less $\varphi(\cdot, t)$ corresponds to such a function, the less $\int_a^b \varphi(s, t)x(s)\, ds$ can be expected to produce a good approximation $x(t)$ for all admissible functions x at the same time. Of course, since ψ_j, $j = 1, \ldots, m$, are functions of t, we will get a new averaging kernel for every value of t.

Description of the Backus–Gilbert Method

The Backus–Gilbert method does not determine *functions* ψ_j, but rather *values* $\psi_j(t)$, $j = 1, \ldots, m$, for some *fixed* parameter $t \in [a, b]$, such that $SKx(t) \approx x(t)$ *at this point* t. The process can then be repeated any number of times to determine values $\psi_j(\bar{t})$ for other parameters $\bar{t} \in [a, b]$ in order to make $SKx(\bar{t}) \approx x(\bar{t})$ hold. So let us now keep $t \in [a, b]$ fixed. To find $v_j := \psi_j(t)$, $j = 1, \ldots, m$ (for fixed t), one solves the minimization problem

$$\text{minimize} \quad \int_a^b (s - t)^2 \varphi(s, t)^2\, ds, \quad \text{where} \quad \varphi(s, t) = \sum_{j=1}^{m} v_j k_j(s), \qquad (2.64)$$

subject to the linear constraint

$$\int_a^b \varphi(s, t)\, ds = \int_a^b \sum_{j=1}^{m} v_j k_j(s)\, ds = 1. \qquad (2.65)$$

The normalization condition (2.65) excludes the trivial solution $v = 0$ of (2.64). Together, (2.64) and (2.65) are a mathematical formulation of the qualities desired for φ, as discussed after (2.62). This minimization problem replaces the vague approximate identity (2.63) and can be written in compact form as

$$\text{minimize} \quad v^\top Q(t)v \quad \text{subject to} \quad v^\top c = 1, \qquad (2.66)$$

where $Q(t) \in \mathbb{R}^{m,m}$ and $c \in \mathbb{R}^m$ have components

$$Q_{ij}(t) = \int_a^b (s-t)^2 k_i(s) k_j(s)\, ds, \quad i, j = 1, \dots, m,$$

$$c_i = \int_a^b k_i(s)\, ds, \quad i = 1, \dots, m. \tag{2.67}$$

If $c \neq 0$ and if the functions k_i are linearly independent, then the matrix $Q(t)$ is positive definite for every $t \in [a, b]$ and problem (2.66) does have a unique solution v, see Theorem 3.31 in [Kir11]. The solution can be computed analytically using the method of Lagrange multipliers. Since minimization of $v^\top Q(t)v$ and $v^\top Q(t)v/2$ is equivalent, $L(v, \mu) = \frac{1}{2}v^\top Q(t)v + \mu v^\top c$ can be used as a Lagrange function. Its gradient with respect to v is $Q(t)v + \mu c$. Setting the gradient to zero gives $v = -\mu Q(t)^{-1}c$. Multiplying this term with $c^\top$ from the left and making use of $c^\top v = 1$ give $\mu = -1/(c^\top Q(t)^{-1}c)$. The solution then reads

$$v = \frac{Q(t)^{-1}c}{c^\top Q(t)^{-1}c}, \tag{2.68}$$

but is *not* to be computed by inverting $Q(t)$, but by solving $Q(t)z = c$ for z. The above derivation also shows that v is determined by the linear system

$$\begin{pmatrix} Q(t) & c \\ c^\top & 0 \end{pmatrix} \begin{pmatrix} v \\ \mu \end{pmatrix} = \begin{pmatrix} 0 \\ 1 \end{pmatrix} \in \mathbb{R}^{m+1}. \tag{2.69}$$

Having determined v and knowing $y := (w(t_1), \dots, w(t_m))$, one computes

$$u_m(t) := \sum_{j=1}^m y_j v_j = \int_a^b \varphi(s, t)u(s)\, ds. \tag{2.70}$$

As already mentioned, the whole process can be repeated for any other $t \in [a, b]$. Every time, a new matrix $Q(t)$ has to be compiled and a new linear system of equations has to be solved. This means a high computational effort but can theoretically produce an approximation u_m of u which is defined on the whole interval $[a, b]$. In [Kir11], Lemma 3.33, it is shown that the function u_m is analytic and therefore infinitely often differentiable.

Application to Model Problem 1.12: Linear Waveform Inversion

Let us consider again the problem of linear waveform inversion in the simplified form presented in Sect. 2.2, where we had

$$T : X \to Y, \quad u \mapsto w, \quad w(t) = -\frac{1}{2\sigma_0} \int_0^{Z_0} g(t - 2s/c_0)u(s)\, ds \qquad (2.29)$$

with $X = L_2(0, Z_0)$ and $Y = L_2(0, T_0)$, and where $T_0 > 0$ and $Z_0 = c_0 T_0/2$ are fixed parameters. As noted before, $g \in H^1(\mathbb{R})$ implies $T(X) \subset C[a, b]$. Assume that $m \in \mathbb{N}$ samples

$$y_i := w(t_i) = -\frac{1}{2\sigma_0} \int_0^{Z_0} g(t_i - 2s/c_0)u(s)\, ds, \quad 0 < t_1 < \ldots < t_m < T_0$$

are given. According to (2.29), $k_i(s) = -g(t_i - 2s/c_0)/(2\sigma_0)$, $i = 1, \ldots, m$. For fixed $t \in [a, b]$, one can compute the point value $u_m(t) = \sum_{j=1}^m v_j y_j$ of an approximation u_m of u^*, where $v \in \mathbb{R}^m$ is determined by (2.68).

Example 2.12 We take up Example 2.7, using the same parameters T_0, a, f_0 to define the Ricker pulse g, using the same parameters σ_0 and c_0, and also using the same functions u^* and w. Let us define $y := (w(t_1), \ldots, w(t_m))$ (this computation was done by exact integration), fix $m = 20$, and let $a = 0$, $b = Z_0 = c_0 T_0/2$ and $k_i(s) = -g(t_i - 2s/c_0)/(2\sigma_0)$, where $t_i = iT_0/m$, $i = 1, \ldots, m$.[3] The exact solution u^* was approximated at $n = 30$ equidistant locations $\tau_j = jZ_0/n$, $j = 1, \ldots, n$, by computing $Q(t) \in \mathbb{R}^{m,m}$ (numerical integration by the trapezoidal rule on a fine grid with mesh size 10^{-4}) and $c \in \mathbb{R}^m$ (exact integration) for every $t = \tau_j$, solving (2.69), and then computing $u_m(t) = \sum_j y_j v_j$. The graph of u_m (with $u_m(0) := 0$) is shown in the left part of Fig. 2.10 (in black) as compared to the graph of u^* (in red). The reconstruction is good except at the right border of the interval $[0, Z_0]$. To see why this is so, note again from considering (2.70) that the averaging kernel $\varphi(s, t) = \sum_j v_j k_j(s)$ is required to be sharply peaked at $s = t$ and to be approximately zero elsewhere, if the reconstruction is to be good at $s = t$. But since all functions $k_j(s)$, $j = 1, \ldots, m$, vanish at $t = Z_0$, we cannot expect that $\varphi(s, t)$ has this quality near $t = Z_0$. As shown in Fig. 2.10 (to the right) the averaging kernel computed for $t = Z_0/2 = 0.25$ (orange line) is peaked as and where it should be, but not so the averaging kernel for $t = Z_0 = 0.5$ (green line).   ◊

---

[3]We cannot use $t_0 = 0$ as a sampling point here, since the function k_0 equals 0, so $Q(t)$ would not be positive definite.

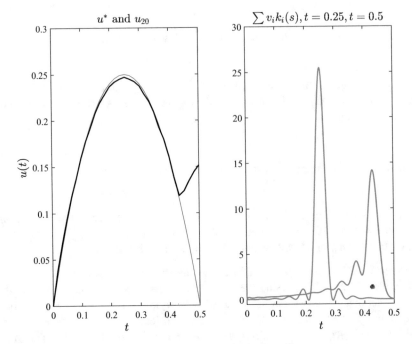

Fig. 2.10 Backus–Gilbert reconstruction and two averaging kernels

Analysis of the Method

Error estimates and convergence results for the Backus–Gilbert method can be found
in Theorems 3.34 and 3.36 of [Kir11]. The approximation error $u^* - u_m$ can only
be bounded if

1. u^* has some degree of smoothness. In Theorem 3.34 of [Kir11], Lipschitz
   continuity is required or, alternatively, $u^* \in H^1(a, b)$.
2. The space

$$X_m = \langle k_1, \dots, k_m \rangle$$

spanned by the Fredholm kernel functions k_i must allow the construction of
"good" averaging kernels. More precisely,

$$e_m(t) := \min \left\{ \int_a^b (s - t)^2 \psi(s)^2 \, ds; \ \psi \in X_m, \ \int_a^b \psi(s) \, ds = 1 \right\}$$

must be bounded (uniformly in t).

The difficulties observed in Example 2.12 can be explained by the fact that near $t \approx Z_0$ the functions $k_1, \ldots, k_m$ cannot be combined to an averaging kernel sharply peaked at t and therefore $e_m(t)$ becomes large for these values of t.

The Approximative Inverse

Assumption 2.10 is still made. The Backus–Gilbert method was already interpreted as a method to approximately invert the operator K from (2.59), but the term "Approximative Inverse" usually is meant to designate a different, although related method. We describe it following [Lou96], restricting Louis' much more general approach to Fredholm integral equations. The same ideas were previously applied in the context of computerized tomography, refer to the very readable presentation in [Nie86]. To start with, define for a fixed parameter $t \in \mathbb{R}$ the function

$$e_\gamma(\cdot, t) : \mathbb{R} \to \mathbb{R}, \quad s \mapsto e_\gamma(s, t) = \begin{cases} 1/\gamma, & |s - t| \le \gamma/2 \\ 0, & \text{else} \end{cases}, \tag{2.71}$$

where $\gamma > 0$ is another parameter. The integral

$$\int_a^b e_\gamma(s, t) u(s)\, ds \quad = \quad \langle u | e_\gamma(\cdot, t) \rangle_{L_2(a,b)},$$

which can be computed for any function $u \in L_2(a, b)$, gives a mean value of u, locally averaged in a neighborhood of t, the size of which is controlled by γ. It defines a continuous function

$$u_\gamma : [a, b] \to \mathbb{R}, \quad t \mapsto \langle u | e_\gamma(., t) \rangle_{L_2(a,b)},$$

which can be considered a smoothed approximation of u. For this reason, e_γ commonly is called a **mollifier**. Further, let us introduce the operator

$$K^* : \mathbb{R}^m \to L_2(a, b), \quad y \mapsto K^* y, \quad K^* y(s) = \sum_{i=1}^m y_i k_i(s), \tag{2.72}$$

which is called **adjoint operator** with respect to K, since

$$\langle Ku | y \rangle \quad = \quad \langle u | K^* y \rangle_{L_2(a,b)} \quad \text{for all} \quad u \in L_2(a, b) \text{ and } y \in \mathbb{R}^m, \tag{2.73}$$

where $\langle \bullet | \bullet \rangle$ on the left-hand side means the Euclidean scalar product. Assume one could find a solution $v_\gamma \in \mathbb{R}^m$ of the equation

$$K^*v = e_\gamma(\cdot, t), \quad v \in \mathbb{R}^m, \tag{2.74}$$

and then, by virtue of (2.73), one would get

$$u_\gamma(t) = \langle u|e_\gamma(\cdot, t)\rangle_{L_2(a,b)} = \langle u|K^*v_\gamma\rangle_{L_2(a,b)} = \langle Ku|v_\gamma\rangle = \langle y|v_\gamma\rangle = \sum_{i=1}^m y_i v_{\gamma,i},$$

whenever $Ku = y$. This means that one could construct a smooth approximation u_γ of an exact solution u^* of $Ku = y$, with $u_\gamma(t) \to u^*(t)$ for $\gamma \to 0$ and $t \in]a, b[$, if u^* is continuous. It is unlikely, however, that a solution of equation (2.74) exists, since the range of K^* is an at most m-dimensional subspace of $L_2(a, b)$. Alternatively, one determines a vector $v = v_\gamma \in \mathbb{R}^m$ such that $\|K^*v - e_\gamma(\cdot, t)\|_{L_2(a,b)}$ becomes minimal. If the functions $k_i \in L_2(a, b)$, $i = 1, \ldots, m$, are linearly independent, then it follows from Theorem B.13 (the projection theorem) that this optimization problem has a unique solution, which can be determined by solving the equations

$$\langle K^*v - e_\gamma(\cdot, t)|k_i\rangle_{L_2(a,b)} = 0, \quad i = 1, \ldots, m.$$

An equivalent formulation is

$$Av = b(t), \tag{2.75}$$

where $A \in \mathbb{R}^{m,m}$ is a positive definite matrix with components

$$a_{ij} = \int_a^b k_i(s)k_j(s)\, ds, \quad i, j = 1, \ldots, m, \tag{2.76}$$

independent of t, and where $b(t) \in \mathbb{R}^m$ has components

$$b_i(t) = \int_a^b e_\gamma(s, t)k_i(s)\, ds, \quad i = 1, \ldots, m. \tag{2.77}$$

Formally, one may define an operator

$$S_\gamma : \mathbb{R}^m \to L_2(a, b), \quad y \mapsto x_\gamma, \quad x_\gamma(t) = \sum_{i=1}^m y_i v_{\gamma,i}, \quad v_\gamma = A^{-1}b(t), \tag{2.78}$$

which is called **approximative inverse** of K. It can be used to compute an approximation

$$\tilde{u}_\gamma = S_\gamma y \approx u_\gamma \approx u^*$$

of a solution u^* of $Ku = y$. A prerequisite for A^{-1} to exist is the linear independence of the functions k_i. Technically, one chooses $\gamma > 0$ and $t \in [a, b]$ and then carries out the following three steps:

(1) Compute the positive definite matrix $A \in \mathbb{R}^{m,m}$ defined in (2.76) and the vector $b(t) \in \mathbb{R}^m$ defined in (2.77).
(2) Find the solution $v_\gamma \in \mathbb{R}^m$ of the linear system (2.75).
(3) Compute $\tilde{u}_\gamma(t) = \sum_{i=1}^m y_i v_{\gamma,i} \approx u_\gamma(t)$.

This procedure can be repeated for any other parameter t, similar to the method of Backus and Gilbert. There are two differences, though. First, one no longer tries to recover u itself, but rather a smoothed (or "mollified") approximation of it. Second, the parameter t *does not enter the matrix A* (which can be computed and factorized once and for all), but only into the right-hand side $b(t)$ of (2.75). Thus, the computational burden is much reduced. The method can be generalized with respect to the choice of e_γ, see [Lou96].

Example 2.13 Linear waveform inversion as in Example 2.12 is reconsidered, with all parameter values retained. Reconstructions were performed at $n = 30$ equidistant locations $t = \tau_j = jZ_0/n$, $j = 0, \ldots, n$, using the approximate inverse as in (2.78), with $\gamma = 0.05$. The matrix A and the right-hand side $b(t)$ were computed by numerical integration (trapezoidal rule on a fine grid with mesh size 10^{-4}). Figure 2.11, to the left, shows the function u^* (in red), its mollified approximation u_γ (in green), and the achieved reconstruction $\tilde{u}_\gamma$ (in black). To see why the reconstruction quality diminishes at the right boundary, look at Eq. (2.74), which the solution v_γ of (2.75) has to satisfy approximately, if $\tilde{u}_\gamma$ is expected to approximate u_γ well. This means that $\sum_i v_{\gamma,i} k_i$ must be a good approximation of $e_\gamma(\cdot, t)$. In the right picture, we illustrate $e_\gamma(\cdot, t)$ (solid black line) and its (rather good) approximation $\sum_i v_{\gamma,i} k_i$ (orange line) for $t = 0.25$. For $t = 0.4667$, near the right border, $e_\gamma(\cdot, t)$ (dashed black line) and $\sum_i v_{\gamma,i} k_i$ (green line) differ grossly. The difficulties caused by Fredholm kernel functions k_i vanishing near the right boundary, which already showed up for the Backus–Gilbert method, persist. ◊

2.5 Discrete Fourier Inversion of Convolutional Equations

In this section, we make use of the definitions and results presented in Appendix C. Many linear inverse problems take the form of convolutional Fredholm equations as introduced—in the one-dimensional case—in (1.10). As a multi-dimensional

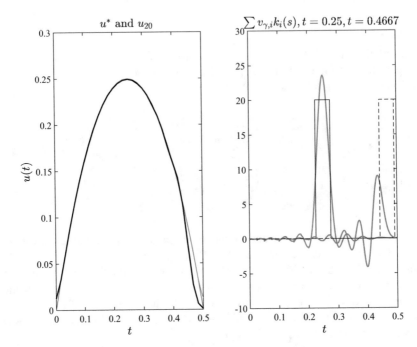

Fig. 2.11 Reconstruction by the approximate inverse and two averaging kernels

example, we consider Problem 1.10 of inverse gravimetry. To recall, one is asked to solve the equation

$$w(x_1, x_2) = \int_{-a}^{a} \int_{-a}^{a} k(x_1 - y_1, x_2 - y_2) u(y_1, y_2), \, dy_2 \, dy_1 \tag{2.79}$$

where $k \in L_2(\mathbb{R}^2)$ is defined by

$$k(u, v) := \left(u^2 + v^2 + u_0^2 \right)^{-3/2}, \quad u, v \in \mathbb{R}, \quad u_0 > 0. \tag{2.80}$$

The continuous function $w : \mathbb{R}^2 \rightarrow \mathbb{R}$ is known for $(x_1, x_2) \in [-b, b]^2$, and the solution function $u^* : [-a, a]^2 \rightarrow \mathbb{R}$ of (2.79) is continuous, too. Any $u \in C([-a, a]^2)$ can be extended to become an $L_2(\mathbb{R}^2)$-function by setting $u(x) := 0$ for $x \notin [-a, a]^2$. With this extension, the domain of integration in (2.79) formally can be extended to $\mathbb{R}^2$, such that (2.79) can be written as a two-dimensional convolution:

$$w(x_1, x_2) = \int_{-\infty}^{\infty} \int_{-\infty}^{\infty} k(x_1 - y_1, x_2 - y_2) u(y_1, y_2) \, dy_2 \, dy_1. \tag{2.81}$$

In the same way, convolutions are defined in the general case of s-variate L_2-functions (i.e., of functions having s arguments), see (C.7), where the usual notation

$$w = k * u, \quad u, k, w \in L_2(\mathbb{R}^s) \tag{2.82}$$

is used to express (2.81) (with $s = 2$). We already saw in (1.73), that Model Problem 1.12 (i.e., linearized waveform inversion) can also be cast in this form (with $s = 1$). From (C.8), one gets for the Fourier transform of w:

$$w(x) = (k * u)(x) \circ\!\!-\!\bullet \widehat{k}(y)\widehat{u}(y) \quad \text{for} \quad k(x) \circ\!\!-\!\bullet \widehat{k}(y), \ u(x) \circ\!\!-\!\bullet \widehat{u}(y).$$

Thus, under the condition that the Fourier transformed kernel function

$$\widehat{k}(y) = \int\limits_{\mathbb{R}^s} k(x)e^{-2\pi i x \cdot y} \, dy$$

has no zeros, a principle possibility to solve (2.82) for u consists in performing the following three steps, which is what is meant by "Fourier inversion of convolutional equations":

(1) Compute the Fourier transform $\widehat{w}$ of w.
(2) Compute $\widehat{u} = \widehat{w}/\widehat{k}$.
(3) Compute the inverse Fourier transform u of $\widehat{u}$.

Fourier inversion is not possible for the linearized model problem of waveform inversion, since the (one-dimensional) Fourier transform of the Ricker pulse $k := g$ with g defined in (2.30) is given by

$$\widehat{k}(y) = \frac{2ay^2}{\sqrt{\pi} f_0^3} e^{-y^2/f_0^2 - 2\pi i y/f_0} \tag{2.83}$$

and *does* have a zero at frequency $y = 0$. The situation is different for the linearized problem of inverse gravimetry: For k as defined in (2.80), one gets the two-dimensional Fourier transform

$$\widehat{k}(y) = \widehat{k}(y_1, y_2) = \frac{2\pi}{u_0} e^{-2\pi u_0 \sqrt{y_1^2 + y_2^2}}, \tag{2.84}$$

which has no zeroes. Nevertheless, the division $\widehat{w}/\widehat{k}$ means to multiply high-frequency components of w by huge values. This will have a disastrous effect when only a noisy approximation $\widetilde{w}$ of w is known, because at high frequencies the spectrum of noisy data usually is dominated by noise. Noise will therefore blow up catastrophically when $\widehat{w}/\widehat{k}$ is computed instead of $\widehat{w}/\widehat{k}$. It is the task of regularization to cope with this problem, as detailed in the next chapter. For the moment, we only ask how Fourier inversion can be discretized.

Description of the Method

With an intended application to the model problem of linearized gravimetry in mind, a discretization of Fourier inversion will only be discussed for bivariate functions. The one-dimensional case is briefly summarized in Appendix C, whereas a generalization to spaces of arbitrary dimension $s \in \mathbb{N}$, albeit possible, is omitted.

According to what was said after (2.79), we think of Fourier inversion (for the linearized gravimetry model problem) as a method to reconstruct a function

$$u \in C(\overline{Q}) \cap L_2(\mathbb{R}^2) \cap L_1(\mathbb{R}^2), \quad Q =]-a, a[^2, \quad a > 0,$$

vanishing outside $\overline{Q}$. Such a function can be approximated by a bilinear spline function. Let $N \in \mathbb{N}$ be an even number and let

$$h := \frac{2a}{N} \quad \text{and} \quad W := \left\{ \alpha \in \mathbb{Z}^2; \ -\frac{N}{2} \leq \alpha_j < \frac{N}{2}, \ j = 1, 2 \right\}. \tag{2.85}$$

For samples of u on an equidistant grid, let us write

$$u_\alpha := u(x_\alpha), \quad \text{where} \quad x_\alpha = (h\alpha_1, h\alpha_2), \ \alpha \in W. \tag{2.86}$$

A bilinear B-spline adapted to the grid points x_α is defined by

$$\Phi(x) := B_2(x_1) \cdot B_2(x_2), \quad x = (x_1, x_2) \in \mathbb{R}^2, \tag{2.87}$$

based in turn on the univariate B-spline B_2 defined by

$$B_2(t) := \begin{cases} t + 1, & -1 \leq t \leq 0 \\ 1 - t, & 0 \leq t \leq 1 \\ 0, & \text{else} \end{cases},$$

as in (C.11). See also Sect. 2.1, where the different notation $N_{i,j,2}$ was used for bilinear B-splines with non-equidistant knots and see Fig. 2.3 for an illustration. A bilinear spline interpolant of u is given by

$$u_N(x) := \sum_{\alpha \in W} u_\alpha \Phi(x/h - \alpha), \quad x \in \mathbb{R}^2. \tag{2.88}$$

This interpolant can be Fourier transformed *exactly*:

$$\widehat{u_N}(y_1, y_2) = h^2 \left(\frac{\sin(\pi h y_1)}{\pi h y_1}\right)^2 \left(\frac{\sin(\pi h y_2)}{\pi h y_2}\right)^2 \cdot \sum_{\alpha \in W} u_\alpha e^{-2\pi i h(y_1\alpha_1 + y_2\alpha_2)}.$$

(2.89)

For $y = \beta/(2a)$, $\beta \in \mathbb{Z}^2$, one gets

$$\widehat{u_N}\left(\frac{\beta}{2a}\right) = \sigma_\beta \cdot \underbrace{\left(\frac{1}{N}\right)^2 \sum_{\alpha \in W} u_\alpha e^{-2\pi i(\alpha_1\beta_1 + \alpha_2\beta_2)/N}}_{=: U_\beta}$$

(2.90)

with data independent so-called **attenuation factors**

$$\sigma_\beta := 4a^2 \left(\frac{\sin(\pi\beta_1/N)}{\pi\beta_1/N}\right)^2 \cdot \left(\frac{\sin(\pi\beta_2/N)}{\pi\beta_2/N}\right)^2, \quad \beta \in \mathbb{Z}^2,$$

(2.91)

and values U_β, which need only be computed for $\beta \in W$ because of their periodicity. In fact, for an arbitrary index $\beta \in \mathbb{Z}^2$, there exists a unique index $\gamma \in W$ such that $\beta = \gamma + N\alpha$ for some $\alpha \in \mathbb{Z}^2$, and $U_\beta = U_\gamma$. The computation of values U_β, $\beta \in W$, from values u_α, $\alpha \in W$, is called (two-dimensional) **discrete Fourier transform (DFT)**. Inversely, the values u_α can be computed from U_β by

$$u_\alpha = \sum_{\beta \in W} U_\beta e^{+2\pi i(\alpha_1\beta_1 + \alpha_2\beta_2)/N}, \quad \alpha \in W,$$

(2.92)

which is the two-dimensional **inverse discrete Fourier transform (IDFT)**. We will use the notation

$$\{u_\alpha\}_{\alpha \in W} \,{\circ}\!\!-\!\!\bullet\, \{U_\beta\}_{\beta \in W} \quad \text{and} \quad \{U_\beta\}_{\beta \in W} \,\bullet\!\!-\!\!{\circ}\, \{u_\alpha\}_{\alpha \in W}$$

to express DFT and IDFT. If N is a power of two, the two-dimensional DFT and IDFT can both be computed with $\mathcal{O}(N^2 \log(N))$ arithmetical operations by the two-dimensional FFT algorithm. Refer to [PTVF92], pp. 521 ff. for a detailed description of this algorithm. In case one does not dispose of equidistant samples of u or in case one wants to compute non-equidistant samples of its Fourier transform $\widehat{u}$, there also exist (more complex) efficient algorithms. Appendix C explains one approach for the one-dimensional case and gives a pointer to the literature for the multi-dimensional case.

Formula (2.90) can also be used "backwards," by which we mean the following: assume that one does not know $u \in C(\overline{Q}) \cap L_2(\mathbb{R}^2) \cap L_1(\mathbb{R}^2)$, but rather its Fourier transform $\widehat{u}$ (which necessarily is a continuous function). Then an approximant u_N of u can be constructed by setting

$$U_\beta := \frac{\hat{u}\,(\beta/2a)}{\sigma_\beta}, \quad \beta \in W, \tag{2.93}$$

and by computing

$$\{U_\beta\}_{\beta \in W} \circ\!\!-\!\!\bullet \{u_\alpha\}_{\alpha \in W}, \quad u_N(x) = \sum_{\alpha \in W} u_\alpha \Phi(x/h - \alpha). \tag{2.94}$$

This amounts to adjusting u_N such that the Fourier interpolation conditions

$$\widehat{u_N}\left(\frac{\beta}{2a}\right) = \hat{u}\left(\frac{\beta}{2a}\right), \quad \beta \in W, \tag{2.95}$$

hold. We remark that because of $\sin(x)/x \geq 2/\pi$ for $|x| \leq \pi/2$, there is no danger of a division by zero in (2.93). We propose the following method for an approximate solution of (2.82).

Approximate Inversion of Convolutional Equation $w = k * u$.

- Compute via DFT:

$$\{w_\alpha\}_{\alpha \in W} \circ\!\!-\!\!\bullet \{W_\beta\}_{\beta \in W} \quad \text{where} \quad w_\alpha := w(x_\alpha), \ \alpha \in W. \tag{2.96}$$

- Compute

$$U_\beta := \frac{W_\beta}{\widehat{k}(\beta/2a)}, \quad \beta \in W. \tag{2.97}$$

- Compute via IDFT:

$$\{U_\beta\}_{\beta \in W} \bullet\!\!-\!\!\circ \{u_\alpha\}_{\alpha \in W} \tag{2.98}$$

and take

$$u_N(x) = \sum_{\alpha \in W} u_\alpha \Phi(x/h - \alpha) \tag{2.99}$$

as an approximant of u.

The same method can be used in the one-dimensional case, using a one-dimensional grid $W = \{-N/2, \ldots, N/2 - 1\}$. The idea behind the first step is to take

$$w_N(x) = \sum_{\alpha \in W} w_\alpha \Phi(x/h - \alpha), \quad w_\alpha = w(x_\alpha), \ \alpha \in W \tag{2.100}$$

as an approximant of w and make use of its Fourier transform values

$$\widehat{w_N}\left(\frac{\beta}{2a}\right) = \sigma_\beta W_\beta \approx \widehat{w}\left(\frac{\beta}{2a}\right). \tag{2.101}$$

The idea behind the last two steps is to determine the coefficients u_α of an approximant

$$u_N(x) = \sum_\alpha u_\alpha \Phi(x/h - \alpha) \approx u(x)$$

such that

$$\widehat{k}\left(\frac{\beta}{2a}\right)\widehat{u}\left(\frac{\beta}{2a}\right) \approx \widehat{k}\left(\frac{\beta}{2a}\right)\widehat{u_N}\left(\frac{\beta}{2a}\right) = \widehat{k}\left(\frac{\beta}{2a}\right)\sigma_\beta U_\beta = \sigma_\beta W_\beta =$$

$$= \widehat{w_N}\left(\frac{\beta}{2a}\right) \approx \widehat{w}\left(\frac{\beta}{2a}\right), \quad \beta \in W.$$

Application to Model Problem 1.10 of Inverse Gravimetry

It was already found that Fourier inversion is (mathematically) viable for Problem 1.10, since the Fourier transform of the kernel function has no zeros, see (2.84). To find an approximant u_N of u as in (2.99), one has to (approximately) know the values

$$\widehat{u}\left(\frac{\beta}{2a}\right) = \left[\widehat{k}\left(\frac{\beta}{2a}\right)\right]^{-1}\widehat{w}\left(\frac{\beta}{2a}\right), \quad \beta \in W. \tag{2.102}$$

Fourier transform values of k can be computed analytically. Approximations of $\widehat{w}(\beta/(2a))$ could be computed according to (2.96) if the values $w(x_\alpha)$, $\alpha \in W$, were known. For the sake of simplicity, we will actually assume that they *are* known, thereby also presuming that $b = a$ in the setting of Problem 1.10. In practice, we cannot rely on our good luck to have just the values $w(x_\alpha)$, $\alpha \in W$, available. It is still possible to efficiently compute $\widehat{w}(\beta/(2a))$, $\beta \in W$, from non-equidistantly sampled function values of w, even if their number is not equal to N^2. One such approach was developed by Fourmont, see [Fou99] and Appendix C.

The factors $1/\widehat{k}(\beta/2a)$ are exponentially growing with $\|\beta\|_2$, and small errors in the computed values $\widehat{w}(\beta/2a)$ will therefore be catastrophically amplified in

(2.102). Without regularization, formula (2.102) is practically useless. For this reason, numerical results for Fourier inversion will only be presented after introducing a regularized reconstruction in Sect. 3.8.

Analysis of the Method

A detailed investigation of the approximation errors associated with the discretized Fourier reconstruction process is given in Appendix C. It turns out that u can be approximated well by u_N if the Fourier interpolation condition (2.95) holds (the corresponding error is estimated in Lemma C.4) and $\widehat{w}$ can also be approximated well by $\widehat{w}_N$ if w_N is chosen as a bilinear spline interpolant of w (the corresponding error is estimated in Lemma C.3). The essential problem with Fourier inversion is the division step (2.97): whenever k is a smooth, rapidly decaying function, then $\widehat{k}$ also decays rapidly for large arguments; even small errors in the computed coefficients W_β are then grossly magnified. This difficulty will be addressed in Sect. 3.8.

2.6   Discretization of Nonlinear Inverse Gravimetry

There are many ways to discretize nonlinear problems, and there is no generally recommendable approach. (Admittedly, even for linear problems, the methods presented in Sects. 2.2 and 2.3 could be judged oversimplified, since by Assumption 2.6 no constraints were considered, nor were adaptive or multiscale discretizations discussed, which are of great practical importance.) In the present section, a specific discretization for the specific Model Problem 1.9 for inverse gravimetry is considered. Model Problem 1.11 for nonlinear waveform inversion will be considered in Sect. 2.7.

The goal is to discretize a nonlinear Fredholm equation of the first kind

$$w(x_1, x_2) = \int_{-a}^{a} \int_{-a}^{a} k(x_1, x_2, y_1, y_2, u(y_1, y_2)) \, dy_2 \, dy_1, \qquad (2.103)$$

where the kernel function is defined by

$$k(x_1, x_2, y_1, y_2, z) := \frac{1}{\sqrt{(x_1 - y_1)^2 + (x_2 - y_2)^2 + z^2}}, \qquad (2.104)$$

for $(x_1, x_2), (y_1, y_2) \in \mathbb{R}^2$ and $z > 0$ and where

$$\mathbb{U} := \left\{ u : [-a, a]^2 \to [b_1, b_2]; \ u \text{ continuous}, \ 0 < b_1 < b_2 \right\}$$

is the set of admissible causes. Here, the parameters h and H from Problem 1.9 were renamed into b_1 and b_2, respectively. The continuous function $w \in C(\mathbb{R}^2)$ is observed for $x \in [-b, b]^2$, and the corresponding exact solution of (2.103) will be denoted by u^*.

A very simple discretization scheme will be used. For $n \in \mathbb{N}$ to be chosen, it is based on equidistant grid points

$$x_\alpha := h\alpha, \quad h := \frac{a}{n}, \quad \alpha \in G_n := \{(\alpha_1, \alpha_2) \in \mathbb{Z}^2; \ -n \leq \alpha_j \leq n, \ j = 1, 2\}.$$
$$(2.105)$$

As an approximant of the unknown function u^*, we use

$$u^*(x) \approx u_n(x) = \sum_{\alpha \in G_n} c_\alpha \Phi(x/h - \alpha), \qquad (2.106)$$

where the parameters c_α are yet unknown and where the bilinear B-spline

$$\Phi(x) := B_2(x_1) \cdot B_2(x_2), \quad x = (x_1, x_2) \in \mathbb{R}^2 \qquad (2.107)$$

is defined exactly as in (2.87). Any approximant u_n of the form (2.106) is continuous, and the constraint $u_n \in \mathbb{U}$ immediately translates into **box constraints**

$$c_\alpha \in [b_1, b_2], \quad \alpha \in G_n, \qquad (2.108)$$

for the parameters c_α. Substituting u_n for u, the integral in (2.103) can be evaluated approximately for any $x = (x_1, x_2)$ by the two-dimensional trapezoidal rule, *based on the same grid points x_α as in (2.105)*, i.e.,

$$w(x) \approx w_n(x) := \sum_{\alpha \in G_n} \omega_\alpha k(x, x_\alpha, c_\alpha) \qquad (2.109)$$

with constant weights

$$\omega_\alpha := \begin{cases} h^2/4, & \text{if } |\alpha_1| = |\alpha_2| = n, \\ h^2/2, & \text{if } |\alpha_1| = n \text{ or } |\alpha_2| = n, \ |\alpha_1| + |\alpha_2| < 2n, \\ h^2, & \text{else.} \end{cases} \qquad (2.110)$$

To arrive at (2.109), one makes use of the identity $u_n(x_\alpha) = c_\alpha$, $\alpha \in G_n$, which is the reason why the trapezoidal rule was based on the grid $\{x_\alpha; \ \alpha \in G_n\}$. More sophisticated quadrature rules exist than the trapezoidal rule, but since we cannot guarantee any smoothness of the integrand beyond continuity, there is no guarantee that these rules produce more accurate results. Approximating (2.103) by (2.109) is known as **Nyborg's method**. Suppose now in accordance with Problem 1.9 that the observation of w consists in sampling values

$$w(\hat{x}_\beta), \quad \hat{x}_\beta \in [-b, b]^2, \quad \beta \in B, \tag{2.111}$$

where the sample points $\hat{x}_\beta$ are pairwise different and where $B \subset \mathbb{Z}^2$ is a finite index set. This leads to the following nonlinear system of equations:

$$y_\beta := w(\hat{x}_\beta) = \sum_{\alpha \in G_n} \omega_\alpha k(\hat{x}_\beta, x_\alpha, c_\alpha), \quad \beta \in B, \tag{2.112}$$

consisting of

$$M := |B| \quad \text{equations for} \quad N := (2n + 1)^2 = |G_n| \quad \text{unknowns } c_\alpha. \tag{2.113}$$

Summarizing observations and unknowns as vectors[4]

$$y = (y_\beta; \ \beta \in B) \in \mathbb{R}^M, \quad c = (c_\alpha; \ \alpha \in G_n) \in \mathbb{R}^N, \tag{2.114}$$

and defining $C := [b_1, b_2]^N$ and a mapping

$$F : C \to \mathbb{R}^M, \quad c \mapsto F(c) = (F_\beta(c); \ \beta \in B)$$

by setting

$$F_\beta(c) = \sum_{\alpha \in G_n} \omega_\alpha k(\hat{x}_\beta, x_\alpha, c_\alpha), \quad \beta \in B, \tag{2.115}$$

this system can be written in condensed form as

$$F(c) = y, \quad c \in C. \tag{2.116}$$

As a consequence of discretization errors, even if exact values $c_\alpha = c_\alpha^* = u^*(x_\alpha)$ were used in (2.115), $w(\hat{x}_\beta) \approx F_\beta(c^*)$ would not be an exact equality. Therefore, a solution of system (2.116) cannot be expected to exist[5] and this system will be replaced by a minimization problem with box constraints

$$\text{minimize} \quad \frac{1}{2}\|y - F(c)\|_2^2, \quad c \in C = [b_1, b_2]^N. \tag{2.117}$$

---

[4] Since two-dimensional index sets can be ordered in multiple ways, the following requires an additional agreement. For example, the vector c consists of elements $c_{(-n,-n)}, \ldots, c_{(n,-n)}, \ldots, c_{(-n,n)}, \ldots, c_{(n,n)}$, and we can agree to list them in precisely that order. But we could agree on any other ordering as well.

[5] Already the basic assumption made in Problem 1.9, namely that a body of constant mass density includes another body of a different, constant mass density, is an idealization and means a modelling error.

This minimization problem will be investigated and solved only in Sect. 4.5. Nonetheless, we already anticipate the following principle problem with the solution of (2.117). The objective function $c \mapsto \frac{1}{2}\|y - F(c)\|_2^2$ exhibits multiple local minima, which prevent efficient iterative solvers to detect a global minimum, unless when started close to one. In the case of (2.117), a good starting point can be found using a "multiscale discretization" of (2.103). This will be explained in the following paragraph.

Multiscale Discretizations

We start with some technical considerations. For $k \in \mathbb{N}_0$, the so-called **scale parameter** or **level parameter**, let

$$n_k := 2^k, \quad h_k := \frac{a}{n_k}, \quad G^{(k)} := \{(\alpha_1, \alpha_2); \; -n_k \leq \alpha_j \leq n_k, \; j = 1, 2\}. \tag{2.118}$$

Further, let

$$x_\alpha^{(k)} := h_k \alpha, \qquad \alpha \in G^{(k)},$$

be the grid points "at level k," distanced from each other in both directions by the mesh size h_k. At level k, we look for an approximation

$$u^{(k)}(x) = \sum_{\alpha \in G^{(k)}} c_\alpha^{(k)} \Phi(x/h_k - \alpha), \quad x \in [-a, a]^2, \tag{2.119}$$

of the unknown function u^* with Φ the bilinear B-spline defined in (2.107). We assemble all (yet unknown) coefficients $c_\alpha^{(k)}$ from (2.119) into a column vector

$$c^{(k)} := (c_\alpha^{(k)}; \; \alpha \in G^{(k)}) \in \mathbb{R}^{N_k}, \quad N_k = (2 \cdot 2^k + 1)^2,$$

by ordering $\alpha \in G^{(k)}$ (in the same manner as for G_n above, say). The box constraints for $u^{(k)}$ read

$$c^{(k)} \in C^{(k)} := [b_1, b_2]^{N_k}.$$

Moreover, let us assume that for some $K \in \mathbb{N}$

$$n = n_K = 2^K,$$

where n is the parameter from (2.105). Within the range $k = 0, \ldots, K$ of levels, the parameter n thus corresponds to the finest grid we are considering (level K). Every spline function $u^{(k)}$ as in (2.119), $k \in \{0, \ldots, K\}$, has a unique representation

$$u^{(k)}(x) = \sum_{\alpha \in G^{(k)}} c_\alpha^{(k)} \Phi(x/h_k - \alpha) = \sum_{\alpha \in G_n} c_\alpha \Phi(x/h - \alpha), \qquad (2.120)$$

as a spline function defined on the finest level. The corresponding coefficients c_α on the right-hand side of (2.120) will be assembled into a column vector

$$c \in C \subset \mathbb{R}^N, \quad N = (2 \cdot 2^K + 1)^2,$$

as we already did in (2.114). The coefficient vectors $c^{(k)}$ and $c = c^{(K)}$ from (2.120) are related by an identity

$$c = I^{(k)} \cdot c^{(k)}. \qquad (2.121)$$

Here, $I^{(k)}$ is a matrix of dimension $(2 \cdot 2^K + 1)^2 \times (2 \cdot 2^k + 1)^2$, which represents the interpolation from the coarse grid at level k to the finest grid at level K.

Seeking an approximation $u^{(k)}$ at level k means to ask for $c^{(k)} \in C^{(k)}$ such that

$$F(I^{(k)} c^{(k)}) = y, \qquad (2.122)$$

where y and F are defined as in (2.114) and in (2.115), respectively. By this construction, we always base our discretization of the integral equation (2.103) on the fine grid but look at approximations $u^{(k)}$ of u^* at different levels. For this reason, $c \in C$ might be called **simulation parameters**, whereas $c^{(k)}$ are called **optimization parameters** at level k, cf. [Cha09]. Of course, with even more reason than for (2.116), we will have to replace (2.122) by a minimization problem. Then, by **multiscale optimization**, we mean the following:

Multiscale Optimization of Nonlinear Inverse Gravimetry. Choose $K \in \mathbb{N}$, set $k = 0$. Choose start value $c^{(-1)} \in C^{(0)}$.

Step 1:   Starting from $c^{(k-1)}$, interpolated to the next level k, find

$$c^{(k)} = \text{argmin} \left\{ \|y - F(I^{(k)}v)\|_2; \ v \in C^{(k)} \right\}.$$

Step 2:   Increment k. If $k = K$, then stop. Otherwise, go to Step 1.

Here, we assume that the minimization problem of Step 1 is solved by some iterative method like the Gauß–Newton method or the Levenberg–Marquardt method, which need to be provided with a good starting point, see Sect. 4.2.

Although multiscale optimization seems to be a good idea to approach a global solution of (2.117), there is no general guarantee that this idea works. This rather depends on the "degree of nonlinearity" of the objective function we ask to minimize—and thus on the nonlinearity of F. We now explain why we can hope that multiscale optimization works in the case of inverse gravimetry, following a geometric reasoning established in Sect. 3.6 of [Cha09].

To start with, let us consider a twice differentiable curve

$$\gamma : [a, b] \to \mathbb{R}^m.$$

The tangent unit vector of this curve is defined by

$$T(t) := \frac{\dot{\gamma}(t)}{\|\dot{\gamma}(t)\|_2},$$

where we have assumed that the regularity condition $\dot{\gamma}(t) \neq 0$ holds. The **curvature** at the point $\gamma(t)$ is defined as

$$\kappa(t) := \left\| \frac{d}{ds} T(t) \right\|_2 = \frac{\|\dot{T}(t)\|_2}{\|\dot{\gamma}(t)\|_2}.$$

Here, d/ds means the derivative with respect to the arc length; this means that one defines the curvature as a measure of how fast $T(t)$ changes *in space*, not in time. The curvature can be considered as a measure of nonlinearity. For linear curves, $\kappa(t) \equiv 0$. By the chain rule, we get

$$\dot{T}(t) = \frac{1}{\|\dot{\gamma}(t)\|_2} \ddot{\gamma}(t) - \frac{\sum_j \dot{\gamma}_j(t) \ddot{\gamma}_j(t)}{(\sum_j \dot{\gamma}_j^2(t))^{3/2}} \dot{\gamma}(t).$$

By the inequality of Cauchy–Schwarz, we get

$$\|\dot{T}(t)\|_2 \leq \frac{\|\ddot{\gamma}(t)\|_2}{\|\dot{\gamma}(t)\|_2} + \frac{\|\dot{\gamma}(t)\|_2 \|\ddot{\gamma}(t)\|_2}{\|\dot{\gamma}(t)\|_2^3} \|\dot{\gamma}(t)\|_2 = 2 \frac{\|\ddot{\gamma}(t)\|_2}{\|\dot{\gamma}(t)\|_2}.$$

We thus have found the following upper bound for the curvature:

$$\kappa(t) \leq 2 \frac{\|\ddot{\gamma}(t)\|_2}{\|\dot{\gamma}(t)\|_2^2}. \tag{2.123}$$

We will now investigate the sensitivity and the nonlinearity of the function F (defined in (2.115)) with respect to its arguments c_α. Because of the validity of the box constraints (2.108), we may choose a reference vector e in the interior of C. For simplicity, let us assume $e = (1, 1, \ldots, 1)$. Let us then consider the curve

$$\gamma : I \to \mathbb{R}^M, \quad t \mapsto F(e + t \cdot e_\alpha) \quad \text{for some fixed value} \quad \alpha \in G_n. \tag{2.124}$$

Here, M is defined in (2.113), I is a (small) interval containing 0, and by e_α we mean the unit vector into the direction of the αth coordinate. One readily computes the derivatives

$$\frac{dF_\beta(e + te_\alpha)}{dt} = -\frac{\omega_\alpha(1 + t)}{\left(\|\hat{x}_\beta - x_\alpha\|_2^2 + (1 + t)^2\right)^{3/2}}, \quad \beta \in B,$$

and

$$\frac{d^2 F_\beta(e + te_\alpha)}{dt^2} = \frac{\omega_\alpha\left(2(1 + t)^2 - \|\hat{x}_\beta - x_\alpha\|_2^2\right)}{\left(\|\hat{x}_\beta - x_\alpha\|_2^2 + (1 + t)^2\right)^{5/2}}, \quad \beta \in B.$$

These derivatives form the components of $\dot{\gamma}(t)$ and $\ddot{\gamma}(t)$ for the curve γ defined by (2.124). Because of (2.110), one deduces

$$S := \|\dot{\gamma}(0)\|_2 = \mathcal{O}(h^2) \tag{2.125}$$

and likewise

$$\|\ddot{\gamma}(0)\|_2 = \mathcal{O}(h^2).$$

From (2.123), we also get the following bound:

$$\kappa(0) \leq \Gamma := 2\frac{\|\ddot{\gamma}(0)\|_2}{\|\dot{\gamma}(0)\|_2^2} = \mathcal{O}(h^{-2}). \tag{2.126}$$

Taking S as a measure of the sensitivity of F at the reference point e with respect to a change in the αth variable and taking Γ as a measure for the corresponding nonlinearity of F, we have the following interpretation of (2.125) and (2.126): *With grid sizes becoming finer and finer (decreasing values of h), the nonlinearity of F increases and the sensitivity decreases. With grid sizes becoming coarser and coarser (increasing values of h), F becomes less nonlinear and more sensitive.* This is just the reason why one may hope that for coarser discretizations, the minimization problem (2.117) will be closer to a *linear* least squares problem and will exhibit less local minima.

Further Reading

For further reading on multiscale and adaptive multiscale methods in general, we recommend [Cha09]. The article [FN18a] especially deals with the problem of inverse gravimetry and also proposes further discretization methods, as do [Mic05], [FM12], and [MS20].

2.7   Discretization of Nonlinear Waveform Inversion

Model Problem 1.11 with additional agreements (1.58) and (1.59) was formulated in Sect. 1.4. To recall, for

$$\mathscr{K} = \left\{ \kappa \in H^1(0, Z_0);\ 0 < \kappa_- \le \kappa(z) \le \kappa_+ < \infty,\ \|\kappa'\|_{L_2(0, Z_0)} \le M \right\},$$

where κ_-, κ_+, and M are constants, an operator $T : \mathscr{K} \to C[0, T_0]$ was defined, which maps an elasticity coefficient $\kappa \in \mathscr{K}$ to an observed seismogram

$$u(0, \bullet) \in C[0, T_0],$$

where u is the (weak) solution of

$$\rho_0 \frac{\partial^2 u(z, t)}{\partial t^2} - \frac{\partial}{\partial z} \left(\kappa(z) \frac{\partial u(z, t)}{\partial z} \right) = 0, \qquad 0 < z < Z_0,\ 0 < t < T_0, \qquad (1.52)$$

$$u(z, t) = 0, \qquad \frac{\partial u(z, t)}{\partial t} = 0, \qquad\qquad 0 < z < Z_0,\ t = 0, \qquad (1.53)$$

$$-\kappa(z) \frac{\partial u(z, t)}{\partial z} = g(t), \qquad\qquad z = 0,\ 0 < t < T_0, \qquad (1.54)$$

$$\kappa(z) \frac{\partial u(z, t)}{\partial z} = -\sqrt{\rho_0 \kappa(z)}\, \frac{\partial u(z, t)}{\partial t}, \qquad z = Z_0,\ 0 < t < T_0. \qquad (1.55)$$

The inverse problem consists in solving the equation

$$T(\kappa) = u(0, \bullet), \qquad u(0, \bullet) \in C[0, T_0], \qquad\qquad (2.127)$$

for $\kappa \in \mathscr{K}$. Equation (2.127) will now be replaced by an equation in finite-dimensional spaces. As a first step, the initial/boundary-value problem (1.52)–(1.55) will be discretized.

Discretization of the Initial/Boundary-Value Problem

We will use standard finite differencing to solve the system of equations (1.52)–(1.55). At first, we choose

$$m, n \in \mathbb{N}, \quad \text{set} \quad \Delta z := Z_0/m, \quad \Delta t := T_0/n, \tag{2.128}$$

and define grids

$$z_i := i \cdot \Delta z, \ i = 0, \ldots, m \quad \text{and} \quad t_j := j \cdot \Delta t, \ j = 0, \ldots, n. \tag{2.129}$$

Functions $\kappa \in \mathcal{K}$ and $g \in C[0, T_0]$ may (approximately) be represented by samples

$$\kappa_i := \kappa(z_i), \quad i = 0, \ldots, m, \quad \text{and} \quad g_j := g(t_j), \quad j = 0, \ldots, n, \tag{2.130}$$

from which one gets spline approximations

$$\kappa^{(m)} := \sum_{i=0}^{m} \kappa_i \cdot N_{i,2} \quad \text{and} \quad g^{(n)} := \sum_{j=0}^{n} g_j \cdot N_{j,2}$$

of κ and g, respectively. To derive a discretized version of T from (2.127), we compute approximate values

$$u_i^j \approx u(z_i, t_j), \quad i = 0, \ldots, m, \quad j = 0, \ldots, n \tag{2.131}$$

of the exact solution u of (1.52)–(1.55). This will be done using finite differencing—a technique that could also be used to solve the acoustic wave equation in more than one space dimension. In the following, the points $(z_i, t_j) \in [0, Z_0] \times [0, T_0]$, $i = 0, \ldots, m$, $j = 0, \ldots, n$, defined by (2.129), are called **grid points**. The points (z_i, t_j) for $i = 1, \ldots, m-1$ and $j = 1, \ldots, n-1$ are called **interior grid points**. The other grid points are called **boundary grid points**.

Discretization of (1.52) At interior grid points, we use

$$\frac{\partial^2 u(z_i, t_j)}{\partial t^2} \approx \frac{u(z_i, t_{j+1}) - 2u(z_i, t_j) + u(z_i, t_{j-1})}{\Delta t^2}$$

to approximate the time derivative in (1.52). To discretize the second term in (1.52), we first approximate the outer derivative

$$\frac{\partial \phi(z_i, t_j)}{\partial z}, \quad \phi(z, t) = \kappa(z) \frac{\partial u(z, t)}{\partial z},$$

by the finite difference

$$\frac{\partial \phi(z_i, t_j)}{\partial z} \approx \frac{\phi\left(z_i + \frac{\Delta z}{2}, t_j\right) - \phi\left(z_i - \frac{\Delta z}{2}, t_j\right)}{\Delta z}.$$

Then, we approximate

$$\phi\left(z_i + \frac{\Delta z}{2}, t_j\right) \approx \kappa\left(z_i + \frac{\Delta z}{2}\right) \cdot \frac{u(z_{i+1}, t_j) - u(z_i, t_j)}{\Delta z}$$

and, similarly,

$$\phi\left(z_i - \frac{\Delta z}{2}, t_j\right) \approx \kappa\left(z_i - \frac{\Delta z}{2}\right) \cdot \frac{u(z_i, t_j) - u(z_{i-1}, t_j)}{\Delta z}.$$

We arrive at

$$\frac{\partial \phi(z_i, t_j)}{\partial z}$$

$$\approx \frac{\kappa\left(z_i + \frac{\Delta z}{2}\right)\left[u(z_{i+1}, t_j) - u(z_i, t_j)\right] - \kappa\left(z_i - \frac{\Delta z}{2}\right)\left[u(z_i, t_j) - u(z_{i-1}, t_j)\right]}{\Delta z^2},$$

but since we want to rely only on the samples (2.130) of κ, we use the further approximations

$$\kappa\left(z_i + \frac{\Delta z}{2}\right) \approx \frac{1}{2}(\kappa_{i+1} + \kappa_i) \quad \text{and} \quad \kappa\left(z_i - \frac{\Delta z}{2}\right) \approx \frac{1}{2}(\kappa_i + \kappa_{i-1}).$$

All in all, we replace the differential equation (1.52) by the finite difference equations

$$\rho_0 \frac{u_i^{j+1} - 2u_i^j + u_i^{j-1}}{\Delta t^2} = \frac{\frac{1}{2}(\kappa_i + \kappa_{i+1})(u_{i+1}^j - u_i^j) - \frac{1}{2}(\kappa_i + \kappa_{i-1})(u_i^j - u_{i-1}^j)}{\Delta z^2},$$

$$(2.132)$$

valid at interior grid points. Using Taylor expansion and assuming u and κ to be smooth enough, one can show that (2.132) is "of second order," which means that (2.132) holds with u_i^j replaced by $u(z_i, t_j)$ (with u the exact solution of (1.52)) up to an error of magnitude $\mathcal{O}(\Delta t^2) + \mathcal{O}(\Delta z^2)$.

Discretization of (1.53) The first of the initial conditions (1.53) obviously is dealt with by setting

$$u_i^0 := 0, \quad i = 0, \ldots, m. \tag{2.133}$$

Further, we use the formal approximation (also of second order) of the other initial condition

$$0 = \frac{\partial u(z_i, 0)}{\partial t} \approx \frac{u_i^1 - u_i^{-1}}{2\Delta t}$$

to define "ghost values" $u_i^{-1} := u_i^1$, $i = 0, \ldots, m$. With this definition and with (2.133), (2.132) may already be used for $j = 0$. We get

$$\rho_0 \frac{2u_i^1}{\Delta t^2} = 0 \quad \Longleftrightarrow \quad u_i^1 = 0, \quad i = 1, \ldots, m. \tag{2.134}$$

Discretization of (1.54) Next, we use the formal approximation (of second order) of the boundary condition (1.54)

$$g_j = -\kappa(0) \frac{\partial u(0, t_j)}{\partial z} \approx -\kappa_0 \frac{u_1^j - u_{-1}^j}{2\Delta z}, \quad j = 1, \ldots, n - 1,$$

which is equivalent to the definition of "ghost values"

$$u_{-1}^j := u_1^j + \frac{2\Delta z}{\kappa_0} g_j, \quad j = 1, \ldots, n - 1.$$

Here, g_j means the samples of the Ricker pulse g as in (2.130). Substituting the values u_{-1}^j in (2.132), thereby making use of the additional approximations

$$\frac{1}{2}(\kappa_0 + \kappa_1) \approx \kappa_0 \approx \frac{1}{2}(\kappa_{-1} + \kappa_0),$$

and setting $u_0^{-1} := 0$, we get

$$\rho_0 \frac{u_0^{j+1} - 2u_0^j + u_0^{j-1}}{\Delta t^2} - \frac{2\kappa_0(u_1^j - u_0^j)}{\Delta z^2} = \frac{2g_j}{\Delta z}, \quad j = 0, \ldots, n - 1. \tag{2.135}$$

Discretization of (1.55) Finally, we approximate the absorbing boundary condition by the formal finite difference equation (again of second order)

$$\sqrt{\rho_0} \cdot \frac{u_m^{j+1} - u_m^{j-1}}{2\Delta t} + \sqrt{\kappa_m} \cdot \frac{u_{m+1}^j - u_{m-1}^j}{2\Delta z} = 0, \quad j = 1, \ldots, n - 1.$$

Once more, this is equivalent to the definition of ghost values

$$u_{m+1}^j := u_{m-1}^j + \frac{\Delta z}{\Delta t} \cdot \sqrt{\frac{\rho_0}{\kappa_m}} \cdot (u_m^{j-1} - u_m^{j+1}), \quad j = 1, \ldots, n - 1.$$

Substituting these values in Eq. (2.132) for $i = m$, thereby making use of the additional approximations

$$\frac{1}{2}\left(\kappa_{m-1} + \kappa_m\right) \approx \kappa_m \approx \frac{1}{2}\left(\kappa_m + \kappa_{m+1}\right),$$

we get (for $j = 1, \ldots, n-1$)

$$u_m^{j+1} - 2u_m^j + u_m^{j-1} = 2\frac{\Delta t^2}{\Delta z^2} \cdot \frac{\kappa_m}{\rho_0} \cdot (u_{m-1}^j - u_m^j) + \frac{\Delta t}{\Delta z} \cdot \sqrt{\frac{\kappa_m}{\rho_0}} \cdot \left(u_m^{j-1} - u_m^{j+1}\right).$$

Using the abbreviation

$$\alpha = \alpha(\kappa_m) := \frac{\Delta t}{\Delta z} \cdot \sqrt{\frac{\kappa_m}{\rho_0}}, \tag{2.136}$$

the latter equations can be reformulated as

$$\frac{1-\alpha}{1+\alpha} \cdot u_m^{j-1} + \left[-\frac{2\alpha^2}{1+\alpha} \cdot u_{m-1}^j + 2(\alpha - 1) \cdot u_m^j\right] + u_m^{j+1} = 0 \tag{2.137}$$

for $j = 1, \ldots, n-1$.

Next, we will summarize Eqs. (2.132)–(2.137) in the form of a single system of linear equations. Let us assemble the approximate values $u_i^j \approx u(z_i, t_j)$, $i = 0, \ldots, m$, $j = 1, \ldots, n$, of the solution u of (1.52)–(1.55) into a vector

$$\mathbf{u} := (u_0^1, \ldots, u_m^1, \ldots, u_0^n, \ldots, u_m^n)^\top \in \mathbb{R}^{(m+1)n}. \tag{2.138}$$

(There is no need to store the zero values u_i^0.) Respecting the same ordering, we form the vector

$$\mathbf{g} := \frac{2}{\Delta z} \cdot (g_0, \underbrace{0, \ldots, 0}_{m}, g_1, \underbrace{0, \ldots, 0}_{m}, \ldots, g_{n-1}, \underbrace{0, \ldots, 0}_{m})^\top \in \mathbb{R}^{(m+1)n}$$

$$\tag{2.139}$$

from the samples g_j, $j = 0, \ldots, n-1$, of the Ricker pulse. Further, we define two diagonal matrices $T_1, T_3 \in \mathbb{R}^{(m+1),(m+1)}$ by

$$T_1 := \begin{pmatrix} \frac{\rho_0}{\Delta t^2} & & & \\ & \frac{\rho_0}{\Delta t^2} & & \mathbf{0} \\ & & \ddots & \\ \mathbf{0} & & & \frac{\rho_0}{\Delta t^2} \\ & & & & 1 \end{pmatrix} \quad \text{and} \quad T_3 := \begin{pmatrix} \frac{\rho_0}{\Delta t^2} & & & \\ & \frac{\rho_0}{\Delta t^2} & & \mathbf{0} \\ & & \ddots & \\ \mathbf{0} & & & \frac{\rho_0}{\Delta t^2} \\ & & & & \frac{1-\alpha}{1+\alpha} \end{pmatrix}.$$

Also, we define the tridiagonal matrix $T_2 \in \mathbb{R}^{(m+1),(m+1)}$ by

$$
T_2 := \begin{pmatrix} -2\frac{\rho_0}{\Delta t^2} & & & \\ & -2\frac{\rho_0}{\Delta t^2} & & 0 \\ & & \ddots & \\ & 0 & & -2\frac{\rho_0}{\Delta t^2} \\ & & & & 0 \end{pmatrix} +
$$

$$
\frac{1}{2\Delta z^2} \cdot \begin{pmatrix} 4\kappa_0 & -4\kappa_0 & & & \\ -(\kappa_0+\kappa_1) & (\kappa_0+2\kappa_1+\kappa_2) & -(\kappa_1+\kappa_2) & & \\ \ddots & \ddots & \ddots & & \\ & & -(\kappa_{m-2}+\kappa_{m-1}) & (\kappa_{m-2}+2\kappa_{m-1}+\kappa_m) & -(\kappa_{m-1}+\kappa_m) \\ & & & -\frac{4\alpha^2\Delta z^2}{1+\alpha} & 4(\alpha-1)\Delta z^2 \end{pmatrix}.
$$

Here, the scaling factor before the second matrix in the sum does not apply to the last row of this matrix.

From T_1, T_2, and T_3, we build a sparse lower triangular matrix

$$
A = A(\mathbf{k}) = \begin{pmatrix} T_1 & & & & \\ T_2 & T_1 & & & \\ T_3 & T_2 & T_1 & & \\ & \ddots & \ddots & \ddots & \\ & & T_3 & T_2 & T_1 \end{pmatrix} \in \mathbb{R}^{n(m+1),n(m+1)}, \tag{2.140}
$$

which depends on

$$
\mathbf{k} := (\kappa_0, \ldots, \kappa_m)^\top \in \mathbb{R}^{m+1}. \tag{2.141}
$$

Solving the discrete equations (2.132)–(2.137) is equivalent to solving the linear system

$$
A \cdot \mathbf{u} = \mathbf{g}, \tag{2.142}
$$

which can very easily be done by simple forward substitution thanks to the triangular structure of the matrix $A = A(\mathbf{k})$. Forward substitution corresponds to a "stepping forward in time," i.e., to the computation of the values u_i^j in ascending order of j.

We have chosen a standard finite differencing scheme, resulting in equations (2.132)–(2.137). More refined methods and their analysis are beyond the scope of this monograph. Nevertheless, we at least have to mention two requirements which must be met for finite differencing to work in practice, cf., e.g., [Tre96]. First, data inaccuracies and rounding errors must be prevented from being magnified in the course of computations. This is the requirement of **numerical stability**. Numerical stability holds, if the condition of Courant–Friedrichs–Lewy (**CFL condition**)

$$\Delta t \leq \frac{\Delta z}{c_{\max}} \tag{2.143}$$

holds, where

$$c_{\max} := \max\left\{ c = c(z) = \sqrt{\kappa(z)/\rho_0}; \ z \in [0, Z_0] \right\}$$

means the maximal wave velocity. Second, discretization leads to an undesired phenomenon known as **numerical dispersion**. This effect is mitigated, if the spatial discretization is chosen fine enough. One often requires

$$\Delta z \leq \frac{c_{\min}}{10 f_{\max}} \tag{2.144}$$

to hold, where

$$c_{\min} = \min\left\{ c = c(z); \ z \in [0, Z_0] \right\} \quad \text{and} \quad f_{\max} \approx 3 \cdot f_0$$

are the minimal wave velocity and the maximal frequency of the Ricker source wavelet with central frequency f_0, respectively.

Relation (2.144) imposes a (high) spatial resolution requirement on the parameter function κ—large value m, small value Δz. But in our case, κ is not directly available. We may be happy to find an approximation of κ resolved on a grid comprising $k + 1 < m + 1$ samples only. This means we are looking for a vector

$$\mathbf{x} \in \mathbb{R}^{k+1}, \quad x_1 \approx \kappa(0), x_2 \approx \kappa(Z_0/k), x_3 \approx \kappa(2Z_0/k), \ldots, x_{k+1} \approx \kappa(Z_0), \tag{2.145}$$

where $k < m$. From samples x_j, we may construct a linear spline approximant

$$\kappa^{(k)} := \sum_{i=0}^{k} x_{i+1} \cdot N_{i,2}$$

and substitute *its* samples

$$\kappa_j := \kappa^{(k)}(j \Delta z), \quad j = 0, \ldots, m,$$

for $\kappa(j \cdot \Delta z)$. It is easily seen that $\mathbf{k}$ relates to $\mathbf{x}$ via

$$\mathbf{k} = G \cdot \mathbf{x}, \tag{2.146}$$

where $G \in \mathbb{R}^{m+1,k+1}$ is an interpolation operator (a matrix). Via G, $A = A(\mathbf{k}) = A(G\mathbf{x})$ depends on $\mathbf{x}$ only, and so does $\mathbf{u}$.

Discretization of the Operator T

An observed seismogram is approximated by the values u_0^j. From (2.138), it becomes clear that a corresponding "discrete observation operator" has to be defined as the linear mapping $\mathbb{R}^{n(m+1)} \to \mathbb{R}^n$, $\mathbf{u} \mapsto M\mathbf{u}$, determined by the matrix

$$
M := \begin{pmatrix}
1, 0, 0, \ldots, 0 & & & \\
& 1, 0, 0, \ldots, 0 & & \\
& & \ddots & \\
& & & 1, 0, 0, \ldots, 0
\end{pmatrix} \in \mathbb{R}^{n, n(m+1)}.
$$

$M\mathbf{u}$ just selects the values u_0^j, $j = 1, \ldots, n$, from the computed discrete solution $\mathbf{u}$ of (2.132)–(2.137). With the $(k+1)$-dimensional interval

$$
D := [\kappa_-, \kappa_+]^{k+1} \subset \mathbb{R}^{k+1}, \tag{2.147}
$$

we can now define the mapping

$$
F : D \to \mathbb{R}^n, \quad \mathbf{x} \mapsto M \cdot [A(G\mathbf{x})]^{-1} \mathbf{g}. \tag{2.148}
$$

This is the discrete approximation of the operator T from Model Problem 1.11. An evaluation $F(\mathbf{x})$ means the following:

(1) Compute $\mathbf{k} = G\mathbf{x}$, a piecewise linear interpolation of $\mathbf{x}$ to a finer grid of meshlength Δz.
(2) Assemble the matrix $A = A(\mathbf{k})$.
(3) Solve the linear system $A\mathbf{u} = \mathbf{g}$.
(4) Compute the discrete approximation $\mathbf{v} = M\mathbf{u}$ of the observed seismogram.

For the moment, we have dropped the restriction $\|\kappa'\|_{L_2(0, z_0)} \leq M$ from the definition of $\mathcal{K}$. This restriction will be reconsidered later in the context of regularization. An actual observation $y \in C[0, T_0]$ will be sampled

$$
y_i := y(t_i), \quad i = 1, \ldots, n, \quad \mathbf{y} := (y_1, \ldots, y_n)^T \in \mathbb{R}^n,
$$

and the equation

$$
F(\mathbf{x}) = \mathbf{y}, \quad \mathbf{x} \in D \tag{2.149}
$$

will have to be solved for $\mathbf{x}$. This is Eq. (2.127) in discretized form.

Discrete Inverse Model Problem

As for inverse gravimetry, the discretized problem (2.149) does not have to have a solution due to discretization and data measurement errors. Consequently, it will be replaced by a nonlinear least squares problem with box constraints

$$\text{minimize} \quad \frac{1}{2}\|\mathbf{y} - F(\mathbf{x})\|_2^2, \quad \mathbf{x} \in D = [\kappa_-, \kappa_+]^{k+1}, \tag{2.150}$$

with F from (2.148). As for its counterpart from inverse gravimetry, the efficient solution of (2.149) by iterative solvers is impeded by the presence of multiple local minima apart from the global minimum. In contrast to inverse gravimetry, this cannot be cured by multiscale optimization. The reason for this will be explained later in Sect. 4.6, where an alternative approach to finding a solution of (2.150) is discussed.

Further Reading

As already mentioned, Model Problem 1.11 only gives a simplified view of full waveform inversion. For a more realistic view, seismic wave propagation would have to be modelled by the elastic (and not the acoustic) wave equation and the three-dimensional case would have to be considered. Then, numerical methods more elaborate than the basic finite differencing method presented above are needed to set up a system of discrete equations like (2.149). A huge amount of research was and is done in this area. Not being a specialist for the numerical solution of PDEs, I only mention [Fic11] as one possible starting point for further reading. Further pointers to the literature will be given in Sect. 4.6, where we consider the solution of (2.150).

Chapter 3
Regularization of Linear Inverse Problems

The discretization of linear identification problems led to linear systems of algebraic equations. In case a solution does not exist, these can be replaced by linear least squares problems, as already done in Sect. 2.3. We will give a detailed sensitivity analysis of linear least squares problems and introduce their **condition number** as a quantitative measure of ill-posedness. If the condition of a problem is too bad, it cannot be solved practically. We introduce and discuss the concept of **regularization** which formally consists in replacing a badly conditioned problem by a better conditioned one. The latter can be solved reliably, but whether its solution is of any relevance depends on how the replacement problem was constructed. We will especially consider Tikhonov regularization and iterative regularization methods. The following discussion is restricted to the Euclidean space over the field of real numbers $\mathbb{K} = \mathbb{R}$, but could easily be extended to complex spaces.

3.1 Linear Least Squares Problems

Solving a system of linear equations

$$Ax = b \quad \text{with} \quad A \in \mathbb{R}^{m,n}, \ x \in \mathbb{R}^n, \ b \in \mathbb{R}^m, \ m \geq n, \tag{3.1}$$

is an inverse problem: b represents an effect, x represents a cause, and "the physical law" is represented by the function $T : \mathbb{R}^n \to \mathbb{R}^m$, $x \mapsto Ax$. The case $m < n$ is excluded in (3.1), because injectivity of the mapping T ("identifiability of a cause") requires $\text{rank}(A) = n$, which is only possible if $m \geq n$. If $m > n$, the system of equations is called **overdetermined**. Inaccuracies in the components of A or b lead to contradictions between the equations of an overdetermined system, such that no exact solution x of (3.1) exists: the **residual**

© The Author(s), under exclusive license to Springer Nature Switzerland AG 2020
M. Richter, *Inverse Problems*, Lecture Notes in Geosystems
Mathematics and Computing, https://doi.org/10.1007/978-3-030-59317-9_3

$$r(x) := b - Ax$$

does not vanish for any $x \in \mathbb{R}^n$. As a substitute for a solution, one can look for a vector x minimizing the residual, for example, with respect to the Euclidean norm:

Find $\hat{x}$ such that $\|r(\hat{x})\|_2 \leq \|r(x)\|_2$ for all $x \in \mathbb{R}^n$.   (3.2)

(3.2) is called **linear least squares problem**. Solving (3.2) means to minimize

$$f(x) = \|r(x)\|_2^2 = r(x)^\top r(x) = x^\top A^\top Ax - 2x^\top A^\top b + b^\top b.$$

The gradient of f at x is $\nabla f(x) = 2A^\top Ax - 2A^\top b$ and since $\nabla f(\hat{x}) = 0$ is a necessary condition for $\hat{x}$ to be a minimizer of f, one gets the so-called **normal equations**

$$A^\top A\hat{x} = A^\top b \quad \Longleftrightarrow \quad A^\top \hat{r} = 0 \quad \text{with} \quad \hat{r} := r(\hat{x}).$$ (3.3)

These conditions are not only necessary but also sufficient for $\hat{x}$ to be a minimizer. To see this, take any $x \in \mathbb{R}^n$ with according residual $r(x) = \hat{r} + A(\hat{x} - x)$. Consequently

$$\|r(x)\|_2^2 = \hat{r}^\top \hat{r} + \underbrace{2(\hat{x} - x)^\top A^\top \hat{r}}_{= 0} + (\hat{x} - x)^\top A^\top A(\hat{x} - x) \geq \|\hat{r}\|_2^2,$$

where $\|A(\hat{x} - x)\|_2 = 0 \Leftrightarrow A(\hat{x} - x) = 0 \Leftrightarrow r(x) = \hat{r}$. This proves

Theorem 3.1 (Existence and Uniqueness of a Solution of the Least Squares Problem) *A necessary and sufficient condition for $\hat{x}$ to be a solution of (3.2) is the compliance with the normal equations (3.3). A unique minimizer $\hat{x}$ exists if and only if all columns of A are linearly independent, i.e., if rank$(A) = n$. The residual $\hat{r}$ is always unique.*

The "full rank condition" rank$(A) = n$ is equivalent to the injectivity of T. As mentioned repeatedly, it would be impossible to identify causes responsible for observed effects, if this condition was violated. Therefore, rank$(A) = n$ will usually be required in the following.

The normal equations have a geometrical interpretation. Since

$$A = \left(\begin{array}{c|c|c|c} & | & | & | \\ a_1 & a_2 & \cdots & a_n \\ & | & | & | \end{array} \right), \quad \text{all } a_j \in \mathbb{R}^m \quad \Longrightarrow \quad A^\top \hat{r} = \begin{pmatrix} a_1^\top \hat{r} \\ \vdots \\ a_n^\top \hat{r} \end{pmatrix},$$

Fig. 3.1 Solution of a linear
least squares problem

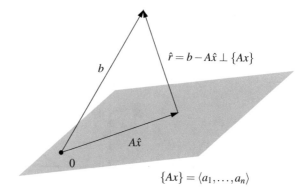

$\hat{r} = b - A\hat{x} \perp \{Ax\}$

b

$A\hat{x}$

0

$\{Ax\} = \langle a_1, \ldots, a_n \rangle$

(3.3) means $\hat{r} \perp \mathscr{R}_A$, i.e. the residual $\hat{r}$ is orthogonal to $\mathscr{R}_A = \operatorname{span}\{a_1, \ldots, a_n\} = \{Ax; \ x \in \mathbb{R}^n\}$, see Fig. 3.1.

Example 3.2 (Straight Line of Best Fit) Assume there is a causal dependence

$$T : \mathbb{R} \to \mathbb{R}, \quad t \mapsto T(t) = \alpha + \beta(t - 4),$$

with unknown parameters $\alpha, \beta \in \mathbb{R}$. Assume further we have the following measurements (taken from [DB74, Example 5.7.3]):

t	1	3	4	6	7
$T(t)$	−2.1	−0.9	−0.6	0.6	0.9

This corresponds to an overdetermined system of linear equations for $x = (\alpha, \beta)^\top$:

$$\underbrace{\begin{pmatrix} 1 & -3 \\ 1 & -1 \\ 1 & 0 \\ 1 & 2 \\ 1 & 3 \end{pmatrix}}_{=: A} \cdot \begin{pmatrix} \alpha \\ \beta \end{pmatrix} = \underbrace{\begin{pmatrix} -2.1 \\ -0.9 \\ -0.6 \\ 0.6 \\ 0.9 \end{pmatrix}}_{=: b}.$$

This system has no solution, since inexact measurements led to deviations from the true values $T(t)$. The normal equations in this example read

$$\begin{pmatrix} 5 & 1 \\ 1 & 23 \end{pmatrix} \begin{pmatrix} \alpha \\ \beta \end{pmatrix} = \begin{pmatrix} -2.1 \\ 11.1 \end{pmatrix}$$

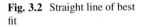

Fig. 3.2 Straight line of best fit

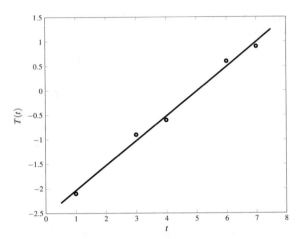

and have a unique solution $\alpha \approx -0.52105$ and $\beta \approx 0.50526$. Figure 3.2 shows the graph of the reconstructed function T, called **straight line of best fit** to the measurements. The measurement values are also marked in the figure.   ◊

Variants of the minimization problem (3.2) can be formulated using norms other than $\| \cdot \|_2$. For example, one could consider the minimization of $\|r(x)\|_1 = |r_1(x)| + \ldots + |r_n(x)|$ or $\|r(x)\|_\infty = \max\{|r_1(x)|, \ldots, |r_n(x)|\}$. The solution of these problems is much more complicated, however, than the minimization of $\|r(x)\|_2$. Under certain conditions concerning the measurement errors, using the Euclidean norm can also be justified by stochastic arguments (the Gauß–Markov theorem).

3.2   Sensitivity Analysis of Linear Least Squares Problems

Under the assumption that $A \in \mathbb{R}^{m,n}$ has full rank $n \leq m$, the linear least squares problem (3.2) has a unique solution $\hat{x} = (A^\top A)^{-1} A^\top b$. Since linear mappings between finite-dimensional spaces are always continuous, (3.2) is a well-posed problem according to Definition 1.5. We will now investigate, *how* sensitively the solution of (3.2) depends on A and b. As a technical tool, we will use the singular value decomposition of A, see (A.1) in Appendix A.

The matrix A and the right-hand side b are input values of the linear least squares problem. The according result $\hat{x}$ is characterized by the inequality

$$\|b - A\hat{x}\|_2 \leq \|b - Ax\|_2 \quad \text{for all } x \in \mathbb{R}^n. \tag{3.4}$$

If the input data are changed toward a matrix $A + \delta A$ and a right-hand side $b + \delta b$, the minimizer changes too, to become $\tilde{x}$. It is characterized by the inequality

$$\|(b + \delta b) - (A + \delta A)\tilde{x}\|_2 \leq \|(b + \delta b) - (A + \delta A)x\|_2 \quad \text{for all } x \in \mathbb{R}^n. \tag{3.5}$$

The following two theorems give bounds of the absolute and relative difference of $\hat{x}$ and $\tilde{x}$ in dependence of absolute and relative differences of input values, respectively.

Theorem 3.3 (Absolute Sensitivity of the Linear Least Squares Problem) *Let $A, \delta A \in \mathbb{R}^{m,n}$ and $b, \delta b \in \mathbb{R}^m$, $m \geq n$. Let A have singular values $\sigma_1 \geq \ldots \geq \sigma_n > 0$ and thus have full rank. Let $\hat{x}$ be the unique solution of (3.2), characterized by (3.4), having residual $\hat{r} := b - A\hat{x}$. Under the condition*

$$\eta := \frac{\|\delta A\|_2}{\sigma_n} < 1, \quad i.e., \quad \|\delta A\|_2 < \sigma_n,$$

there exists a unique $\tilde{x} \in \mathbb{R}^n$ with property (3.5). For $\delta x := \tilde{x} - \hat{x}$,

$$\|\delta x\|_2 \leq \frac{1}{\sigma_n(1 - \eta)} \cdot \left(\|\delta b\|_2 + \|\delta A\|_2 \|\hat{x}\|_2\right) + \frac{1}{\sigma_n^2(1 - \eta)^2} \cdot \|\delta A\|_2 \cdot \|\hat{r}\|_2 \quad (3.6)$$

holds.

Proof The matrix $A + \delta A$ has full rank, since

$$(A + \delta A)x = 0 \Rightarrow Ax = -\delta Ax \Rightarrow \sigma_n \|x\|_2 \leq \|Ax\|_2 \leq \|\delta A\|_2 \|x\|_2$$

$$\Rightarrow \underbrace{(\sigma_n - \|\delta A\|_2)}_{> 0} \|x\|_2 \leq 0,$$

which is only possible in case $x = 0$. According to Theorem 3.1, $\tilde{x}$ is uniquely determined.

Setting $\tilde{A} = A + \delta A$ and $\tilde{b} = b + \delta b$, one concludes from Theorem 3.1 that $\tilde{x} = M\tilde{A}^\top \tilde{b}$, where $M := (\tilde{A}^\top \tilde{A})^{-1}$ (normal equations). Consequently, we have

$$\delta x = \tilde{x} - \hat{x} = M\tilde{A}^\top \tilde{b} - \hat{x} = M\tilde{A}^\top(\tilde{b} - \tilde{A}\hat{x})$$

$$= M\tilde{A}^\top(b - A\hat{x}) + M\tilde{A}^\top(\delta b - \delta A\hat{x})$$

$$= M(\delta A)^\top \hat{r} + M\tilde{A}^\top(\delta b - \delta A\hat{x}) \quad [(A + \delta A)^\top \hat{r} = (\delta A)^\top \hat{r} \text{ from (3.3)}]$$

and from this, we get

$$\|\delta x\|_2 \leq \|M\|_2 \|\delta A\|_2 \|\hat{r}\|_2 + \|M\tilde{A}^\top\|_2 \left(\|\delta b\|_2 + \|\delta A\|_2 \|\hat{x}\|_2\right).$$

Everything is settled if $\|M\|_2 \leq 1/(\sigma_n^2(1 - \eta)^2)$ and $\|M\tilde{A}^\top\|_2 \leq 1/(\sigma_n(1 - \eta))$. From an SVD $\tilde{A} = \tilde{U}\tilde{\Sigma}\tilde{V}^\top$, we get

$$M = \tilde{V}(\tilde{\Sigma}^\top\tilde{\Sigma})^{-1}\tilde{V}^\top = \tilde{V}\text{diag}\left(1/\tilde{\sigma}_1^2, \ldots, 1/\tilde{\sigma}_n^2\right)\tilde{V}^\top$$

and from this, we get $M\tilde{A}^\top = \tilde{V}\tilde{\Sigma}^+\tilde{U}^\top$, where $\tilde{\Sigma}^+ = \text{diag}(1/\tilde{\sigma}_1, \ldots, 1/\tilde{\sigma}_n) \in \mathbb{R}^{n,m}$. We conclude $\|M\|_2 = 1/\tilde{\sigma}_n^2$ as well as $\|M\tilde{A}^\top\|_2 = 1/\tilde{\sigma}_n$. From Theorem A.3 it follows $|\sigma_n - \tilde{\sigma}_n| \leq \|\delta A\|_2$, so that $\tilde{\sigma}_n \geq \sigma_n - \|\delta A\|_2 = \sigma_n(1 - \eta)$ and thus $\|M\tilde{A}^\top\|_2 \leq 1/(\sigma_n(1 - \eta))$. An analogous argument holds for $\|M\|_2$. $\square$

The bound given for $\|\delta x\|_2$ in the above theorem is "nearly sharp": [Bjö96, p. 29], gives an example where it is approximately attained.

Theorem 3.4 (Relative Sensitivity of the Linear Least Squares Problem) *For $m \geq n$, let $A, \delta A \in \mathbb{R}^{m,n}$ and $b, \delta b \in \mathbb{R}^m$. Let A have singular values $\sigma_1 \geq \cdots \geq \sigma_n > 0$ and*

$$\kappa_2(A) := \frac{\sigma_1}{\sigma_n} . \tag{3.7}$$

Let $\hat{x} \neq 0$ be the solution of (3.2), characterized by (3.4) and having residual $\hat{r} := b - A\hat{x}$. Further assume that for some $\varepsilon > 0$,

$$\|\delta A\|_2 \leq \varepsilon\|A\|_2, \quad \|\delta b\|_2 \leq \varepsilon\|b\|_2, \quad \text{and} \quad \kappa_2(A)\varepsilon < 1 \tag{3.8}$$

holds. Then, (3.5) uniquely characterizes $\tilde{x} \in \mathbb{R}^n$ and for $\delta x := \tilde{x} - \hat{x}$,

$$\frac{\|\delta x\|_2}{\|\hat{x}\|_2} \leq \frac{\kappa_2(A)\varepsilon}{1 - \kappa_2(A)\varepsilon}\left(2 + \left(\frac{\kappa_2(A)}{1 - \kappa_2(A)\varepsilon} + 1\right)\frac{\|\hat{r}\|_2}{\|A\|_2\|\hat{x}\|_2}\right) \tag{3.9}$$

holds.

Proof From an SVD of A, one gets $\sigma_1 = \|A\|_2$. Therefore, $\|\delta A\|_2 \leq \varepsilon\|A\|_2$ means that $\eta = \|\delta A\|_2/\sigma_n \leq \kappa_2(A)\varepsilon$. Under the condition $\kappa_2(A)\varepsilon < 1$, the condition $\eta < 1$ from Theorem 3.3 is fulfilled a fortiori and the estimate (3.6) remains valid, if η is replaced by $\kappa_2(A)\varepsilon$. Division by $\|\hat{x}\|_2$ shows

$$\frac{\|\delta x\|_2}{\|\hat{x}\|_2} \leq \frac{1}{\sigma_n(1 - \kappa_2(A)\varepsilon)}\left(\frac{\|\delta b\|_2}{\|\hat{x}\|_2} + \|\delta A\|_2\right) + \frac{1}{(\sigma_n(1 - \kappa_2(A)\varepsilon))^2}\cdot\frac{\|\delta A\|_2\|\hat{r}\|_2}{\|\hat{x}\|_2} .$$

Because of $\|\delta A\|_2 \leq \varepsilon\|A\|_2$, $\|\delta b\|_2 \leq \varepsilon\|b\|_2$, and $\|A\|_2/\sigma_n = \kappa_2(A)$, we get

$$\frac{\|\delta x\|_2}{\|\hat{x}\|_2} \leq \frac{\kappa_2(A)\varepsilon}{1 - \kappa_2(A)\varepsilon}\left(\frac{\|b\|_2}{\|A\|_2\|\hat{x}\|_2} + 1\right) + \frac{\kappa_2(A)^2\varepsilon}{(1 - \kappa_2(A)\varepsilon)^2}\cdot\frac{\|\hat{r}\|_2}{\|A\|_2\|\hat{x}\|_2} .$$

Equation (3.9) finally follows from $\|b\|_2 \leq \|b - A\hat{x}\|_2 + \|A\hat{x}\|_2 \leq \|\hat{r}\|_2 + \|A\|_2\|\hat{x}\|_2$. $\square$

Theorems 3.3 and 3.4 give a measure of how sensitively the solution of the least squares problem depends on perturbations of the input data A and b in the worst case. Evidently, $\delta x \to 0$ for $\delta A \to 0$ and $\delta b \to 0$, so we can state once more that according to Definition 1.5, the linear least squares problem formally is well-posed.

In practice, however, we cannot hope that ε tends to zero. It rather remains a finite, positive value, the size of which is limited by the finite precision of measurement devices. It is important to note that neither the input error size ε nor the output error size $\|\delta x\|_2/\|\hat{x}\|_2$ has anything to do with the finite precision of computer arithmetic. These errors would even show up if computations could be carried out exactly.

Both theorems contain statements about linear systems of equations as a special case: if A is non-singular, the solution $\hat{x}$ of (3.1) is unique and coincides with the solution of (3.2). Theorems 3.3 and 3.4 can then be used with $\hat{r} = 0$. It turns out that the number $\kappa_2(A)\varepsilon$ is decisive for the worst case error in the solution of $Ax = b$. This number clearly has to be much smaller than 1 to guarantee a meaningful solution. If $\varepsilon \ll \kappa_2(A)$, one has $\kappa_2(A)/(1 - \kappa_2(A)\varepsilon) \approx \kappa_2(A)$. Then, $\kappa_2(A)$ can be interpreted as a factor, by which relative input errors of size ε may be amplified in the result. For linear least squares problems, an additional (approximate) amplification factor $\kappa_2(A)^2\|\hat{r}\|_2$ appears on the right-hand side of (3.9), meaning that in this case the worst case error also depends on b. If b is close to $\mathscr{R}_A$, i.e. if the system $Ax = b$ is nearly consistent ("nearly solvable"), then $\|\hat{r}\|_2$ is small. Otherwise, the solution of a least squares problem may depend more sensitively on A and b than does the solution of a linear system of equations.

An inverse problem of form (3.1) or (3.2) is called **well conditioned**, if small perturbations in the data A and b do only lead to small perturbations in the solution. It is called **badly conditioned**, if small perturbations in the data A and b may lead to large perturbations in the solution. The **condition number** of the inverse problem is a factor by which relative perturbations in the input data A or b might be amplified in the result *in the worst case*. The condition number of (3.1) is approximately given by $\kappa_2(A)$, which is called **condition number of matrix** A. The condition number of (3.2) is approximately given by the maximum of $\kappa_2(A)$ and $\kappa_2(A)^2\|\hat{r}\|_2/(\|A\|_2\|\hat{x}\|_2)$.

Example 3.5 (Condition Numbers for the Collocation Method) Reconsider Example 2.11 from Sect. 2.3. The linearized model problem of waveform inversion was discretized by the collocation method. This led to an overdetermined system $Ax = b$, which was replaced by the corresponding least squares problem. In this case, we suppose the system $Ax = b$ to be nearly consistent. It is to be expected, therefore, that $\kappa_2(A)\varepsilon$ bounds the relative error in the least squares solution (disregarding computing errors due to the finite precision computer arithmetic). In Fig. 3.3, the condition number $\kappa_2(A)$ is shown (on a logarithmic scale) for matrix A resulting from various discretization levels n. Visibly, $\kappa_2(A)$ is growing quickly with n. For

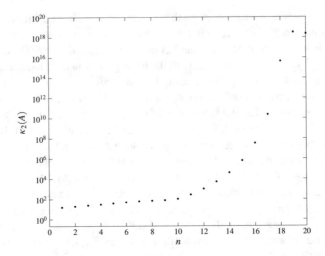

Fig. 3.3 Condition numbers $\kappa_2(A)$ as a function of n

computers with "double precision arithmetic,"[1] an input error size $\varepsilon \geq 10^{-16}$ has
to be assumed, which would result from the conversion of measurement data to
machine numbers—even in the purely hypothetical case of exact measurements.
According to Fig. 3.3, a discretization level $n > 18$ may then already lead to a
meaningless result. In Example 2.11, we actually had $\varepsilon \approx 10^{-5}$ (compare definition
of w^δ and w). Thus, even a condition number $\kappa_2(A) > 10^5$ will have a disastrous
effect and a discretization level $n \geq 15$ should therefore be avoided.   ◊

Matrices with Deficient Rank

In case of a rank deficient matrix A, the smallest singular value vanishes: $\sigma_n = 0$.
One may then formally set

$$\kappa_2(A) = \sigma_1/0 = \infty$$

and call problem (3.1) or (3.2) "infinitely badly conditioned." This is in accordance
with the fact that no unique solution exists anymore. The uniqueness can be re-

---

[1]Such computers use 53 bits to represent the mantissae of floating point numbers. Then, the
rounding error committed when converting a real to a machine number is bounded by $2^{-53} \approx 10^{-16}$.

established if an *additional requirement* is added.[2] In case of (3.2), we might ask for the solution with smallest Euclidean norm. This means that (3.2) is changed to become a *new problem*, namely

$$\text{Find } \hat{x} \text{ such that } \|\hat{x}\|_2 \leq \|x\|_2 \text{ for all } x \in M := \arg\min\{\|b - Ax\|_2\}, \qquad (3.10)$$

where $\arg\min\{\|b - Ax\|_2\}$ is the set of all vectors solving (3.2). A solution of (3.10) is called **minimum-norm solution** of the least squares problem (3.2). In case $\text{rank}(A) = n$, the unique solution $\hat{x}$ of (3.2) is also necessarily the unique solution of (3.10). The following theorem shows that (3.10) has a unique solution for any matrix A. It also shows that (3.10) may be better conditioned than (3.2).

Theorem 3.6 (Linear Least Squares Problem for General Matrices) *Let $A \in \mathbb{R}^{m,n}$ have rank $r \leq \min\{m, n\}$ and have an SVD*

$$A = U\Sigma V^\top = (\ \underbrace{U_1}_{r} \quad \underbrace{U_2}_{m-r}\ ) \left. \begin{pmatrix} \Sigma_1 & 0 \\ 0 & 0 \end{pmatrix} \right\} \begin{matrix} r \\ m\text{-}r \end{matrix} \ (\ \underbrace{V_1}_{r} \quad \underbrace{V_2}_{n-r}\ )^\top$$

$$\underbrace{}_{r} \underbrace{}_{n-r}$$

$$= U_1 \Sigma_1 V_1^\top .$$

(a) All solutions of (3.2) have a representation of the form

$$x = V_1 \Sigma_1^{-1} U_1^\top b + V_2 z, \quad z \in \mathbb{R}^{n-r}. \qquad (3.11)$$

(b) Among all solutions of (3.2), there is a unique one having minimal Euclidean norm, i.e., there is a unique solution of (3.10). This solution is given by

$$x = V_1 \Sigma_1^{-1} U_1^\top b .$$

Its norm is bounded by $\|x\|_2 \leq \|b\|_2 / \sigma_r$.

(c) If the right-hand side b is perturbed to become $b + \delta b$, then the according (unique) solution of (3.10) becomes $x + \delta x$ and

$$\|\delta x\|_2 \leq \frac{\|\delta b\|_2}{\sigma_r}$$

holds.

---

[2]The same kind of idea was used to force the inverse gravimetry problem into having a unique solution.

Proof Part (a):

$$\|b - Ax\|_2^2 = \left\| U^\top b - \begin{pmatrix} U_1^\top \\ U_2^\top \end{pmatrix} U_1 \Sigma_1 V_1^\top x \right\|_2^2 = \left\| \begin{pmatrix} U_1^\top b - \Sigma_1 V_1^\top x \\ U_2^\top b \end{pmatrix} \right\|_2^2$$
$$= \| U_1^\top b - \Sigma_1 V_1^\top x \|_2^2 + \| U_2^\top b \|_2^2$$

becomes minimal if and only if

$$\Sigma_1 V_1^\top x = U_1^\top b \quad \Longleftrightarrow \quad x = V_1 \Sigma_1^{-1} U_1^\top b + V_2 z$$

for arbitrary $z \in \mathbb{R}^{n-r}$, since

$$\mathscr{N}_{V_1^\top} = \mathscr{R}_{V_1}^\perp = \{V_2 z; \ z \in \mathbb{R}^{n-r}\}.$$

Part (b): Since V_1 and V_2 are orthogonal, from (3.11) and Pythagoras' theorem, we get

$$\|x\|_2^2 = \| V_1 \Sigma^{-1} U_1^\top b \|_2^2 + \| V_2 z \|_2^2.$$

The right-hand side becomes minimal if and only if $V_2 z = 0$, i.e., if and only if $z = 0$. For $z = 0$, we get

$$\|x\|_2^2 = \| V_1 \Sigma_1^{-1} U_1^\top b \|_2^2 = \left\| \begin{pmatrix} u_1^\top b/\sigma_1 \\ \vdots \\ u_r^\top b/\sigma_r \end{pmatrix} \right\|_2^2 \le \frac{1}{\sigma_r^2} \sum_{j=1}^r |u_j^\top b|^2 \le \frac{\|b\|_2^2}{\sigma_r^2}.$$

Part (c): Replacing b by $b + \delta b$ in part (b), we get the minimum-norm solution $x + \delta x = V_1 \Sigma_1^{-1} U_1^\top (b + \delta b)$. Thus, $\delta x = V_1 \Sigma_1^{-1} U_1^\top \delta b$, and we can directly use the last estimate from part (b) to end the proof.

$\square$

The sensitivity of the minimum-norm solution with respect to perturbations of b is determined by the size of the smallest non-vanishing singular value σ_r of A. More generally, one can also consider perturbations of the elements of A. Bounds for $\|\delta x\|_2$ do exist also in this general case, which are analogous to the one given in Theorem 3.3 with σ_n replaced by σ_r, see, e.g., Theorem 1.4.6 in [Bjö96].

Definition 3.7 (Pseudoinverse) Using the notation of Theorem 3.6, define

$$A^+ := V_1 \Sigma_1^{-1} U_1^\top.$$

This matrix is called the **pseudoinverse** of A.

According to Theorem 3.6, the unique minimum-norm solution of (3.2) can formally be written as

$$x = A^+ b.$$

In case $\text{rank}(A) = n$, a unique solution of (3.2) exists and $A^+ = (A^T A)^{-1} A^T$. In case $\text{rank}(A) = n$ *and* $m = n$, a unique solution of (3.1) exists and $A^+ = A^{-1}$.

Example 3.8 Let $\varepsilon > 0$ and

$$A = \begin{pmatrix} 1 & 0 \\ 0 & \varepsilon \end{pmatrix}, \quad b = \begin{pmatrix} b_1 \\ b_2 \end{pmatrix} \quad \Longrightarrow \quad A^+ = \begin{pmatrix} 1 & 0 \\ 0 & \varepsilon^{-1} \end{pmatrix}, \quad A^+ b = \begin{pmatrix} b_1 \\ b_2/\varepsilon \end{pmatrix}.$$

The errors in component b_2 of the right-hand side are amplified by a factor ε^{-1}. Since $\kappa_2(A) = \varepsilon^{-1}$, this is in accordance with Theorem 3.4. The condition of problem (3.2) becomes arbitrarily bad with $\varepsilon \to 0$. Consider, on the other hand,

$$B = \begin{pmatrix} 1 & 0 \\ 0 & 0 \end{pmatrix} = \underbrace{\begin{pmatrix} 1 & 0 \\ 0 & 1 \end{pmatrix}}_{=\,U} \underbrace{\begin{pmatrix} 1 & 0 \\ 0 & 0 \end{pmatrix}}_{=\,\Sigma} \underbrace{\begin{pmatrix} 1 & 0 \\ 0 & 1 \end{pmatrix}^{\mathsf{T}}}_{=\,V^{\mathsf{T}}} = \begin{pmatrix} 1 \\ 0 \end{pmatrix} 1 \begin{pmatrix} 1 \\ 0 \end{pmatrix}^{\mathsf{T}} \quad \Longrightarrow \quad B^+ = B.$$

Matrix B has rank 1 with smallest non-vanishing singular value $\sigma_1 = 1$. The solution $B^+ b = (b_1, 0)^{\mathsf{T}}$ of problem (3.10) is perfectly well conditioned with respect to the data b. $\diamond$

Theorem 3.6 and the above example show that the presence of small singular values can be more annoying than a deficient rank.

Generalized Least Squares Problems

Even if the matrix A is rank deficient, the minimum-norm problem (3.10) still has a unique solution. In this paragraph, a more general additional requirement is added to the least squares problem than the minimum-norm criterion. This will be of use in the context of regularization. To start with, a characterization of the pseudoinverse will be given, which differs from Definition 3.7.

Let $S \subset \mathbb{R}^n$ be a linear subspace of $\mathbb{R}^n$ with orthonormal basis $\{s_1, \ldots, s_k\}$. Define the matrices

$$C := (s_1 | \cdots | s_k) \in \mathbb{R}^{n,k}, \quad P_S := CC^T, \quad \text{and} \quad P_{S^\perp} := I_n - CC^T. \qquad (3.12)$$

It can easily be seen that P_S defines a mapping $\mathbb{R}^n \to S$, $y \mapsto P_S y$, where $\|y - P_S y\|_2 \leq \|y - x\|_2$ for all vectors $x \in S$. For this reason, P_S is called an

orthogonal projector onto the subspace S of $\mathbb{R}^n$. Likewise, $P_{S^\perp}$ is an orthogonal projector onto the orthogonal complement $S^\perp$ of S. Any vector $x \in \mathbb{R}^n$ has a unique decomposition $x = u + v$ with $u = P_S x \in S$ and $v = P_{S^\perp} x \in S^\perp$. From Theorem 3.6 and Definition 3.7, it can readily be seen that

$$P_{\mathscr{R}_A} = AA^+ = U_1 U_1^\top, \; P_{\mathscr{R}_A^\perp} = I_m - AA^+ = U_2 U_2^\top$$
$$P_{\mathscr{N}_A^\perp} = A^+ A = V_1 V_1^\top, \; P_{\mathscr{N}_A} = I_n - A^+ A = V_2 V_2^\top. \tag{3.13}$$

Using these relations, the pseudoinverse can be interpreted as follows.

Lemma 3.9 *Under the conditions of Theorem 3.6, one has*

$$A^+ = B \circ P_{\mathscr{R}_A},$$

where B is the inverse of the bijective mapping

$$\tilde{A} = A_{|\mathscr{N}_A^\perp} : \mathscr{N}_A^\perp \to \mathscr{R}_A.$$

Further on, the pseudoinverse A^+ is the unique matrix $X \in \mathbb{R}^{n,m}$ having the following four properties:

$$AXA = A, \; (AX)^\top = AX,$$
$$XAX = X, \; (XA)^\top = XA, \tag{3.14}$$

which are called **Moore–Penrose axioms**.

Proof Reusing the notation from Theorem 3.6, any $y \in \mathbb{R}^m$ can uniquely be written in the form $y = U_1 z_1 + U_2 z_2$, where $z_1 \in \mathbb{R}^r$ and $z_2 \in \mathbb{R}^{m-r}$. One therefore gets $A^+ y = V_1 \Sigma_1^{-1} z_1$. On the other hand, one also has

$$B \circ P_{\mathscr{R}_A} y = A^{-1}(U_1 z_1) = V_1 \Sigma_1^{-1} z_1,$$

since $V_1 \Sigma_1^{-1} z_1 \in \mathscr{N}_A^\perp$ and $A(V_1 \Sigma_1^{-1} z_1) = U_1 \Sigma_1 V_1^\top V_1 \Sigma_1^{-1} z_1 = U_1 z_1$. This proves the first part of the lemma. Property (3.14) follows immediately from Theorem 3.6 and Definition 3.7 for $X = A^+$. Knowing that $X = A^+$ satisfies (3.14), assume that X_1 and X_2 both satisfy (3.14). Then,

$$AX_1 = (AX_1)^\top = X_1^\top A^\top = X_1^\top (AX_2 A)^\top = (AX_1)^\top (AX_2)^\top = AX_1 AX_2 = AX_2.$$

In the same way, $X_1 A = X_2 A$. Therefore,

$$X_1 = X_1 AX_1 = X_1 AX_2 = X_2 AX_2 = X_2,$$

meaning that *only* $X = A^+$ can satisfy (3.14): the Moore–Penrose axioms *determine* A^+. □

Theorem 3.10 (Generalized Linear Least Squares Problem) *Let all conditions of Theorem 3.6 hold and $L \in \mathbb{R}^{p,n}$. There exists a unique solution of*

$$\min_{x \in M} \|Lx\|_2, \qquad M := \arg\min\{\|b - Ax\|_2\} \tag{3.15}$$

if and only if

$$\mathcal{N}_A \cap \mathcal{N}_L = \{0\}. \tag{3.16}$$

If (3.16) holds, this minimizer is given by

$$x = (I_n - (LP_{\mathcal{N}_A})^+ L)A^+ b. \tag{3.17}$$

Proof The necessity of (3.16) for the uniqueness of a minimizer is clear, since for every $z \in \mathcal{N}_A \cap \mathcal{N}_L$ and for all $x \in \mathbb{R}^n$, one has $L(x + z) = Lx$ and $A(x + z) - b = Ax - b$. So, let now (3.16) hold. From Theorem 3.6, it is known that all elements of M have the form

$$x = A^+ b + P_{\mathcal{N}_A} z, \quad z \in \mathbb{R}^n,$$

so the minimization problem (3.15) can be written in the form

$$\min_{z \in \mathbb{R}^n} \|LP_{\mathcal{N}_A} z + LA^+ b\|_2$$

which is a standard linear least squares problem. Again from Theorem 3.6, one concludes that the general solution of *this* problem has the form

$$z = -(LP_{\mathcal{N}_A})^+ LA^+ b + (I_n - (LP_{\mathcal{N}_A})^+ LP_{\mathcal{N}_A})y, \quad y \in \mathbb{R}^n, \tag{3.18}$$

where (3.13) was used to express the projection onto the nullspace of $LP_{\mathcal{N}_A}$. Thus, the general solution of (3.15) has the form

$$x = A^+ b + P_{\mathcal{N}_A} z, \quad z \text{ from (3.18)}. \tag{3.19}$$

Observe now that

$$LP_{\mathcal{N}_A} x = 0 \iff \underbrace{P_{\mathcal{N}_A} x}_{\in \mathcal{N}_A} \in \mathcal{N}_L \overset{\mathcal{N}_A \cap \mathcal{N}_L = \{0\}}{\iff} P_{\mathcal{N}_A} x = 0 \iff x \in \mathcal{N}_A^\perp,$$

$$\tag{3.20}$$

such that, by (3.13), $(I_n - (LP_{\mathcal{N}_A})^+ LP_{\mathcal{N}_A})y \in \mathcal{N}_A^\perp$ for all $y \in \mathbb{R}^n$. This shows that

$$P_{\mathcal{N}_A}(I_n - (LP_{\mathcal{N}_A})^+ LP_{\mathcal{N}_A})y = 0$$

for all $y \in \mathbb{R}^n$. Using this and (3.18) in (3.19), we have proven that there is a unique solution

$$x = A^+ b - P_{\mathcal{N}_A}(LP_{\mathcal{N}_A})^+ LA^+ b = (I_n - P_{\mathcal{N}_A}(LP_{\mathcal{N}_A})^+ L)A^+ b$$

of (3.15). From this, (3.17) follows, if $P_{\mathcal{N}_A}(LP_{\mathcal{N}_A})^+ = (LP_{\mathcal{N}_A})^+$. The latter identity can be verified as follows:

$$\mathcal{R}_{(LP_{\mathcal{N}_A})^+} \overset{\text{Lemma (3.9)}}{=} \mathcal{N}_{LP_{\mathcal{N}_A}}^\perp \overset{(3.20)}{=} \left(\mathcal{N}_A^\perp\right)^\perp = \mathcal{N}_A,$$

and this ends the proof. $\qquad\qquad\qquad\qquad\qquad\qquad\qquad\qquad\qquad\qquad\qquad\square$

The above theorem gives rise to the following definition.

Definition 3.11 (Weighted Pseudoinverse) Let all conditions of Theorem 3.6 hold, $L \in \mathbb{R}^{p,n}$, and (3.16) hold. Then, the matrix

$$A_L^+ := (I_n - (LP_{\mathcal{N}_A})^+ L)A^+ \in \mathbb{R}^{n,m}$$

is called L-**weighted pseudoinverse** of A. Analogously, the matrix

$$L_A^+ := (I_n - (AP_{\mathcal{N}_L})^+ A)L^+ \in \mathbb{R}^{n,p}$$

is called A-**weighted pseudoinverse** of L.

Whereas A^+ was introduced as a mapping relating $b \in \mathbb{R}^m$ to the unique least squares solution of $Ax = b$ having minimal norm, A_L^+ is defined as the mapping relating b to the unique least squares solution of $Ax = b$ which minimizes the norm of Lx. But the L-weighted pseudoinverse of A has another interpretation similar to the one given for A^+ in Lemma 3.9. To see this, introduce the scalar product on $\mathbb{R}^n$ defined by

$$\langle x|\bar{x}\rangle_* := \langle Ax|A\bar{x}\rangle + \langle Lx|L\bar{x}\rangle, \quad x, \bar{x} \in \mathbb{R}^n \qquad (3.21)$$

with standard Euclidean scalar products $\langle \bullet|\bullet\rangle$ on the right-hand side. Note that (3.16) is needed for positive definiteness to hold. Let us write

$$x \perp_* y = 0 \quad :\Longleftrightarrow \quad \langle x|y\rangle_* = 0.$$

The orthogonal complement of $\mathcal{N}_A$ with respect to $\langle \bullet|\bullet\rangle_*$ is given by

$$\mathcal{N}_A^{\perp_*} = \{x \in \mathbb{R}^n; \ L^\top Lx \perp \mathcal{N}_A\}, \qquad (3.22)$$

which can be verified easily:

$$x \in \mathcal{N}_A^{\perp *} \iff \langle x | \bar{x} \rangle_* = \langle Ax | A\bar{x} \rangle + \langle Lx | L\bar{x} \rangle = 0 \text{ for all } \bar{x} \in \mathcal{N}_A$$

$$\iff \langle Lx | L\bar{x} \rangle = \langle L^{\top} Lx | \bar{x} \rangle = 0 \text{ for all } \bar{x} \in \mathcal{N}_A.$$

Lemma 3.12 *Under the conditions of Definition 3.11, one has*

$$A_L^+ = B_* \circ P_{\mathcal{R}_A},$$

where B_* is the inverse of the bijective mapping

$$A_* = A_{|\mathcal{N}_A^{\perp *}} : \mathcal{N}_A^{\perp *} \to \mathcal{R}_A.$$

Proof It has to be shown that for every $b \in \mathbb{R}^m$, $B_*(P_{\mathcal{R}_A} b)$ is a least squares solution of $Ax = b$ and minimizes $\|Lx\|_2$ among all least squares solutions. Let $x_0 := B_*(P_{\mathcal{R}_A} b)$ and $x_1 := A^+ b$. Then,

$$A(x_0 - x_1) = AA^{-1}(P_{\mathcal{R}_A} b) - AA^{-1}(P_{\mathcal{R}_A} b) = 0,$$

showing that $x_0 - x_1 \in \mathcal{N}_A$. This shows that x_0 is a least squares solution, as is x_1. Next, let x_2 be any least squares solution. Then,

$$\|Lx_2\|_2^2 = \|Lx_0 + L(x_2 - x_0)\|_2^2 = \|Lx_0\|_2^2 + \|L(x_2 - x_0)\|_2^2 + 2(x_2 - x_0)^{\top} L^{\top} Lx_0$$

$$= \|Lx_0\|_2^2 + \|L(x_2 - x_0)\|_2^2,$$

since $x_2 - x_0 \in \mathcal{N}_A$, $x_0 = B_*(P_{\mathcal{R}_A} b) \in \mathcal{N}_A^{\perp *}$, and (3.22) holds. The term $\|L(x_2 - x_0)\|_2^2$ vanishes if and only if $x_2 - x_0 \in \mathcal{N}_A \cap \mathcal{N}_L \overset{(3.16)}{=} \{0\}$. Thus, x_0 is the only least squares solution minimizing the norm of Lx. $\qquad\square$

3.3 The Concept of Regularization

To explain what is meant by "regularization," we first turn back to linear parameter identification problems in their general form. This means we consider the solution of an equation $Tu = w$, where $T : X \to \mathbb{W} \subseteq Y$ is linear and bijective, where $(X, \|\bullet\|_X)$ and $(Y, \|\bullet\|_Y)$ are normed spaces and $w \in \mathbb{W}$ is fixed. A unique solution u^* of this problem is assumed to exist, but usually two difficulties impede its practical computation.

1. Only an approximation $w^{\delta} \in Y$ of w is known. At best, one can estimate the data error $w - w^{\delta}$ by $\|w - w^{\delta}\|_Y \leq \delta$ for some $\delta > 0$, but one can generally *not* assume $w^{\delta} \in \mathbb{W}$.
2. The inverse $T^{-1} : \mathbb{W} \to X$ is possibly not continuous. Then, even if one knew a good approximation $\tilde{w} \in \mathbb{W}$ of w, $T^{-1}(\tilde{w})$ could be far away from u^*.

Regularization is an attempt to bypass both difficulties in order to find a good approximation of u^*. Before giving the usual, abstract definition of regularization, we tentatively define it to be a set $\{R_t;\ t \in I\}$ of *continuous* operators $R_t : Y \to X$, such that

$$\|R_t(w) - T^{-1}(w)\|_X \overset{t \to 0}{\longrightarrow} 0 \quad \text{for all} \quad w \in \mathbb{W}. \tag{3.23}$$

The index set I may be any (possibly uncountable) set $I \subseteq \mathbb{R}_0^+$ having 0 as a limit point, so that (3.23) makes sense. If all operators R_t are linear, then the regularization is called linear. Given below are two essential points:

1. The operators R_t are defined on the whole space Y and thus can be applied to any $w^\delta \in Y$, even if $w^\delta \notin \mathbb{W}$.
2. The operators R_t are continuous approximants of the possibly discontinuous operator $T^{-1} : \mathbb{W} \to X$, converging pointwise to T^{-1} on $\mathbb{W}$.

The next example shows that discretization already means regularization (in the sense of (3.23)).

Example 3.13 (Regularization by the Least Squares Discretization Method) This method was introduced in Sect. 2.2. We chose a subspace $X_n = \langle \varphi_1, \ldots, \varphi_{d_n} \rangle \subset X$ and defined an approximation $u_n = \sum_{j=1}^{d_n} x_j \varphi_j$ of the true solution u^*. The approximation was formalized by operators

$$R_n : Y \to X_n \subseteq X, \quad y \mapsto u_n,$$

defined in (2.32), see Theorem 2.8. These operators are continuous for all $n \in \mathbb{N}$. Let us rename R_n to $R_{1/n}$, $n \in \mathbb{N}$. Theorem 2.9 states that under conditions (2.40) and (2.41), convergence $\|R_{1/n}(w) - T^{-1}(w)\|_X \to 0$ holds for $n \to \infty$ and for all $w \in T(X) = \mathbb{W}$. Thus, the operator set $\{R_t;\ t \in I\}$ with $I = \{1/n;\ n \in \mathbb{N}\}$ is a regularization in the sense of (3.23). $\diamond$

The above tentative definition is not sufficient, since perturbed data are not considered. Assume some $w^\delta \in Y$ (possibly not contained in $\mathbb{W}$) was given and assume knowledge of $\delta > 0$ such that $\|w^\delta - w\|_Y \le \delta$. If the regularization is linear, then one can estimate, making use of $Tu^* = w \in \mathbb{W}$,

$$\|R_t w^\delta - u^*\|_X \le \|R_t(w^\delta - w)\|_X + \|R_t w - u^*\|_X$$

$$\le \|R_t\|\delta + \|R_t w - u^*\|_X, \tag{3.24}$$

which corresponds to estimate (2.34) from Theorem 2.8. The total reconstruction error is bounded by the sum of two terms, with the second term $\|R_t w - u^*\|_X = \|R_t w - T^{-1}(w)\|_X$ tending to zero for $t \to 0$, if (3.23) holds. But at the same time, the operator norm $\|R_t\|$ grows beyond all bounds for $t \to 0$, whenever T^{-1} is not continuous.[3] This was shown in the special case of regularization by discretization (by the linear least squares method) in (2.44) and (2.45). For a finite value $\delta > 0$ and for given data w^δ, one would like to choose an optimal parameter t^* such that $\|R_{t^*}(w^\delta) - u^*\|_X$ becomes as small as possible. In the special case of Example 3.13 (regularization by discretization), this would mean to select an optimal dicretization level, which is not possible in practice, since u^* is not known. For general operators R_t, an optimal choice $t = t^*$ is even less possible. Anyway, some rule $\rho(\delta, w^\delta) = t$ is required in order to choose a parameter t in dependence of given data w^δ and given data error magnitude δ. Since an optimal parameter selection is out of reach, one more modestly demands the total error $\|R_{\rho(\delta, w^\delta)}(w^\delta) - T^{-1}(w)\|_X$ to tend to zero for all $w \in \mathbb{W}$, if $\delta \to 0$. This convergence shall be uniform for all w^δ in the neighborhood of w. These considerations lead to the following definition.

Definition 3.14 (Regularization) Let $(X, \| \bullet \|_X)$ and $(Y, \| \bullet \|_Y)$ be normed linear spaces and $T : X \to \mathbb{W} \subseteq Y$ be linear, continuous, and bijective. Let $I \subset \mathbb{R}_0^+$ be an index set having 0 as a limit point and $\{R_t;\ t \in I\}$ be a set of continuous operators $R_t : Y \to X$. If a mapping $\rho : \mathbb{R}^+ \times Y \to \mathbb{R}_0^+$ exists such that for all $w \in \mathbb{W}$

$$\sup\{\rho(\delta, w^\delta);\ w^\delta \in Y, \|w - w^\delta\|_Y \le \delta\} \xrightarrow{\delta \to 0} 0 \qquad (3.25)$$

and

$$\sup\{\|R_{\rho(\delta, w^\delta)}(w^\delta) - T^{-1}(w)\|_X;\ w^\delta \in Y,\ \|w - w^\delta\|_Y \le \delta\} \xrightarrow{\delta \to 0} 0, \qquad (3.26)$$

then the pair $(\{R_t;\ t \in I\}, \rho)$ is called **regularization** of T^{-1}. If all operators R_t are linear, the regularization is called linear. Each number $\rho(\delta, w^\delta)$ is called a **regularization parameter**. The mapping ρ is called **parameter choice**. If a parameter choice does only depend on δ but not on the data w^δ, then it is called a **parameter choice a priori**. Otherwise, it is called a **parameter choice a posteriori**.

From (3.26), it follows that

$$\|R_{\rho(\delta, w)}(w) - T^{-1}(w)\|_X \xrightarrow{\delta \to 0} 0 \quad \text{for all} \quad w \in \mathbb{W},$$

meaning that the above definition generalizes the previous, tentative one. From (3.24), it can be seen that for linear regularizations with (3.23), an a priori parameter choice $\rho = \rho(\delta)$ is admissible if

$$\rho(\delta) \xrightarrow{\delta \to 0} 0 \quad \text{and} \quad \|R_{\rho(\delta)}\| \delta \xrightarrow{\delta \to 0} 0. \qquad (3.27)$$

---

[3]Assume, on the contrary, that $\|R_t\| \le C$ for some constant C and for $t \to 0$. Then, $\|R_t y\|_Y \le C\|y\|_Y$ for all $y \in T(X)$ and $t \to 0$. Since $R_t y \to T^{-1}y$, it follows that $\|T^{-1}y\|_Y \le C\|y\|_Y$ for all $y \in T(X)$, a contradiction to the unboundedness of T^{-1}.

Example 3.15 (Regularization by the Least Squares Discretization Method) Let us look again at discretization by the least squares method. With the renaming introduced in Example 3.15, the continuous discretization operators form a set $\{R_t; \ t \in I\}, I = \{1/n; \ n \in \mathbb{N}\}$. From Theorem 2.9, it follows that under conditions (2.40) and (2.41), the convergence property $\|R_t w - T^{-1}(w)\|_X \to 0$ holds for $t \to 0$. From Theorem 2.8, estimate (2.34), it can be seen that a parameter choice satisfying (3.27) makes discretization by the least squares method a regularization in the sense of Definition 3.14 (if all conditions (2.40), (2.41), and (3.27) hold). $\lozenge$

Discretization of a parameter identification problem $Tu = w$ is a kind of a regularization. This could be verified in the above examples for the least squares method, but it is also true for the collocation method. In both cases, discretization leads to a system of equations $Ax = b$ or, more generally, to a least squares problem

$$\text{minimize} \quad \|b - Ax\|_2 \quad \text{for} \quad A \in \mathbb{R}^{m,n} \text{ and } b \in \mathbb{R}^m, \tag{3.28}$$

where, in practice, only an approximation b^δ of b is known. But (3.28) is a parameter identification problem by itself. In case $\text{rank}(A) = n$, it has a unique solution $x^* = A^+ b$, the pseudoinverse A^+ now playing the role of T^{-1}. The mapping $b \mapsto A^+ b$ formally is continuous (as a linear mapping between finite-dimensional spaces), but A may be badly conditioned, so that $A^+ b^\delta$ may be far away from x^*, even if b^δ is close to b. As confirmed by the examples of Chap. 2, a bad condition of A results from choosing too fine a discretization—which is likely to happen, since we have no way of choosing an optimal one. So, there may well be a need to regularize (3.28)—not in the sense of approximating a discontinuous operator by a continuous one (as in Definition 3.14), but in the sense of approximating a badly conditioned operator (matrix) by a better conditioned one. This cannot be achieved by discretization again. Also, there is no point in just replacing A^+ by a better conditioned operator (matrix) M, if Mb^δ is not close to x^*. We can only hope to make Mb^δ be close to x^*, *if we have some additional information about x^** beyond the knowledge $x^* = A^+ b$, which cannot be exploited, since b is not known.

"Additional information" can take various forms. We now present one possibility, assuming the following.

Assumption 3.16 (Additional Information) *Let Assumption 2.6 hold and u^* be the solution of the corresponding inverse problem $Tu = w$. Let $X_0 \subseteq X$ be a subspace of X and let $\| \bullet \|_0$ be a semi-norm on X_0. Assume that the "additional information"*

$$u^* \in X_0 \quad and \quad \|u^*\|_0 \leq S, \tag{3.29}$$

is available about u^*.

Remark Often, $\| \bullet \|_0$ is required to be a *norm*, not only a semi-norm (see Appendix B for the difference between both), and that, moreover, $\| \bullet \|_0$ is stronger than the norm $\| \bullet \|_X$ on X, i.e., that

$$\|u\|_X \leq C\|u\|_0 \quad \text{for all } u \in X_0 \tag{3.30}$$

holds for some constant C, compare Definition 1.18 in [Kir11]. These stronger requirements on $\| \bullet \|_0$ are needed to develop a theory of "worst case errors" and "regularizations of optimal order," i.e., regularizations for which the convergence (3.26) is of highest possible order, compare Lemma 1.19, Theorem 1.21, and Theorems 2.9 and 2.12 in [Kir11]. We will not go into these abstract concepts, because we are rather interested in achieving a small error

$$\|R_{\rho(\delta,w^\delta)}(w^\delta) - T^{-1}(w)\|_X$$

for a finite value $\delta > 0$ than achieving optimal convergence order for $\delta \to 0$. Also, "optimal order of convergence" is not a helpful concept in the finite-dimensional setting, where the inversion operator A^+ is continuous, see the highlighted comments after Theorem 3.26. Requiring $\| \bullet \|_0$ to be only a semi-norm will prove practical in the examples of the following section.

Example 3.17 (Fredholm Integral Equation) Assume we want to solve a linear Fredholm integral equation of the first kind. This problem is defined by a mapping

$$T : L_2(a,b) \to L_2(a,b), \quad u \mapsto w, \quad w(t) = \int_a^b k(s,t)u(s)\,ds.$$

Here, $X = L_2(a,b)$ with norm $\| \bullet \|_X = \| \bullet \|_{L_2(a,b)}$. Let $X_0 := H^1(a,b) \subset L_2(a,b)$ with norm $\| \bullet \|_0 = \| \bullet \|_{H^1(a,b)}$ (in this case, even (3.30) holds). Functions contained in X_0 are distinguished from L_2 functions by some guaranteed extra smoothness. Additional information $u^* \in X_0$ means that we have an a priori knowledge about the smoothness of u^*. $\diamond$

If additional information about u^* as in (3.29) is available, it is natural to require (3.29) to hold also for an approximant. This will be described in the following Problem 3.18, which is not yet a concrete problem, but rather a guideline to set up one.

Problem 3.18 (Regularized Linear Least Squares Problem) Let Assumption 3.16 hold, especially let $Tu = w$ be an identification problem with unique exact solution u^*, about which we have the additional information

$$u^* \in X_0 \quad \text{and} \quad \|u^*\|_0 \leq S$$

(continued)

Problem 3.18 (continued)
as in (3.29). As with standard discretization, we choose an n-dimensional
subspace $X_n \subset X$, but now under the additional requirement

$$\varphi_j \in X_0, \quad j = 1, \ldots, n, \tag{3.31}$$

for the basis vectors φ_j of X_n. Approximants $u_n \in X_n \subset X_0$ of u^* are rated
in the form

$$u_n = \sum_{j=1}^{n} x_j \varphi_j, \quad x_j \in \mathbb{R}, \ j = 1, \ldots, n,$$

as in Sects. 2.2 and 2.3 (for simplicity, let $d_n = n$ here). The coefficients x_j
are to be found by solving a system $Ax = b$ defined in (2.25) for the least
squares method and in (2.51) for the collocation method. It will be assumed
that rank$(A) = n$. Let only perturbed data $b^\delta \in \mathbb{R}^m$ be given such that
$\|b - b^\delta\|_2 \leq \delta$ for some $\delta > 0$. Based on these perturbed data compute
an approximation $u_n^\delta = \sum_{j=1}^{n} x_j^\delta \varphi_j$ of u^*, with coefficients x_j^δ defined as a
solution of the problem

$$\text{minimize} \quad \|b^\delta - Ax\|_2 \quad \text{under the constraint} \quad \left\| \sum_{j=1}^{n} x_j \varphi_j \right\|_0 \leq S. \tag{3.32}$$

An investigation of why (3.32) is indeed a regularization of (3.28) will be
postponed to Sect. 3.4. The following example shows, how adding information
works in practice.

Example 3.19 (Numerical Differentiation) Let $w \in C^1[a, b]$ with $w(a) = 0$. The
derivative $u^* = w'$ of w is the solution of the Volterra integral equation $Tu = w$
defined by the mapping

$$T : C[a, b] \to C^1[a, b], \quad u \mapsto Tu = w, \quad w(t) = \int_a^t u(s) \, ds, \quad a \leq t \leq b.$$

We have already seen that $Tu = w$ is ill-posed, if the norm $\| \bullet \|_{C[a,b]}$ is used for
both, $X = C[a, b]$ and $Y = C^1[a, b]$. Assume we have the following additional
information: $w \in H^2(a, b)$ and $\|w''\|_{L_2(a,b)} \leq S$. Consequently, $u^* \in H^1(a, b)$ and
$\|(u^*)'\|_{L_2(a,b)} \leq S$. The normed linear space $(X, \| \bullet \|_X)$ given by $X = C[a, b]$ and
$\| \bullet \|_X = \| \bullet \|_{C[a,b]}$ contains $X_0 := H^1(a, b)$ as a subspace. X_0 can be equipped
with the semi-norm

$$\|\bullet\|_0 : X_0 \to \mathbb{R}, \quad x \mapsto \|x\|_0 := \|x'\|_{L_2(a,b)}.$$

(All constant functions x are contained in X_0 and have $\|x\|_0 = 0$ even if $x \neq 0$. Therefore, $\|\bullet\|_0$ is a semi-norm only). Thus, we have an additional information of the form

$$u^* \in X_0 \quad \text{and} \quad \|u^*\|_0 = \|(u^*)'\|_{L_2(a,b)} \leq S,$$

which corresponds to (3.29). The inverse problem will be discretized by the collocation method. To find a spline approximant u_n of u^*, $n \in \mathbb{N}$ with $n \geq 2$, let $h := (b-a)/(n-1)$, and $t_i := a + (i-1)h, i = 1, \ldots, n$. Every $u_n \in \mathscr{S}_2(t_1, \ldots, t_n)$ can be written as

$$u_n = \sum_{j=1}^{n} x_j N_{j,2} \quad \Longrightarrow \quad \|u_n\|_0^2 = \sum_{j=1}^{n-1} h \left(\frac{x_{j+1} - x_j}{h} \right)^2 = \frac{1}{h} \|Lx\|_2^2, \tag{3.33}$$

where L is the matrix defined by

$$L = \begin{pmatrix} -1 & 1 & 0 & 0 & 0 & 0 & 0 & \cdots & 0 \\ 0 & -1 & 1 & 0 & 0 & 0 & 0 & \cdots & 0 \\ \vdots & & & \ddots & & & & & \vdots \\ 0 & \cdots & 0 & 0 & 0 & 0 & -1 & 1 & 0 \\ 0 & \cdots & 0 & 0 & 0 & 0 & 0 & -1 & 1 \end{pmatrix} \in \mathbb{R}^{n-1,n}. \tag{3.34}$$

By construction, $u_n \in H^1(a, b)$. The semi-norm condition for u_n directly translates into a semi-norm condition for the coefficient vector x, see (3.33). To make computations easy, assume that samples of $w(t) = \int_a^t u(s)\,ds$ are given at t_i, $i = 2, \ldots, n$, and additionally at $t_{i-1/2} := t_i - h/2, i = 2, \ldots, n$. Then, (2.51) takes the form of the following equations:

$$w(t_i) = h \left(\frac{x_1}{2} + x_2 + \ldots + x_{i-1} + \frac{x_i}{2} \right), \quad i = 2, \ldots, n,$$

$$w(t_{1.5}) = \frac{3}{8} x_1 + \frac{1}{8} x_2, \quad \text{and}$$

$$w(t_{i-1/2}) = h \left(\frac{x_1}{2} + x_2 + \ldots + x_{i-2} + \frac{7x_{i-1}}{8} + \frac{x_i}{8} \right), \quad i = 3, \ldots, n.$$

Setting

$$\beta := \begin{pmatrix} w(t_{1.5}) \\ w(t_2) \\ w(t_{2.5}) \\ w(t_3) \\ \vdots \\ w(t_{n-0.5}) \\ w(t_n) \end{pmatrix} \quad \text{and } A := h \begin{pmatrix} 0.375 & 0.125 \\ 0.5 & 0.5 \\ 0.5 & 0.875 & 0.125 \\ 0.5 & 1 & 0.5 \\ \vdots & \vdots & \ddots & \ddots \\ 0.5 & 1 & \cdots & 1 & 0.875 & 0.125 \\ 0.5 & 1 & \cdots & 1 & 1 & 0.5 \end{pmatrix} \in \mathbb{R}^{2n-2,n}$$

and assuming that only perturbed data β^δ are available instead of β, the minimization problem (3.32) takes the form

$$\text{minimize } \|\beta^\delta - Ax\|_2 \text{ under the constraint } \|Lx\|_2 \le \sqrt{h}S. \tag{3.35}$$

For a numerical example, let $a = 0$, $b = 1$, and $u^*(t) = t(1-t) - 1/6$ such that $\|(u^*)'\|_{L_2(a,b)} = 1/\sqrt{3} =: S$. Choose $n = 101$. In Fig. 3.4 (left), samples β of the exact effect w corresponding to u^* are shown (in red) as well as perturbed values β^δ (in black), which were obtained from β by adding to each component a random number drawn from a normal distribution with a mean value of 0 and a standard deviation of 10^{-3}. To the right, we show the exact solution u^* (in red) as well as the approximation $u_n^\delta = \sum_{j=1}^n x_j^\delta N_{j,2}$ obtained from minimizing $\|\beta^\delta - Ax\|_2$

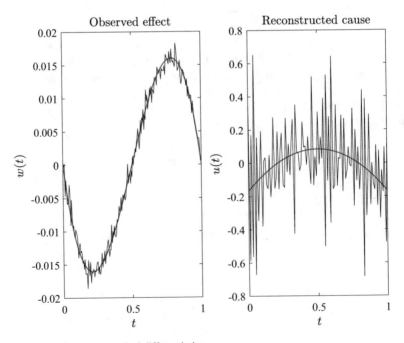

Fig. 3.4 Unregularized numerical differentiation

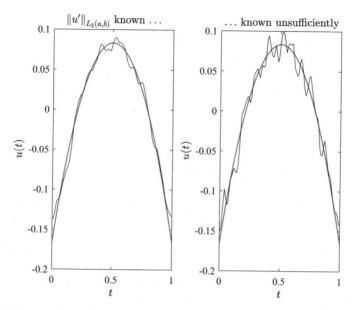

Fig. 3.5 Numerical differentiation regularized by using additional information

(in black). In Fig. 3.5 (left), we show (in black) the approximation u_n^δ obtained from minimizing $\|\beta^\delta - Ax\|_2$ under the constraint $\|Lx\|_2 \le \sqrt{h}S$, as suggested by (3.35). For comparison, the exact solution is again shown in red. A practical solution method for the constrained minimization problem will only be discussed in Sect. 3.4. To the right, we show an approximation u_n^δ obtained using the less sharp constraint $\|(u_n^\delta)'\|_2 \le 1 \cdot \sqrt{h}$. This shows that if the available information about u^* is of unsufficient quality, then the reconstruction quality might suffer accordingly. $\Diamond$

3.4 Tikhonov Regularization

Motivated by Sect. 3.3 and Example 3.19, we will investigate

Problem 3.20 (Linear Least Squares Problem with Quadratic Constraints) Let

$$A \in \mathbb{R}^{m,n}, \quad L \in \mathbb{R}^{p,n}, \quad b \in \mathbb{R}^m, \quad \text{and} \quad S > 0. \tag{3.36}$$

(continued)

Problem 3.20 (continued)
Solve the constrained minimization problem

$$\text{minimize } \|b - Ax\|_2 \text{ under the constraint } \|Lx\|_2 \leq S. \qquad (3.37)$$

An equivalent formulation of (3.36) is

$$\text{min.} \quad f(x) := \|b - Ax\|_2^2 \quad \text{such that} \quad h(x) := \|Lx\|_2^2 - S^2 \leq 0. \qquad (3.38)$$

From the formulation of Problem 3.18, it becomes clear, how Problem 3.20 relates to a linear parameter identification problem $Tu = w$, discretized by the method of least squares or by the collocation method.

Replacing the linear least squares problem (3.28) by (3.37) is called **Tikhonov regularization** and goes back to the Russian mathematician Tikhonov and the American mathematician Phillips. An often considered case is $L = I_n \in \mathbb{R}^{n,n}$ (unity matrix of dimension $n \times n$), the so-called **standard case**. Matrix $L \neq I_n$ is used to formulate constraints on the discrete version of derivatives, as with L defined in (3.34). Likewise, one could choose

$$L = \begin{pmatrix} -1 & 2 & -1 & 0 & 0 & 0 & 0 & \cdots & 0 \\ 0 & -1 & 2 & -1 & 0 & 0 & 0 & \cdots & 0 \\ \vdots & & & \ddots & & & & & \vdots \\ 0 & \cdots & 0 & 0 & 0 & -1 & 2 & -1 & 0 \\ 0 & \cdots & 0 & 0 & 0 & 0 & -1 & 2 & -1 \end{pmatrix} \in \mathbb{R}^{n-2,n}, \qquad (3.39)$$

to formulate a constraint on second derivatives. Note that $x \mapsto \|Lx\|_2$ only is a semi-norm if $p < n$, since in that case $\|Lx\|_2 = 0$ is possible for $x \neq 0$. This is why we only required $\| \bullet \|_0$ to be a semi-norm in Assumption 3.16. The (quadratic) function f in (3.38) is called **objective function**, and the set $N := \{x \in \mathbb{R}^n; \ h(x) \leq 0\}$ is called **feasible region**. The objective function f is continuous and convex, and N is convex, closed, and nonempty (since $0 \in N$). This ensures that a solution of (3.37) always exists. The constraint $h(x) \leq 0$ is called **binding**, if

$$h(\hat{x}) > 0 \quad \text{for all} \quad \hat{x} \in M := \{x \in \mathbb{R}^n; \ f(x) = \min\}. \qquad (3.40)$$

In this case, no element of M can be a solution of (3.37). Put differently, if x^* *is a* solution of (3.37), then $f(\hat{x}) < f(x^*)$ for all $\hat{x} \in M$. Because of the convexity of f and the convexity and closedness of N, this does mean that x^* lies necessarily on the boundary of N whenever (3.40) is valid.

Theorem 3.21 (Linear Least Squares Problem with Quadratic Constraints)
Let A, L, b, and S be defined as in (3.36) and f and h be defined as in (3.38).

Also, assume that

$$\text{rank} \begin{pmatrix} A \\ L \end{pmatrix} = n. \tag{3.41}$$

Then, the linear system of equations

$$(A^\top A + \lambda L^\top L)x = A^\top b \tag{3.42}$$

has a unique solution x_λ for every $\lambda > 0$. If (3.40) holds, i.e., if the constraint of the minimization problem (3.37) is binding, then there exists a unique $\lambda > 0$ such that

$$\|Lx_\lambda\|_2 = S \tag{3.43}$$

and the corresponding x_λ is the unique solution of (3.37).

Remarks Condition (3.41) can equivalently be written in the form

$$\mathcal{N}_A \cap \mathcal{N}_L = \{0\}. \tag{3.44}$$

We usually require $\text{rank}(A) = n$, in which case, condition (3.41) is automatically fulfilled. This also implies the existence of a unique minimizer $\hat{x}$ of $\|b - Ax\|_2$. Either $h(\hat{x}) \leq 0$, such that $\hat{x}$ also is the unique minimizer of (3.37), or $h(\hat{x}) > 0$, meaning that (3.40) holds, since $\hat{x}$ is the only element of M.

Proof Let $\lambda > 0$. When (3.41) holds, then

$$\text{rank} \begin{pmatrix} A \\ tL \end{pmatrix} = n \quad \text{for} \quad t = \sqrt{\lambda} > 0.$$

According to Theorem 3.1, a unique solution of the linear least squares problems

$$\min_{x \in \mathbb{R}^n} \left\{ \left\| \begin{pmatrix} A \\ \sqrt{\lambda}L \end{pmatrix} x - \begin{pmatrix} b \\ 0 \end{pmatrix} \right\|_2 \right\}, \quad \lambda > 0, \tag{3.45}$$

exists, which is defined as the unique solution of the corresponding normal equations, namely (3.42). From now on let (3.40) hold. As already seen, this implies that any solution of (3.37) is necessarily located on the boundary of N. Problem (3.37), then, is equivalent to

$$\text{minimize} \quad f(x) \quad \text{under the constraint} \quad h(x) = 0.$$

It is easy to verify that for the gradient ∇h of h,

$$\nabla h(x) = 0 \quad \Longleftrightarrow \quad L^\top Lx = 0 \quad \Longleftrightarrow \quad \|Lx\|_2 = 0$$

holds. From this, one concludes that problem (3.38) is **regular**, i.e., the gradient of the constraint function does not vanish at an optimal point. In fact, if x^* is an optimal point, then $h(x^*) = \|Lx^*\|_2^2 - S^2 = 0$ and from $S > 0$ follows $\|Lx^*\|_2 \neq 0$, thus $\nabla h(x^*) \neq 0$. The well-known theorem on Lagrange multipliers—applicable since (3.38) is regular—tells us a scalar λ exists for which

$$0 = \nabla f(x^*) + \lambda \nabla h(x^*) = 2(A^\top A + \lambda L^\top L)x^* - 2A^\top b, \qquad (3.46)$$

where x^* still is a solution of (3.38). It is known that the gradient of a function points into the direction of ascending function values. Therefore, $\nabla h(x^*)$ must point outwards of N, x^* being a boundary point. But then λ cannot be negative: if it was, then we would conclude from (3.46) that moving from x^* into the (feasible) interior of N, values of f would diminish, a contradiction to the optimality of x^*. If λ would vanish, then (3.46) would tell us $\nabla f(x^*) = 0$, meaning that x^* would be a minimizer of (the convex function) f, which would contradict (3.40). The only remaining possibility is $\lambda > 0$. But in this case, as we have already seen, (3.46) has a unique solution $x^* = x_\lambda$, for which $h(x_\lambda) = 0$, i.e., for which (3.43) must hold. It remains to show that λ is unique. To do so, define $J, E : \mathbb{R}^+ \to \mathbb{R}_0^+$ by setting

$$J(\lambda) := f(x_\lambda) = \|b - Ax_\lambda\|_2^2 \quad \text{and} \quad E(\lambda) := \|Lx_\lambda\|_2^2. \qquad (3.47)$$

The proof is accomplished if we can show that E is strictly monotonic decreasing. Since $f, h : \mathbb{R}^n \to \mathbb{R}$ both are convex, so is $f + \lambda h : \mathbb{R}^n \to \mathbb{R}$ for every $\lambda > 0$. This function is minimized where its gradient $\nabla f(x) + \lambda \nabla h(x)$—compare (3.46)—vanishes. As we have seen, for every $\lambda > 0$, a unique $x = x_\lambda$ exists where this happens. Consequently, for two values $0 < \lambda_1 < \lambda_2$, we must have

$$J(\lambda_1) + \lambda_1 E(\lambda_1) < J(\lambda_2) + \lambda_1 E(\lambda_2)$$

as well as

$$J(\lambda_2) + \lambda_2 E(\lambda_2) < J(\lambda_1) + \lambda_2 E(\lambda_1).$$

Adding both inequalities shows

$$(\lambda_1 - \lambda_2)(E(\lambda_1) - E(\lambda_2)) < 0,$$

from which $E(\lambda_1) > E(\lambda_2)$ follows. Likewise, one can show that J is strictly monotonic increasing (divide the first inequality by λ_1, the second by λ_2, and add them afterwards).   □

The proof shows how Problem 3.20 can be solved in principle, at least in case $\text{rank}(A) = n$ (in this case, condition (3.40) is easily checked).

Solution Method for Problem 3.20

(1) Compute the unique minimizer x_0 of $\|b - Ax\|_2$ and check, whether $\|Lx_0\|_2 \leq S$. If this is the case, x_0 is the sought-after solution. Otherwise,
(2) Solve $E(\lambda) - S^2 = 0$ using some numerical method (for example, Newton's method) and exploiting the fact that the function E is strictly monotonic decreasing. Each evaluation of E requires the computation of a solution x_λ of the *unconstrained* linear least squares problem (3.45), which can be found by solving the normal equations (3.42). The zero λ^* of $E - S^2$ determines the solution $x^* = x_{\lambda^*}$ of (3.37).

Before Problem 3.20 is analyzed and solved, we turn to a closely related, "symmetric" problem, which will come into play in Sect. 3.5 and which will be relevant, if additional information of the form $\|Lx\|_2 \leq S$ is not available.

Problem 3.22 (Linear Least Squares Problem with Quadratic Constraints) Let

$$A \in \mathbb{R}^{m,n}, \quad L \in \mathbb{R}^{p,n}, \quad b \in \mathbb{R}^m, \quad \text{and} \quad \delta > 0. \tag{3.48}$$

Solve the constrained minimization problem

$$\text{minimize } \|Lx\|_2 \text{ under the constraint } \|Ax - b\|_2 \leq \delta. \tag{3.49}$$

An equivalent formulation of (3.48) is

$$\text{min.} \quad f(x) := \|Lx\|_2^2 \quad \text{such that} \quad h(x) := \|Ax - b\|_2^2 - \delta^2 \leq 0. \tag{3.50}$$

Theorem 3.23 (Linear Least Squares Problems with Quadratic Constraints)
Let A, L, b, and δ be defined as in (3.48) and f and h be defined as in (3.50). Let $\mathcal{N}_A \cap \mathcal{N}_L = \{0\}$, and let $x_0 \in \mathbb{R}^n$ be a minimizer of $\|Ax - b\|_2$ (without constraint). Assume that

$$\|Ax_0 - b\|_2 < \delta < \|b\|_2. \tag{3.51}$$

The linear system of equations

$$(A^\top A + \lambda L^\top L)x = A^\top b \tag{3.52}$$

has a unique solution x_λ *for every* $\lambda > 0$. *If the constraint of* $h(x) \leq 0$ *is binding, then there exists a unique* $\lambda > 0$ *such that*

$$\|Ax_\lambda - b\|_2 = \delta \qquad (3.53)$$

and the corresponding x_λ *is the unique solution of (3.49).*

Remark The first inequality in (3.51) is needed for regularity of Problem 3.22, see the proof below. The second inequality in (3.51) excludes the trivial solution $x = 0$ and is necessary for the constraint to be binding.

Proof The existence of a solution x^* of Problem 3.22 is proven as for Problem 3.20. The existence of a unique solution of (3.52) follows from $\mathcal{N}_A \cap \mathcal{N}_L = \{0\}$ as in Theorem 3.21. From now on, let the constraint be binding, i.e., let $h(x) > 0$ for any minimizer of $\|Lx\|_2$ (for any $x \in \mathcal{N}_L$). In this case, any solution x^* of Problem 3.22 lies necessarily on the boundary of the feasible region, such that Problem 3.22 can be replaced equivalently by

$$\text{minimize} \quad f(x) \quad \text{under the constraint} \quad h(x) = 0. \qquad (3.54)$$

For a solution x^* of (3.54), one must have $h(x^*) = 0$. One easily verifies

$$\nabla h(x^*) = 0 \quad \Longleftrightarrow \quad A^\top A x^* = A^\top b,$$

which is equivalent to x^* being a minimizer of $h(x)$. But in this case, $h(x^*) = h(x_0) < 0$ according to (3.51). Consequently, $\nabla h(x^*) \neq 0$ for a solution of (3.54), which is the regularity condition. From the Lagrange multiplier theorem, one deduces the existence of $\mu \in \mathbb{R}$ such that

$$\nabla f(x^*) + \mu \nabla h(x^*) = 0. \qquad (3.55)$$

As in the proof of Theorem 3.21, it can be shown that necessarily $\mu > 0$. But then (3.55) is equivalent to (3.52), with $\lambda = 1/\mu$. The rest of the proof is equal to that of Theorem 3.21. $\qquad\qquad\qquad\qquad\qquad\qquad\qquad\qquad\qquad\qquad\qquad\qquad\qquad\qquad \square$

Analysis of Problem 3.20 and Its Practical Solution

An analysis of (3.37) often is based on the "Generalized Singular Value Decomposition (GSVD)." This is not necessary. The following alternative approach, which goes back to Reinsch [Rei67], has the advantage that a very efficient practical method to compute the solution x_{λ^*} of (3.37) can directly be derived from it. *The analysis could be carried out under condition (3.41), but to make things easier, we will more specifically assume that* $\text{rank}(A) = n \leq m$. In this case, the normal equations (3.42) already have a unique solution x_0 for $\lambda = 0$. If $h(x_0) \leq 0$, we are done, since then x_0 is the solution of (3.37). So, let $h(x_0) > 0$ from now on.

Since rank$(A) = n$, the matrix $A^\top A$ is positive definite. In addition, $L^\top L$ is positive semi-definite. According to (A.3), a non-singular matrix $V = (v_1|v_2|\ldots|v_n) \in \mathbb{R}^{n,n}$ exists such that

$$V^\top A^\top AV = \mathrm{diag}(1, \ldots, 1) \quad \text{and} \quad V^\top L^\top LV = \mathrm{diag}(\kappa_1, \ldots, \kappa_n), \tag{3.56}$$

where $\kappa_i \geq 0$ for all i. This can be put differently by saying

$$v_i^\top A^\top A v_j = \begin{cases} 1, & i = j \\ 0, & \text{else} \end{cases} \quad \text{and} \quad v_i^\top L^\top L v_j = \begin{cases} \kappa_i, & i = j \\ 0, & \text{else.} \end{cases}$$

Choose the numbering such that

$$\kappa_1 \geq \ldots \geq \kappa_r > 0 \quad \text{and} \quad \kappa_{r+1} = \ldots = \kappa_n = 0 \quad \text{where} \quad r := \mathrm{rank}(L).$$

From (3.56), one gets

$$A^\top A v_i = \frac{1}{\kappa_i} L^\top L v_i, \quad i = 1, \ldots, r.$$

For fixed $\lambda \geq 0$, the solution x_λ of (3.42) can be written in the form $x_\lambda = \sum_{i=1}^n \tau_i v_i$. Substituting this into (3.42), one gets

$$\sum_{i=1}^n \tau_i (A^\top A v_i + \lambda L^\top L v_i) = \sum_{i=1}^r \tau_i \left(\frac{1}{\kappa_i} + \lambda \right) L^\top L v_i + \sum_{i=r+1}^n \tau_i A^\top A v_i = A^\top b.$$

To determine the unknowns τ_i, multiply these identities from the left by $v_1^\top, \ldots, v_n^\top$ to get

$$x_\lambda = \sum_{i=1}^n \left(\frac{\gamma_i}{1 + \lambda \kappa_i} \right) v_i, \quad \gamma_i = v_i^\top A^\top b, \tag{3.57}$$

which is valid for $\lambda \geq 0$. One also gets

$$E(\lambda) = \|Lx_\lambda\|_2^2 = \sum_{i=1}^n \left(\frac{\gamma_i}{1 + \lambda \kappa_i} \right)^2 \kappa_i . \tag{3.58}$$

This shows that under the conditions rank$(A) = n$ and $E(0) = \|Lx_0\|_2^2 > S^2$, the function $E : \mathbb{R}_0^+ \to \mathbb{R}_0^+$ is not only positive and strictly monotonic decreasing from $E(0) > S^2$ to 0, but that it is also convex. As a consequence, if we use Newton's method to find the solution λ^* of $E(\lambda) - S^2 = 0$ and if we start it with $\lambda_0 = 0$ (or with some positive λ_0 to the left of λ^*), then it will produce a sequence $(\lambda_k)_{k \in \mathbb{N}_0}$ converging monotonously to λ^*. Using the abbreviations $E_k := E(\lambda_k)$ and $E'_k := E'(\lambda_k)$, the sequence $(\lambda_k)_{k \in \mathbb{N}_0}$ is recursively defined by Newton's method

$$\lambda_{k+1} = \lambda_k - \frac{E(\lambda_k) - S^2}{E'(\lambda_k)} = \lambda_k - \frac{E_k - S^2}{E'_k}, \quad k = 0, 1, 2, \ldots \quad (3.59)$$

and requires the computation of derivatives $E'(\lambda)$:

$$E'(\lambda) = \frac{d}{d\lambda}\left(x_\lambda^\top L^\top L x_\lambda\right) = 2x_\lambda^\top L^\top L x_\lambda', \quad x_\lambda' = \frac{d}{d\lambda} x_\lambda.$$

Implicit differentiation of identity (3.42) with respect to λ shows

$$L^\top L x_\lambda + (A^\top A + \lambda L^\top L)x_\lambda' = 0,$$

therefore, we get the explicit formula

$$E'(\lambda)/2 = -x_\lambda^\top L^\top L(A^\top A + \lambda L^\top L)^{-1} L L x_\lambda. \quad (3.60)$$

Note that if we compute x_λ from (3.42) via a Cholesky factorization $A^\top A + \lambda L^\top L = R^\top R$, then

$$E'(\lambda)/2 = -\|R^{-\top} L^\top L x_\lambda\|_2^2$$

only needs the computation of $z = R^{-\top} L^\top L x_\lambda$, which can be done very cheaply by solving $R^\top z = L^\top L x_\lambda$. The need to compute derivatives often is a disadvantage of Newton's method, but in the present case, this computation means nearly no effort at all.

From [Rei67], we take the recommendation to replace $E(\lambda) - S^2 = 0$ by the equivalent equation

$$G(\lambda) - \frac{1}{S} = 0, \quad G(\lambda) = \frac{1}{\sqrt{E(\lambda)}}, \quad (3.61)$$

and use Newton's method to solve the latter. Since E is positive and strictly monotonic decreasing, G is positive and strictly monotonic increasing. From $E(0) > S^2$ and $E(\lambda) \to 0$ for $\lambda \to \infty$, we get $G(0) < S^{-1}$ and $G(\lambda) \to \infty$ for $\lambda \to \infty$, so there is a unique solution of (3.61). Using $G'(\lambda) = -(1/2)E(\lambda)^{-3/2}E'(\lambda)$ (which can be computed as efficiently as $E'(\lambda)$), Newton's method applied to (3.61) will produce a sequence

$$\lambda_{k+1} = \lambda_k + 2\frac{E_k^{3/2}}{E'_k} \cdot (E_k^{-1/2} - S^{-1}), \quad k = 0, 1, 2, \ldots \quad (3.62)$$

(the same abbreviations are used as for (3.59)). The recursion will be started again at $\lambda_0 = 0$. Since G is concave (which can be shown with some effort), (3.62) again is a monotonously converging sequence. The advantage of using (3.62) instead of (3.59) lies in its faster convergence. To see this, consider the ratio of the respective increments $(\lambda_{k+1} - \lambda_k)$ which both methods take when started at a common reference point λ_k:

$$v_k := 2 \frac{E_k^{3/2}(E_k^{-1/2} - S^{-1})}{E_k'} \cdot \frac{-E_k'}{E_k - S^2} = \frac{2}{q_k + \sqrt{q_k}}$$

with $q_k := S^2/E_k$. Since the Newton's iteration is started at $\lambda_0 = 0$, to the left of the solution λ^* of $E(\lambda) - S^2$, we have $0 < q_k < 1$. Consequently, $v_k > 1$. In a computer implementation, the iteration will be stopped as soon as $\lambda_{k+1} \leq \lambda_k$ is observed, since this can only happen due to the finite precision arithmetic. Before the iteration is started, the condition $\|Lx_0\|_2 > S$ has to be checked.

As an example for regularization as in Problem 3.20, we have already considered numerical differentiation (Example 3.19). It is a good idea to use additional information as formulated in Assumption 3.16, but unluckily, such kind of information often is not available. In Example 3.19, it turned out that if $\|(u^*)'\|_{L_2(a,b)}$ is not exactly known, then the reconstruction quality of u^* suffers. We are now looking for an alternative approach to regularization.

Generalization of Tikhonov Regularization

When information as in Assumption 3.16 is not available, heuristics come into play. Our starting point for going into this is a modification of Problem 3.20.

Problem 3.24 (Linear Least Squares Problem with Lagrange Parameter)
Let

$$A \in \mathbb{R}^{m,n}, \quad L \in \mathbb{R}^{p,n}, \quad \text{and} \quad b \in \mathbb{R}^m. \tag{3.63}$$

Let rank$(A) = n$. Find the minimizer $x_\lambda \in \mathbb{R}^n$ of

$$\min_x \left\{ \|b - Ax\|_2^2 + \lambda \|Lx\|_2^2 \right\} = \min_x \left\{ \left\| \begin{pmatrix} A \\ \sqrt{\lambda}L \end{pmatrix} x - \begin{pmatrix} b \\ 0 \end{pmatrix} \right\|_2^2 \right\}, \tag{3.64}$$

where the parameter $\lambda \geq 0$ has to be chosen.

We know that (3.64) has a unique minimizer x_λ for each $\lambda \geq 0$. If λ is chosen as the Lagrange multiplier according to Theorem 3.21, then we will get the same result as before. By choosing $\lambda \in [0, \infty[$ freely, the minimizer x_λ can be shaped to some extent. To see how this works, remember the functions $J, E : \mathbb{R}_0^+ \to \mathbb{R}_0^+$ defined by

$$J(\lambda) = \|b - Ax_\lambda\|_2^2 \quad \text{and} \quad E(\lambda) = \|Lx_\lambda\|_2^2 \tag{3.65}$$

in (3.47). We found that J is strictly monotonic increasing, whereas E is strictly monotonic decreasing with $E(0) = \|Lx_0\|_2^2$ and $\lim_{\lambda \to \infty} E(\lambda) = 0$. The choice of the parameter λ determines a trade-off between the conflicting goals of minimizing J and minimizing E. In the limit $\lambda \to 0$, we will get the minimizer x_0 of $\|b - Ax\|_2$. Among all x_λ, $\lambda \geq 0$, this is the one which satisfies best the system of equations $Ax = b$ and thus is "closest to the data" b. Therefore, $J(\lambda)$ can be considered an inverse measure of the data fit of x_λ. In the other limit case $\lambda \to \infty$, we can derive the identity

$$x_\infty = \sum_{i=r+1}^n \frac{v_i^\top A^\top b}{v_i^\top A^\top A v_i} v_i \in \mathcal{N}_L = \langle v_{r+1}, \ldots, v_n \rangle$$

from (3.57). This is the unique minimizer of $\|b - Ax\|_2$ in the subspace $\mathcal{N}_L$.[4] In Example 3.19, function $u_n = \sum_{j=1}^n x_{\infty,j} N_{j,2}$ is the constant function, which fits best the data. Since the matrices L used in (3.34) and (3.39) are related to (discretized) derivatives of function $u_n = \sum x_{\lambda,j} \varphi_j$, one may consider $E(\lambda)$ an inverse measure of smoothness. The trade-off between the minimization of J and the minimization of E, which is determined by the choice of λ, thus turns out to be a trade-off between the fidelity of the minimizer to the data and the smoothness of the minimizer. In contrast to the objective way in which additional information is used according to Assumption 3.16, a free choice of λ corresponds to a more subjective belief of how smooth u_n should be versus how well it should reproduce the data. Choosing λ subjectively can be quite dangerous and drive an approximation u_n into a wrong direction, if one does not really know that the exact solution u^* actually *has* some degree of smoothness. Still, we could replace (3.64) by the more general minimization problem

$$\min_{\alpha \in D} \left\{ \|b - Ax\|_2^2 + \lambda G(x) \right\}, \tag{3.66}$$

where G is some positive function incorporating a priori information about u^*. The term $G(x)$ should measure (inversely), to what extent $u_n = \sum_j x_j \varphi_j$ has a quality that we know u^* also has. Conversely, the value $G(x)$ should become

---

[4]To see this, compute the gradient of the convex function $h(\tau_{r+1}, \ldots, \tau_n) = \|b - A(\sum_{i=r+1}^n \tau_i v_i)\|_2^2$ considering $v_i^\top A^\top A v_j = 0$ for $i \neq j$.

large, whenever $u_n = \sum_j x_j \varphi_j$ does not have this desired quality. For this reason, G is called a **penalty function**. For example, if we have a priori information that $\hat{u} = \sum_{j=1}^{n} \hat{x}_j \varphi_j$ approximates u^* well, then we could choose

$$G(x) := \|x - \hat{x}\|_2^2. \tag{3.67}$$

If we know that $u^* : [a, b] \to \mathbb{R}$ is not oscillating, and if $u_n = \sum x_j \varphi_j$ is a polygonal line with vertices at equidistant abscissae and at ordinates x_j, then we could choose

$$G(x) := \sum_{j=1}^{n-1} |x_{j+1} - x_j|, \tag{3.68}$$

which measures the total variation of u_n. Choosing a subset $D \subseteq \mathbb{R}^n$ in (3.66) is another way to incorporate information about u^*. For example, if we know that u^* is positive, monotone, or convex, then D should be chosen such that for $x \in D$ a corresponding approximant $u_n = \sum x_j \varphi_j$ is also positive, monotone, or convex, respectively. We will come back to generalized Tikhonov regularization of the form (3.66) in Sect. 4.6.

Before we discuss a method to choose λ in Problem 3.24, we show that (3.64) is a regularized version of the linear least squares problem. To make the analysis easier, we specialize on the ("norm") case $L = I_n \in \mathbb{R}^{n,n}$. This is not an important restriction, since any problem of the form (3.64) can be transformed to a standard problem with $L = I$, see Sect. 3.6.

Theorem 3.25 (Approximation Power of Tikhonov Regularization) *Let $A \in \mathbb{R}^{m,n}$ with $rank(A) = n \leq m$ and singular values $\sigma_1 \geq \ldots \geq \sigma_n > 0$. Let $L = I_n$, $b, b^\delta \in \mathbb{R}^m$, and $\|b - b^\delta\|_2 \leq \delta$ for $\delta > 0$. Let $\lambda \geq 0$ and denote by*

$\hat{x}$ *the unique solution of $\|b - Ax\|_2 = $ min! (exact data) and by*

x_λ *the unique solution of (3.64) for b replaced by b^δ (perturbed data).*

Then, the following estimates hold

(1) $\|\hat{x} - x_0\|_2 \leq \dfrac{\delta}{\sigma_n}$ and

(2) $\|\hat{x} - x_\lambda\|_2 \leq \dfrac{\sqrt{\lambda}}{2\sigma_n} \|\hat{x}\|_2 + \dfrac{\delta}{2\sqrt{\lambda}}$ for $\lambda > 0$.

From (2), one concludes that if λ is chosen such that for $\delta \to 0$

$$\lambda \to 0 \quad and \quad \frac{\delta^2}{\lambda} \to 0,$$

then (3.64) is a regularization of the linear least squares problem according to Definition 3.14.

Proof Part (1) was already proven in Theorem 3.6. Part (2): Using the SVD $A = U \Sigma V^\top$, we get

$$V^\top A^\top A V = \mathrm{diag}(\sigma_1^2, \ldots, \sigma_n^2) \quad \text{and} \quad V^\top I_n V = I_n$$

as a special case of (3.56). Observing $v_i^\top A^\top = \sigma_i u_i^\top$, (3.57) becomes

$$x_\lambda = \sum_{i=1}^n \frac{\sigma_i}{\sigma_i^2 + \lambda} \cdot (u_i^\top b^\delta) \cdot v_i . \tag{3.69}$$

For each $\lambda \geq 0$, we can define an operator

$$A_\lambda^+ : \mathbb{R}^m \to \mathbb{R}^n, \quad y \mapsto A_\lambda^+ y := \sum_{i=1}^n \frac{\sigma_i}{\sigma_i^2 + \lambda} \cdot (u_i^\top y) \cdot v_i,$$

such that $x_\lambda = A_\lambda^+ b^\delta$. Since $\hat{x} = A^+ b$, we get

$$\|\hat{x} - x_\lambda\|_2 \leq \|A^+ b - A_\lambda^+ b\|_2 + \|A_\lambda^+ b - A_\lambda^+ b^\delta\|_2. \tag{3.70}$$

Both terms on the right-hand side will be estimated separately. Since $A^+ = A_0^+$, we get

$$A^+ b - A_\lambda^+ b = \sum_{i=1}^n \frac{\lambda}{\sigma_i^2 + \lambda} \frac{1}{\sigma_i} (u_i^\top b) v_i.$$

It is easy to check that

$$\frac{\sigma^2}{\sigma^2 + \lambda} \leq \frac{\sigma}{2\sqrt{\lambda}} \quad \text{for} \quad \sigma, \lambda > 0, \tag{3.71}$$

from which $\lambda/(\sigma_i^2 + \lambda) \leq \sqrt{\lambda}/(2\sigma_n)$ follows, such that

$$\|A^+ b - A_\lambda^+ b\|_2^2 \leq \frac{\lambda}{4\sigma_n^2} \cdot \sum_{i=1}^n \frac{1}{\sigma_i^2} \left| u_i^\top b \right|^2 = \frac{\lambda}{4\sigma_n^2} \|A^+ b\|_2^2.$$

This shows that the first term on the right-hand side of (3.70) has $(\sqrt{\lambda}/(2\sigma_n)) \cdot \|\hat{x}\|_2$ as an upper bound. To estimate the second term, (3.71) is reused:

$$\|A_\lambda^+ b - A_\lambda^+ b^\delta\|_2^2 = \sum_{i=1}^n \underbrace{\left(\frac{\sigma_i^2}{\sigma_i^2 + \lambda} \cdot \frac{1}{\sigma_i}\right)^2}_{\leq 1/(4\lambda)} \left|u_i^\top (b - b^\delta)\right|^2 \leq \frac{1}{4\lambda}\|b - b^\delta\|_2^2.$$

By the way, this gives an upper bound

$$\|A_\lambda^+\|_2 \leq \frac{1}{2\sqrt{\lambda}} \tag{3.72}$$

for the spectral norm of the operator A_λ^+, which regularizes A^+. $\square$

3.5 Discrepancy Principle

We have to set the parameter λ in Problem 3.24. This can be done according to the following rule known as **Morozov's discrepancy principle**. It is assumed that only an approximation b^δ to the data b is at hand, and that its error can be bounded in the form $\|b - b^\delta\|_2 \leq \delta$ with $\delta > 0$ known.

Morozov's discrepancy principle for choosing the regularization parameter λ in (3.64), with b replaced by b^δ.

- Choose $\lambda = \infty$, if

$$\|b^\delta - Ax_\infty\|_2 \leq \delta. \tag{3.73}$$

- Choose $\lambda = 0$, if

$$\|b^\delta - Ax_0\|_2 > \delta. \tag{3.74}$$

- Otherwise choose the unique value λ, for which

$$\|b^\delta - Ax_\lambda\|_2 = \delta. \tag{3.75}$$

Note that $\lambda = \lambda(\delta, b^\delta)$ is chosen depending on the error level δ as well as on the data b^δ, so this is a parameter choice a posteriori. The following are the two basic ideas behind the discrepancy principle.

1. One does not strive to achieve a better data fit than $\|b^\delta - Ax_\lambda\|_2 = \delta$, since b^δ itself does not approximate the exact data b any better.

2. Since $J(\lambda) = \|b^\delta - Ax_\lambda\|_2$ is strictly monotonic increasing, the largest possible λ is chosen such that $\|b^\delta - Ax_\lambda\|_2 \leq \delta$. (In the limiting cases (3.73) and (3.74) all, or no, x_λ satisfy $\|b^\delta - Ax_\lambda\|_2 \leq \delta$).

From Theorem 3.23, it can be seen that *Tikhonov regularization with parameter choice according to the discrepancy principle solves Problem 3.22, if a solution exists.* An a priori information of the form $\|Lx\|_2 \leq S$ about the true solution, as it was used in Problem 3.20, is replaced by an a priori information of the form $\|b - b^\delta\|_2 \leq \delta$ about the data error. The discrepancy principle has to be accompanied by two warnings. *First*, and most importantly, choosing λ as large as possible seems justified *if one knows that u^* has the smoothness property, which is inversely measured by E,* but it can also be misleading, if u^* does not have this property. We cannot make a totally subjective choice of E and then hope to miraculously detect u^* from the data b^δ alone. For example, if u^* is sharply peaked (the location of the peak might be an important information), but E is chosen such that large derivatives are penalized, then regularization will carry us away from the sought-after solution. *Second*, assuming that $u_n^* = \sum x_j^* \varphi_j$ is an optimal approximant of the exact solution u^* (according to some norm), then it may happen that $\|b^\delta - Ax^*\|_2 < \delta$. In this case, the discrepancy principle will choose too large a parameter λ. The discrepancy principle is then said to "over-regularize."

By the strict monotonicity of J, the uniqueness of λ with (3.75) is guaranteed. Note that with $\hat{x} = A^+ b$ we have

$$\|b^\delta - Ax_0\|_2 \leq \|b^\delta - A\hat{x}\|_2 \leq \|b^\delta - b\|_2 + \|b - A\hat{x}\|_2, \qquad (3.76)$$

where the second term on the right-hand side should tend to 0 with the discretization getting finer and finer. Therefore, if we encounter case (3.74) for some finite value δ, this is a hint that the discretization was chosen too coarse. We can only find some x with $\|b^\delta - Ax\|_2 \leq \delta$ if δ is an upper bound for data error *plus* discretization error.

The following theorem formally shows that the discrepancy principle makes Tikhonov regularization a regularization according to Definition 3.14.

Theorem 3.26 (Tikhonov Regularization and Discrepancy Principle) *Let $A \in \mathbb{R}^{m,n}$ with $rank(A) = n \leq m$ and with singular values $\sigma_1 \geq \ldots \geq \sigma_n > 0$ and $L \in \mathbb{R}^{p,n}$. Let $b, b^\delta \in \mathbb{R}^m$. Let $\lambda \geq 0$ and denote by*

$\hat{x}$ *the unique solution of $\|b - Ax\|_2 = $ min! (exact data) and by*

x_λ *the unique solution of (3.64) for b replaced by b^δ (perturbed data).*

Assume that for some known $\delta > 0$

$$\|b - b^\delta\|_2 + \|b - A\hat{x}\|_2 \leq \delta. \tag{3.77}$$

Determine λ according to the discrepancy principle. Then case (3.74) cannot occur and the estimate

$$\|\hat{x} - x_\lambda\|_2 \leq 2\frac{\delta}{\sigma_n} \tag{3.78}$$

holds.

Proof Because of (3.76) and (3.77), case (3.74) cannot occur. The discrepancy principle will choose $0 \leq \lambda \leq \infty$ and a corresponding x_λ. For the residual $r_\lambda = b^\delta - Ax_\lambda$, one has $\|r_\lambda\|_2 \leq \delta$. From

$$A^\top r_\lambda = A^\top b^\delta - A^\top Ax_\lambda = \underbrace{A^\top b}_{} - A^\top(b - b^\delta) - A^\top Ax_\lambda$$
$$= A^\top A\hat{x}$$
$$= A^\top A(\hat{x} - x_\lambda) - A^\top(b - b^\delta),$$

we get

$$\hat{x} - x_\lambda = (A^T A)^{-1} A^T r_\lambda + (A^\top A)^{-1} A^\top (b - b^\delta).$$

From $\|r_\lambda\|_2 \leq \delta$, from $\|b - b^\delta\|_2 \leq \delta$, and from $\|(A^\top A)^{-1}A^\top\|_2 \leq \sigma_n^{-1}$ (using an SVD of A), the estimate (3.78) follows. $\qquad\square$

In functional analytic regularization theory for operator inversion, regularization methods (including parameter choices) as in Definition 3.14 are compared by their "order of convergence." This is the rate at which the error defined in (3.26) tends to zero for $\delta \to 0$ (which does not tell much about the actual error magnitude achieved for some given, finite value $\delta > 0$). Best possible rates can be achieved under specific assumptions on the exact solution u^* of $Tu = w$. In the infinite-dimensional case, one can generally not hope that this error goes to zero faster than $\mathcal{O}(\delta^\mu)$, where $0 < \mu < 1$ is a parameter depending on u^*, see Sect. 3.2 of [EHN96]. In the finite-dimensional case, however, where the "operator" A^+ is continuous, part (c) of Theorem 3.6 shows $\|A^+b - x_0\|_2 \leq \delta/\sigma_n$ for the *unregularized* solution $x_0 = A^+b^\delta$, which is an even better bound than the one given for $\|A^+b - x_\lambda\|_2$ in (3.78). Theorem 3.26, therefore, cannot justify the need for regularization of finite-dimensional problems (which are well-posed in the sense of Definition 1.5) and just shows formally, that the combination of Tikhonov regularization and discrepancy principles fulfills the requirements

(continued)

of Definition 3.14. The benefit of regularizing a finite-dimensional least squares problem can rather be seen from (the proof of) Theorem 3.25: comparing the formulae for x_0 and (3.69) for x_λ, namely

$$x_0 = \sum_{i=1}^{n} \frac{1}{\sigma_i} \cdot (u_i^\top b^\delta) \cdot v_i \quad \text{and} \quad x_\lambda = \sum_{i=1}^{n} \frac{\sigma_i}{\sigma_i^2 + \lambda} \cdot (u_i^\top b^\delta) \cdot v_i,$$

one notes that the error-amplifying effect of small singular values is moderated by choosing a regularization parameter $\lambda > 0$.

In an implementation of the discrepancy principle one first has to check for the cases (3.73) and (3.74). If these can be excluded, then there exists a unique $\lambda = \lambda^*$ fulfilling (3.75). In the special case $\|b^\delta - Ax_0\|_2 = \delta$, the value $\lambda^* = 0$ will be selected. Otherwise, the unique solution $\lambda^* > 0$ of the equation

$$J(\lambda) - \delta^2 = 0,$$

can be found by Newton's method. The required derivatives can be found analogously to (3.60) from

$$J'(\lambda) = -\lambda E'(\lambda).$$

However, the function $J - \delta^2$ is not convex. It is advantageous to consider the equivalent equation

$$I(\lambda) := J\left(\frac{1}{\lambda}\right) - \delta^2 = 0 \tag{3.79}$$

instead, as will be seen now. Since J is strictly monotonic increasing, I is strictly monotonic decreasing. We compute

$$I'(\lambda) = \frac{1}{\lambda^3} E'\left(\frac{1}{\lambda}\right) \quad \Longrightarrow \quad I''(\lambda) = -\frac{3}{\lambda^4} E'\left(\frac{1}{\lambda}\right) - \frac{1}{\lambda^5} E''\left(\frac{1}{\lambda}\right).$$

From (3.58), one gets, with $r = \mathrm{rank}(L)$

$$E'(\lambda) = \sum_{i=1}^{r} \frac{-2\kappa_i^2 \gamma_i^2}{(1 + \lambda\kappa_i)^3} \quad \Rightarrow \quad -\frac{3}{\lambda^4} E'\left(\frac{1}{\lambda}\right) = \sum_{i=1}^{r} \frac{6\kappa_i^2 \gamma_i^2}{\lambda(\lambda + \kappa_i)^3}$$

$$E''(\lambda) = \sum_{i=1}^{r} \frac{+6\kappa_i^3 \gamma_i^2}{(1 + \lambda\kappa_i)^4} \quad \Rightarrow \quad -\frac{1}{\lambda^5} E''\left(\frac{1}{\lambda}\right) = \sum_{i=1}^{r} \frac{-6\kappa_i^3 \gamma_i^2}{\lambda(\lambda + \kappa_i)^4}$$

and from this

$$I''(\lambda) = \sum_{i=1}^{r} \underbrace{\frac{6\kappa_i^2 \gamma_i^2}{\lambda(\lambda + \kappa_i)^3}}_{> 0} \underbrace{\left[1 - \frac{\kappa_i}{\lambda + \kappa_i}\right]}_{> 0} > 0.$$

This shows that the strictly monotonic decreasing function $I :]0, \infty[\to \mathbb{R}$, which has a unique zero λ^*, unless $\|b^\delta - Ax_0\|_2 \geq \delta$ or $\|b^\delta - Ax_\infty\|_2 \leq \delta$, is also strictly convex. Newton's method will, therefore, produce a sequence

$$\lambda_{k+1} = \lambda_k - \frac{I(\lambda_k)}{I'(\lambda_k)}, \quad I'(\lambda) = -\frac{2}{\lambda^3} x_{1/\lambda}^\top L^\top L (A^\top A + \frac{1}{\lambda} L^\top L)^{-1} L^\top s L x_{1/\lambda},$$

converging monotonously to this zero, when started with a small, positive value λ_0. The iteration will be stopped as soon as $\lambda_{k+1} \leq \lambda_k$, which can only happen due to the finite precision arithmetic.

Random Data Perturbations

Often data perturbations are modelled by random variables and cannot be bounded in the form $\|b - b^\delta\|_2 \leq \delta$. We will only consider a very special case, where each component b_i^δ is considered as a realization of a random variable B_i with normal distribution having mean value b_i and standard deviation δb_i. Moreover, $B_1, \ldots, B_m$ are assumed to be stochastically independent. Then,

$$Z_i := \frac{B_i - b_i}{\delta b_i} \quad i = 1, \ldots, m,$$

are stochastically independent random variables with standard normal distribution (mean value of 0 and standard deviation of 1), and therefore

$$X := \sum_{i=1}^{m} Z_i^2$$

is a random variable with χ^2 distribution and m degrees of freedom. Consequently, X has mean value m and standard deviation $\sqrt{2m}$. Setting

$$W := \begin{pmatrix} 1/\delta b_1 & 0 & \cdots & 0 \\ 0 & 1/\delta b_2 & & \vdots \\ \vdots & & \ddots & 0 \\ 0 & \cdots & 0 & 1/\delta b_m \end{pmatrix},$$

the number

$$\|W(b^\delta - b)\|_2^2 = \sum_{i=1}^{m} \left(\frac{b_i^\delta - b_i}{\delta b_i}\right)^2$$

can be considered a realization of the random variable X. In this case, we compute x_λ as minimizer of

$$\|W(b^\delta - Ax)\|_2^2 + \lambda \|Lx\|_2^2 \tag{3.80}$$

instead of solving (3.64), thereby determining λ according to the discrepancy principle with $\delta = \sqrt{m}$. This means we find λ such that the corresponding solution x_λ of (3.80) has the property

$$\|W(b^\delta - Ax_\lambda)\|_2 = \sqrt{m} . \tag{3.81}$$

To do so, we need to know or estimate the standard deviation δb_i associated with each measurement.

Examples

Example 3.27 (Numerical Differentiation) Let us reconsider Example 3.19. Discrete measurement values $b \in \mathbb{R}^m$, $m = 2n - 2$, $n = 101$, are perturbed by adding to each component a random number drawn from a normal distribution with a mean value of 0 and a standard deviation of $\sigma = 10^{-3}$. The term $\|b - b^\delta\|_2^2$ then has an expected value of $m\sigma^2$, see above. We apply the discrepancy principle with $\delta := \sqrt{m}\sigma \approx \|b - b^\delta\|_2$. The results for two randomly generated data sets are shown in Fig. 3.6 (exact u^* in red and reconstruction u_n in black). The parameter λ was determined numerically by finding a zero of function I defined in (3.79), using Newton's method. The quality of the reconstruction evidently depends on the data b^δ. The reconstructed functions u_n flatten at the boundaries of their domains of definition, where the reconstruction error becomes the largest. On the one hand, this is due to using $\|Lx\|_2$ as an inverse measure of smoothness, because this measure penalizes functions having steep slopes. On the other hand, it is inherently due to the boundary position: the difference in the derivatives of u_n and u^* would finally lead to increasing differences in function values and consequently to increasing data errors $Ax - b^\delta$, which would be inhibited by our minimization, but this effect cannot show up at the boundary.   ◊

Example 3.28 (Linearized Waveform Inversion) Let us apply Tikhonov regularization and discrepancy principle to Problem 1.12 of linearized waveform inversion,

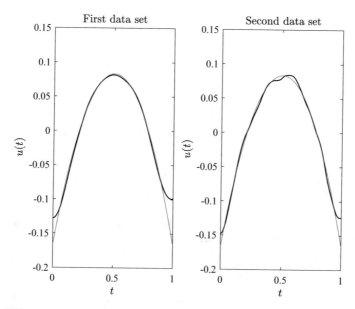

Fig. 3.6 Tikhonov regularization and discrepancy principle

discretized by the collocation method as detailed in Sect. 2.3. We repeat the main points. We want to solve the convolution equation

$$w(t) = -\frac{1}{2\sigma_0} \int_0^{Z_0} g(t - 2s/c_0)u(s) \, ds, \tag{2.29}$$

where g is the Ricker pulse function defined in (2.30). All parameters are chosen as in Example 2.7: $Z_0 = c_0 T_0/2$, $T_0 = 1$, $a = 1$, $f_0 = 5$, $\sigma_0 = \frac{1}{2}$, and $c_0 = 1$. The function w is chosen such that the quadratic polynomial

$$u^* : \left[0, \frac{1}{2}\right] \to \mathbb{R}, \quad t \mapsto 2t(1 - 2t) \tag{2.31}$$

is the exact solution of (2.29). We assume that for collocation points $t_i := i T_0/m$, $i = 0, \ldots, m$, $m \in \mathbb{N}$,

$$\text{measurements} \quad b_i^\delta \quad \text{of the true sample values} \quad b_i := w(t_i) \tag{3.82}$$

are available. We choose $n \in \mathbb{N}$ and define $h := Z_0/n$ and $\tau_j := jh$, $j = 0, \ldots, n$. An approximant u_n of u^* is sought in the space

$$X_{n+1} := \mathscr{S}_2(\tau_0, \ldots, \tau_n) = \langle N_{0,2}, \ldots, N_{n,2} \rangle,$$

i.e., having the form

$$u_n = \sum_{j=0}^{n} x_j \varphi_j, \quad \varphi_j = N_{j,2}, \ j = 0, \ldots, n.$$

This leads to a system of linear equations $Ax = b^\delta$ with $A \in \mathbb{R}^{m+1, n+1}$ having components

$$A_{i,j} = -\frac{1}{2\sigma_0} \int_0^{z_0} g(t_i - 2s/c_0) N_{j,2}(s) \, ds, \quad i = 0, \ldots, m, \ j = 0, \ldots, n,$$

(2.58)

which were computed by exact integration. The values of b_i^δ were simulated by incrementing b_i by random numbers drawn from a normal distribution with zero mean and standard deviation $\sigma = 10^{-5}$, leading to $\|b - b^\delta\|_2 \approx \sqrt{m+1} \cdot \sigma =: \delta$. We found x^δ and a corresponding approximant

$$u_n^\delta = \sum_{j=0}^{n} x_j^\delta N_{j,2}$$

by solving the regularized linear least squares problem (3.64), determining λ according to the discrepancy principle with $\delta = \sqrt{m+1} \cdot \sigma$. In Fig. 3.7, we show the results (u^* in red and u_n^δ in black) obtained for $n = m = 100$ and for two choices of L, namely $L = I_{n+1}$ (identity matrix) and $L = L_2$ as in (3.39), with n replaced by $n + 1$ ("second derivatives"). The second choice is the better one in this case and demonstrates the potential of regularization. It also shows that care must be taken concerning a good choice of the regularization term, since this will drive the reconstruction into a specific direction. Computing an unregularized solution is completely out of question here, the condition number of A being way too large ($\kappa_2(A) > 10^{20}$).   ◇

Example 3.29 (Linearized Waveform Inversion) Example 3.28 is repeated, but this time with a (much less smooth) true solution $u^* : [0, \frac{1}{2}] \to \mathbb{R}$, defined as the step function shown in red in the right parts of Figs. 3.8 and 3.9. Samples b_i were perturbed by adding realizations of independent normally distributed random variables with zero mean and standard deviation σ. The reconstruction was performed using Tikhonov regularization with the regularization term based on $L = L_2$ defined by (3.39) as in Example 3.28. The regularization parameter was determined by the discrepancy principle. Figure 3.8 shows the exact and perturbed seismogram (in red and in black, respectively, in the left picture) and the exact and the reconstructed solution (in red and in black, respectively, in the right picture), where $\sigma = \delta/\sqrt{m+1}$ was chosen such that $\delta = 0.01 \cdot \|b\|_2$ (1% error). Figure 3.9 shows the same for $\delta = 0.1 \cdot \|b\|_2$ (10% error). In the presence of jumps in the true solution function, it is often recommended to base regularization on the

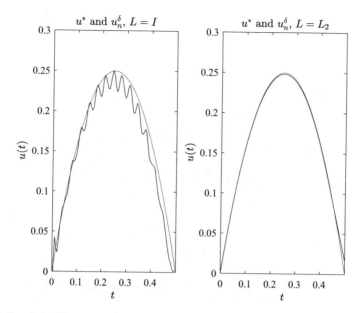

Fig. 3.7 Regularized linear waveform inversion, collocation method

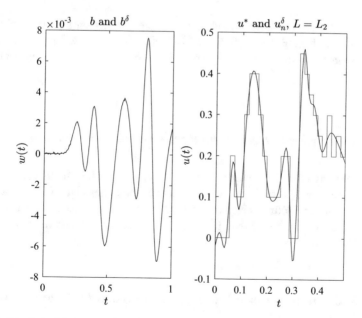

Fig. 3.8 Regularized linear waveform inversion, collocation method, 1% data error

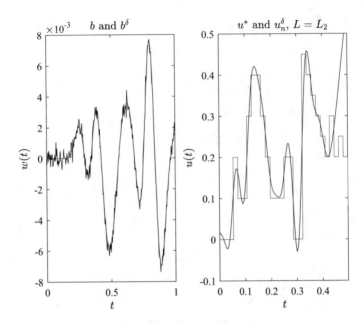

Fig. 3.9 Regularized linear waveform inversion, collocation method, 10% data error

total variation as an inverse smoothness measure and not on the norm of second derivatives, as formulated in (3.66) and (3.68). Since this leads to a nonlinear optimization problem, we will only come back to this choice in the next chapter. ◊

Other Heuristics to Determine Regularization Parameters

There are many other heuristics and recommendations of how to choose a regularization parameter λ in (3.64). Examples include the **generalized cross validation** [GHW79] or the **L-curve criterion** [Han92]. For each of these criteria, one can find examples where they deliver better results than their competitors, but no one is best in all cases.

3.6 Reduction of Least Squares Regularization to Standard Form

Tikhonov regularization as considered in Problem 3.24 with parameter λ determined by Morozov's discrepancy principle was seen to solve the constrained minimization problem formulated in Problem 3.22:

$$\min_{x \in M} \| Lx \|_2, \quad M := \{ x \in \mathbb{R}^n ; \ \| Ax - b \|_2 \le \delta \}, \tag{3.83}$$

where $\delta > 0$. Under the assumptions of Theorem 3.23, especially if the constraint $\| Ax - b \|_2 \le \delta$ is binding, a unique solution of (3.83) exists. It is determined as the unique solution x_λ of $(A^\top A + \lambda L^\top L)x = A^\top b^\delta$ with $\| b^\delta - Ax_\lambda \|_2 = \delta$. As shown by Eldén [Eld82], (3.83) can be transformed to an equivalent problem with $L = I_n$. In the following theorem, the weighted pseudoinverse L_A^+ is used, which was introduced in Definition 3.11.

Theorem 3.30 *Let $A \in \mathbb{R}^{m,n}$ and $L \in \mathbb{R}^{p,n}$ with $p \le n \le m$ and*

$$\mathcal{N}_A \cap \mathcal{N}_L = \{0\}. \tag{3.84}$$

Define

$$\begin{aligned}
B &= AL_A^+, & P &= P_{\mathcal{N}_L} = I_n - L^+ L, \\
Q &= I_m - (AP)(AP)^+, & c &= Qb,
\end{aligned} \tag{3.85}$$

and assume that

$$\| (I_m - AA^+)b \|_2 \le \delta < \| Qb \|_2 \tag{3.86}$$

holds. Then problem (3.83) has a unique solution of the form

$$x^* = L_A^+ \hat{x} + (AP)^+ b, \tag{3.87}$$

where $\hat{x} \in \mathbb{R}^p$ is the unique solution of the constrained least squares problem

$$\min_{z \in N} \| z \|_2, \quad N := \{ z \in \mathbb{R}^p ; \ \| Bz - c \|_2 = \delta \}. \tag{3.88}$$

Conversely, if $x^* \in \mathbb{R}^n$ is the unique solution of (3.83), then $z = Lx^*$ is the unique solution of (3.88).

Proof Any vector $x \in \mathbb{R}^n$ can be decomposed into its projections on $\mathcal{N}_L^\perp$ and $\mathcal{N}_L$, namely

$$x = L^+ Lx + Px = L^+ z + Px, \quad \text{with } z := Lx. \tag{3.89}$$

Q as in (3.85) is the orthogonal projector from $\mathbb{R}^m$ onto $\mathcal{R}_{AP}^\perp$ and $\tilde{Q} := I_m - Q$ is the orthogonal projector onto $\mathcal{R}_{AP}$. From Pythagoras' theorem, one gets

$$\| Ax - b \|_2^2 = \underbrace{\| Q(Ax - b) \|_2^2}_{=: \, \tau_1} + \underbrace{\| \tilde{Q}(Ax - b) \|_2^2}_{=: \, \tau_2}.$$

Since $QAP = AP - AP(AP)^+ AP = 0$ (by virtue of the Moore–Penrose axioms), one concludes from (3.89) and Definition 3.11 that

$$\tau_1 = \|QAL^+z + QAPx - Qb\|_2^2$$
$$= \|A(I_n - P(AP)^+A)L^+z - Qb\|_2^2 = \|AL_A^+z - Qb\|_2^2 \qquad (3.90)$$

(using the identity $P(AP)^+ = (AP)^+$, which follows similarly as the symmetric identity $P_{\mathcal{N}_A}(LP_{\mathcal{N}_A})^+ = (LP_{\mathcal{N}_A})^+$ at the end of the proof of Theorem 3.10). Since $\tilde{Q}AP = (AP)(AP)^+AP = AP$ (by virtue of the Moore–Penrose axioms) and since $P = P^2$, one concludes from (3.89) that

$$\tau_2 = \|AP(Px) - \tilde{Q}(b - AL^+z)\|_2^2. \qquad (3.91)$$

Further on,

$$\|Lx\|_2^2 \overset{(3.89)}{=} \|LL^+Lx + LPx\|_2^2 = \|z\|_2^2,$$

where the Moore–Penrose axioms were used once more. Observe from formulae (3.90) and (3.91) that τ_1 depends on the component L^+z of x alone but not on Px, whereas τ_2 depends on both orthogonal components L^+z and Px of x. We have found that problem (3.83) is equivalent to

$$\min_{z \in \mathbb{R}^p} \|z\|_2 \quad \text{subject to} \quad \tau_1 = \|Bz - c\|_2^2 \le \delta^2 - \tau_2, \qquad (3.92)$$

where equivalence precisely means the following. If x is a solution of (3.83), then $z = Lx$ is a solution of (3.92), the constraint being also determined by Px. If, conversely, z and Px are chosen such that $\|z\|_2$ becomes minimal under the constraints in (3.92), then $x = L^+z + Px$ is a solution of (3.83). Because of (3.84),

$$Px = (AP)^+(b - AL^+z) \qquad (3.93)$$

is the unique choice of Px that leads to $\tau_2 = 0$ (verified by direct computation). Since $\min \|z\|_2$ is decreasing with the right-hand side of the inequality constraint in (3.92) growing, one can state that (3.83) is equivalent to (3.92) with $\tau_2 = 0$.

Since $\|AA^+b - b\|_2 \le \|Ax - b\|_2$ for all $x \in \mathbb{R}^n$, requiring $\|(I_m - AA^+)b\|_2 \le \delta$ is necessary for a solution of (3.83) to exist at all. Requiring $\delta < \|Qb\|$ means that the constraint in (3.83) is binding, since for every minimizer of $\|Lx\|_2$, i.e., for every $x \in \mathcal{N}_L$, one has

$$\min_{x \in \mathcal{N}_L} \{\|Ax - b\|_2\} = \min_{x \in \mathbb{R}^n} \{\|APx - b\|_2\} = \|AP(AP)^+b - b\|_2 = \|Qb\|_2.$$

Thus, from (3.86), it follows by Theorem 3.23 that a unique solution x^* of (3.83) exists, which is located at the boundary of the feasible region, i.e., $\|Ax^* - b\|_2 = \delta$ holds. Since x^* is unique, so must be the solution of (3.92) with $\tau_2 = 0$ and with the constraint in (3.92) replaced by $\|Bz - c\|_2 = \delta$.

From (3.89), (3.93), and Definition 3.11, one concludes that

$$x^* = L^+ z + (AP)^+ (b - AL^+ z) = (I_n - (AP)^+ A) L^+ z + (AP)^+ b = L_A^+ z + (AP)^+ z,$$

where z is the solution of (3.88). $\qquad\square$

By virtue of Theorem 3.30, the solution of (3.83) can be found by solving problem (3.88), which is in standard form. This requires computation of L_A^+, AL_A^+, $(AP)^+$, and Q. In [Eld77], a method is described to achieve this, if one assumes

$$\operatorname{rank}(L) = p, \tag{3.94}$$

i.e., that matrix L has full rank. If $\operatorname{rank}(L) = p = n$, we get $L^+ = L^{-1}$ and $P = 0$, such that

$$L_A^+ = L^{-1}, \quad B = AL^{-1}, \quad (AP)^+ = 0, \quad \text{and} \quad I_m - (AP)(AP)^+ = I_m \tag{3.95}$$

in Theorem 3.30. If $\operatorname{rank}(L) = p < n$, one starts with a QR decomposition

$$L^{\mathsf{T}} = V \left(\begin{array}{c} R \\ 0 \end{array} \right) \begin{array}{l} \}\,p \\ \}\,n-p \end{array}, \qquad V = (\underbrace{V_1}_{p} \;\; \underbrace{V_2}_{n-p}) \in \mathbb{R}^{n,n}, \tag{3.96}$$

where V is orthogonal and R is upper triangular and non-singular because of (3.94). One then computes a QR decomposition of AV_2:

$$AV_2 = Q \left(\begin{array}{c} U \\ 0 \end{array} \right) \begin{array}{l} \}\,n-p \\ \}\,m-n+p \end{array}, \qquad Q = (\underbrace{Q_1}_{n-p} \;\; \underbrace{Q_2}_{m-n+p}) \in \mathbb{R}^{m,m}, \tag{3.97}$$

where Q is orthogonal (this is not the same matrix Q as in Theorem 3.30) and $U \in \mathbb{R}^{n-p,n-p}$ is upper triangular and non-singular, because otherwise AV_2 could not have full rank $n - p$. (To see that AV_2 has full rank, assume $AV_2 x = 0$ for some x. Then, $V_2 x \in \mathcal{N}_A$, but at the same time $V_2 x \in \mathcal{N}_L$ by (3.96). This implies $V_2 x = 0$, since $\mathcal{N}_A \cap \mathcal{N}_L = \{0\}$, and then implies $x = 0$, since V_2 has full rank.)

Lemma 3.31 *Let $A \in \mathbb{R}^{m,n}$ and $L \in \mathbb{R}^{p,n}$ with $p < n \le m$, and let (3.84) and (3.94) hold. Let P be defined according to (3.85), and let V, R, Q, and U be computed according to (3.96) and (3.97). Then,*

$$
\begin{aligned}
L_A^+ &= (V_1 - V_2 U^{-1} Q_1^\top A V_1) R^{-\top}, \\
A L_A^+ &= Q_2 Q_2^\top A V_1 R^{-\top}, \\
(AP)^+ &= V_2 U^{-1} Q_1^\top, \text{ and} \\
I_m - (AP)(AP)^+ &= Q_2 Q_2^\top.
\end{aligned}
$$

If $rank(L) = p = n$, *then the above matrices are given by (3.95).*

Proof From Definition 3.11, one knows that

$$
L_A^+ = (I_n - (AP)^+ A) L^+, \qquad P = P_{\mathcal{N}_L}.
$$

We compute the individual terms in this expression. First, from (3.96), one deduces

$$
L = R^\top V_1^\top \quad \Longrightarrow \quad L^+ = V_1 R^{-\top}
$$

by the Moore–Penrose axioms. Second, $P = V_2 V_2^\top$ and $(AP)^+ = (A V_2 V_2^\top)^+ = V_2 (A V_2)^+$, again easily verified by the Moore–Penrose axioms. Third,

$$
A V_2 = Q_1 U \quad \Longrightarrow \quad (A V_2)^+ = U^{-1} Q_1^\top,
$$

once more by the Moore–Penrose axioms. This proves the first and, by the way, the third equality. Furthermore, $(AP)(AP)^+ = Q_1 Q_1^\top$, which shows the last equality. Concerning the second equality, observe that

$$
A L_A^+ = (A - A V_2 U^{-1} Q_1^\top A) V_1 R^{-\top} = (I_m - Q_1 Q_1^\top) A V_1 R^{-\top},
$$

and that $I_m - Q_1 Q_1^\top = Q_2 Q_2^\top$. $\qquad\qquad\qquad\qquad\qquad\qquad\qquad\qquad\square$

In practice, L often is a sparse, banded matrix, like in (3.34) and (3.39). Moreover, $n - p$ is usually a small number. Then, (3.96) and (3.97) can be computed efficiently. The decompositions (3.96) and (3.97) can also be used to transform (3.64) for any $\lambda > 0$ into standard form, as formulated in the following.

Theorem 3.32 *Let* $A \in \mathbb{R}^{m,n}$ *and* $L \in \mathbb{R}^{p,n}$ *with* $p < n \le m$, *and let (3.84) and (3.94) hold. Let* P *be defined according to (3.85), and let* $V = (V_1, V_2)$, R, $Q = (Q_1, Q_2)$, *and* U *be computed according to (3.96) and (3.97). Let* $\lambda > 0$ *be arbitrary and* x_λ *be the unique minimizer of*

$$
\min \left\{ \|b - Ax\|_2^2 + \lambda \|Lx\|_2^2; \ x \in \mathbb{R}^n \right\}. \tag{3.98}
$$

Then, $z_\lambda = L x_\lambda$ *is the unique minimizer of*

$$
\min \left\{ \|\tilde{b} - \tilde{A} z\|_2^2 + \lambda \|z\|_2^2; \ z \in \mathbb{R}^p \right\}, \tag{3.99}
$$

where

$$\tilde{A} := Q_2^\top A V_1 R^{-\top} \quad and \quad \tilde{b} = Q_2^\top b. \tag{3.100}$$

Conversely, if z_λ is the unique solution of (3.99), then

$$x_\lambda = (V_1 - V_2 U^{-1} Q_1^\top A V_1) R^{-\top} z_\lambda + V_2 U^{-1} Q_1^\top b \tag{3.101}$$

is the unique solution of (3.98). If $\operatorname{rank}(L) = p = n$, the same equivalences hold, but with $\tilde{A} = AL^{-1}$ and $\tilde{b} = b$ in (3.100) and with $x_\lambda = L^{-1} z_\lambda$ in (3.101).

Proof The proof is taken from [Eld77]. Only the non-trivial case $p < n$ is considered. Problem (3.98) is equivalent to minimizing $\|r\|_2$, where

$$r = \begin{pmatrix} A \\ \sqrt{\lambda}L \end{pmatrix} x - \begin{pmatrix} b \\ 0 \end{pmatrix}.$$

With (3.96) and the change of variables

$$x = Vy = V_1 y_1 + V_2 y_2, \tag{3.102}$$

the vector r becomes

$$r = \begin{pmatrix} A V_1 & A V_2 \\ \sqrt{\lambda} R^\top & 0 \end{pmatrix} \begin{pmatrix} y_1 \\ y_2 \end{pmatrix} - \begin{pmatrix} b \\ 0 \end{pmatrix}.$$

The first m components of r are multiplied by $Q^\top$ from (3.97). This defines a vector

$$\tilde{r} = \begin{pmatrix} \tilde{r}_1 \\ \tilde{r}_2 \\ \tilde{r}_3 \end{pmatrix} = \begin{pmatrix} Q_1^\top A V_1 & U \\ Q_2^\top A V_1 & 0 \\ \sqrt{\lambda} R^\top & 0 \end{pmatrix} \begin{pmatrix} y_1 \\ y_2 \end{pmatrix} - \begin{pmatrix} Q_1^\top b \\ Q_2^\top b \\ 0 \end{pmatrix}.$$

with $\|r\|_2 = \|\tilde{r}\|_2$. The components $\tilde{r}_2$ and $\tilde{r}_3$ are independent of y_2, which can always be chosen such that $\tilde{r}_1 = 0$. This means that (3.98) is equivalent to

$$\min \left\{ \left\| \begin{pmatrix} Q_2^\top A V_1 \\ \sqrt{\lambda} R^\top \end{pmatrix} y_1 - \begin{pmatrix} Q_2^\top b \\ 0 \end{pmatrix} \right\|_2 ; \; y_1 \in \mathbb{R}^p \right\} \tag{3.103}$$

and

$$y_2 = U^{-1} Q_1^\top (b - A V_1 y_1). \tag{3.104}$$

By one more change of variables

$$z = R^\top y_1, \tag{3.105}$$

(3.103) takes the standard form

$$\min\{\|\tilde{A}z - \tilde{b}\|_2^2 + \lambda\|z\|_2^2;\ z \in \mathbb{R}^p\}$$

with $\tilde{A}$ and $\tilde{b}$ defined in (3.100).   □

From Lemma 3.31, the following relations between the matrices B and $\tilde{A}$ and between the vectors c and $\tilde{b}$ defined in Theorems 3.30 and 3.32 are immediate:

$$B = Q_2\tilde{A} \quad \text{and} \quad c = Q_2\tilde{b}.$$

Since the columns of Q_2 are orthonormal, from Pythagoras' theorem it follows that

$$\|Bz - c\|_2 = \|\tilde{A}z - \tilde{b}\|_2 \quad \text{for all} \quad z \in \mathbb{R}^p,$$

so we may deliberately replace B by $\tilde{A}$ and c by $\tilde{b}$ in Theorem 3.30. Lemma 3.31 also shows that the transform (3.101) defined in Theorem 3.32 can be written in the form $x_\lambda = L_A^+ z_\lambda + (AP)^+b$, as in Theorem 3.30.

Summary

Let all conditions of Theorems 3.30 and 3.32 hold and assume moreover, that rank$(A) = n$, such that a unique minimizer $\hat{x}$ of $\|Ax - b\|_2$ exists. In this case, one also has rank$(\tilde{A}) = p$,[5] such that a unique minimizer $\hat{z}$ of $\|\tilde{A}z - \tilde{b}\|_2$ exists.

It was seen that the function $x \mapsto z = Lx$ maps the solution of (3.98) to the solution of (3.99) *and also* maps the solution of (3.83) to the solution of (3.88). Conversely, for any $\lambda > 0$, the affine function

$$F_b : \mathbb{R}^p \to \mathbb{R}^n, \quad z \mapsto x = L_A^+ z + (AP)^+b$$

---

[5] Assume that $\tilde{A}x = Q_2^\top AV_1 R^{-\top}x = 0$ for some x. This means that $AV_1 R^{-\top}x$ belongs to the space spanned by the columns of Q_1, which is the space spanned by the columns of AV_2 [see (3.97)]. Therefore, there is some y such that $AV_1 R^{-\top}x = AV_2y$. Since A has full rank, this means $V_1 R^{-\top}x = V_2y$, but the spaces spanned by the columns of V_1 and V_2 are orthogonal. This means that $V_1 R^{-\top}x = 0$, and since V_1 and $R^{-\top}$ have full rank, this implies $x = 0$.

maps the unique solution of (3.99) to the unique solution of (3.98). Since rank$(A) =$ n, the latter must also hold in the limit for $\lambda \to 0$, i.e., the unique minimizer $\hat{z}$ of $\|\tilde{A}z - \tilde{b}\|_2$ is mapped to the unique minimizer $\hat{x} = F_b(\hat{z})$ of $\|Ax - b\|_2$. *The same function F_b also maps the unique solution of (3.88) to the unique solution of (3.83).*

3.7   Regularization of the Backus–Gilbert Method

The method of Backus–Gilbert was introduced in Sect. 2.4 in the context of solving a Fredholm integral equation

$$w(t) = \int_a^b k(t, s)u(s)\, ds$$

under Assumption 2.10, when samples $y_i = w(t_i)$, $i = 1, \ldots, m$ are available. Technically, a parameter t has to be fixed, the matrix $Q(t)$ and the vector c defined in (2.67) have to be computed, the vector $v \in \mathbb{R}^m$ (dependent on t) has to be found by solving the system (2.69), and a pointwise approximation $u_m(t) = \sum_{j=1}^m y_j v_j$ of the solution u^* at t has to be computed according to (2.70). The process has to be repeated for every t, where a reconstruction is desired. We have already seen in Sect. 2.4 that the method will fail, if the averaging kernel

$$\varphi(s, t) = \sum_{i=1}^m v_i k_i(s) \quad \text{with} \quad k_i(s) = k(t_i, s)$$

fails to be a function (of s) sharply peaked at t and approximately zero elsewhere. Another concern is the ill-conditioning of $Q(t)$ which will show up for growing values of m and which requires an adequate regularization when solving (2.69). To describe a possible regularization, let us replace the linear system (2.69) by the equivalent least squares problem

$$\|Ax - b\|_2^2 = \min!, \quad A = A(t) = \begin{pmatrix} Q(t) & c \\ c^\top & 0 \end{pmatrix}, \quad x = \begin{pmatrix} v \\ \mu \end{pmatrix}, \quad b = \begin{pmatrix} 0 \\ 1 \end{pmatrix}$$

$$(3.106)$$

with $A(t) \in \mathbb{R}^{m+1,m+1}$ and $x, b \in \mathbb{R}^{m+1}$ (b has its m first components equal to 0). To derive a regularization term, let us assume that instead of $y_i = w(t_i)$ we will observe perturbed samples

$$\tilde{y}_i = y_i + n_i = w(t_i) + n_i, \quad i = 1, \ldots, m,$$

where n_i are realizations of normally distributed random variables N_i with mean 0 and variance $\sigma_{ii} > 0$. Double indices are used because we will designate the covariance of N_i and N_j by σ_{ij}. Based on perturbed data $\tilde{y}_i$, the reconstructed value becomes

$$\tilde{u}_m(t) := \sum_{i=1}^{m} v_i \tilde{y}_i$$

and can be interpreted as realization of a random variable with variance

$$v^\top S v, \quad S := (\sigma_{ij})_{i,j=1,\dots,m} \in \mathbb{R}^{m,m}. \tag{3.107}$$

The so-called **covariance matrix** S is positive definite (it is a diagonal matrix if the random variables N_i are uncorrelated) and the variance $v^\top S v$ is a measure *not* of the error $\tilde{u}_m(t) - u^*(t)$, but of how much the reconstructed value $\tilde{u}_m(t)$ is subject to change, if measurement data are perturbed. It can therefore be considered an inverse measure of the **stability** of the reconstruction. In contrast, the value $v^\top Q v$ is an inverse measure of the "peakedness" of the averaging kernel $\varphi(s, t)$ at t and thus assesses the resolution achieved by the Backus–Gilbert reconstruction. One approach to a regularized method of Backus–Gilbert consists in minimizing

$$v^\top Q(t) v + \lambda v^\top S v \text{ under the constraint } v^\top c = 1, \tag{3.108}$$

for some suitable parameter λ. An explicit solution of (3.108) can be computed as in (2.68), with Q replaced by $Q + \lambda S$. An alternative approach is Tikhonov regularization of the linear system $Ax = b$ from (3.106). To this end, determine a Cholesky factorization $S = L_1^\top L_1$, where L_1 is a lower triangular matrix—in case of uncorrelated random variables N_i, this is just a diagonal matrix with entries $\sqrt{\sigma_{ii}}$ on the diagonal. Then, define

$$L := \begin{pmatrix} L_1 & 0 \end{pmatrix} \in \mathbb{R}^{m,m+1} \quad \Longrightarrow \quad \|Lx\|_2^2 = v^\top S v \quad \text{for} \quad x = \begin{pmatrix} v \\ \mu \end{pmatrix}.$$

Finally, define a regularized vector v as the first m components of a solution of

$$\min \left\{ \|b - Ax\|_2^2 + \lambda \|Lx\|_2^2; \ x \in \mathbb{R}^{m+1} \right\}. \tag{3.109}$$

To determine the parameter λ, one could devise some bound Δ and require $\|Lx\|_2 \leq \Delta$. This would mean that in fact we have to solve a constrained optimization problem like Problem 3.20:

$$\min_x \left\{ \|b - Ax\|_2^2 \right\} \quad \text{under the constraint} \quad \|Lx\|_2 \leq \Delta, \tag{3.110}$$

which was discussed in Sect. 3.4. Note that it is well possible to choose $\Delta = \Delta(t)$ in dependence of t.

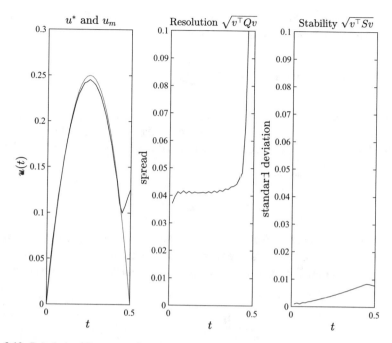

Fig. 3.10 Regularized linear waveform inversion, Backus–Gilbert method

Example 3.33 (Linearized Waveform Inversion) We take up linearized waveform inversion as in Example 2.12, but now we choose $m = 40$ sampling points and use perturbed values $\tilde{y}_i$ obtained from exact measurement values $y_i = w(t_i)$, $i = 1, \ldots, m$, by adding realizations n_i of independent normal random variables with zero mean and standard deviation $\sigma = 10^{-5}$. In Fig. 3.10, we show the results obtained by the regularized Backus–Gilbert method (3.109) for a fixed value $\lambda = 10^{-6}$ (u^* in red and u_m in black). We also show the measures for the achieved resolution and stability. Regularization can compensate the ill-conditioning of the matrix $Q(t)$ (without regularization, u_m would nowhere be an acceptable approximation of u^*). At the right boundary, the approximation quality remains poor however. As explained in Example 2.12, the low resolution achieved by the kernel functions k_i at the right boundary is responsible for this.   ◊

3.8   Regularization of Fourier Inversion

A discrete Fourier inversion of convolutional, s-variate Fredholm equations

$$w = k * u, \quad u, k, w \in L_2(\mathbb{R}^s) \tag{2.82}$$

with exact solution u^* was presented in Sect. 2.5. Reusing the notation from that section, the method can be summarized in one line:

$$\{w_\alpha = w(x_\alpha)\}_\alpha \multimap \{W_\beta\}_\beta, \quad \left\{U_\beta = \frac{W_\beta}{\hat{k}(\beta/2a)}\right\}_\beta, \quad \{U_\beta\}_\beta \multimapdotinv \{u_\alpha\}_\alpha, \quad (3.111)$$

where $u_\alpha = u_N(x_\alpha) \approx u(x_\alpha)$ and where $\alpha, \beta \in W := \{-N/2, \ldots, N/2 - 1\}^s$ for an even integer $N \geq 2$. With an application to the linearized problem of inverse gravimetry in mind, we will focus on the bivariate case $s = 2$ from now on. Let us define

$$K_\beta := \hat{k}\left(\frac{\beta}{2a}\right), \quad \beta \in \mathbb{Z}^2.$$

The smoother the function k, the faster the values $|K_\beta|$ will decay to 0 for $|\beta| \to \infty$. In practice, we will not dispose of exact measurement values (samples) $w_\alpha = w(x_\alpha), \alpha \in W$ but will have to work with perturbed data $w_\alpha^\delta, \alpha \in W$, assuming that for some $\delta > 0$

$$\sum_{\alpha \in W} \left|w_\alpha - w_\alpha^\delta\right|^2 = \delta^2. \quad (3.112)$$

Accordingly, the first step in (3.111) will produce perturbed values W_β^δ instead of W_β. Then, for $|K_\beta|$ close to 0, $U_\beta^\delta = W_\beta^\delta/K_\beta$ will contain a huge error $U_\beta^\delta - U_\beta$, such that carrying out the last step of (3.111) will produce a useless result. Tikhonov regularization suggests to avoid the division step in (3.111) and determine U_β^δ by minimizing

$$\sum_{\beta \in W} \left|W_\beta^\delta - U_\beta^\delta K_\beta\right|^2 + \lambda \sum_{\beta \in W} \left|U_\beta^\delta\right|^2$$

for some value $\lambda \geq 0$ yet to be determined. The minimization can be carried out for each index β separately. A short calculation shows

$$\left|W_\beta^\delta - U_\beta^\delta K_k\right|^2 + \lambda \left|U_\beta^\delta\right|^2 =$$

$$= \left(\lambda + |K_\beta|^2\right)\left|U_\beta^\delta - \frac{W_\beta^\delta \overline{K_\beta}}{\lambda + |K_\beta|^2}\right|^2 + \left|W_\beta^\delta\right|^2 - \frac{\left|W_\beta^\delta \overline{K_\beta}\right|^2}{\lambda + |K_\beta|^2}$$

and this term becomes minimal for the choice

$$U_\beta^\delta = \frac{W_\beta^\delta \overline{K_\beta}}{\lambda + |K_\beta|^2}, \quad \beta \in W. \quad (3.113)$$

Note that for $\lambda = 0$ ("no regularization"), this formula becomes $U_\beta^\delta = W_\beta^\delta/K_\beta$, which is exactly the division step from (3.111). Choosing a value $\lambda > 0$, one can avoid division by frequencies close to zero. A discrete inverse Fourier transform of U_β^δ (the last step of (3.111)) will give coefficients u_α^δ, $\alpha \in W$, of a B-spline approximant u^δ of $u*$.[6] One can insert u^δ into (2.82) and sample *its* output—this gives the samples of a "simulated effect" $\tilde{w}_\alpha^\delta$. The discrepancy principle then suggests choosing a maximal value λ, such that

$$\sum_{\alpha \in W} |w_\alpha^\delta - \tilde{w}_\alpha^\delta|^2 \le \delta^2.$$

To apply this criterion directly in the frequency domain, observe that, starting from (3.112), one gets (with $h = 2a/N$):

$$\delta^2 = \frac{1}{h^2} \cdot h^2 \cdot \sum_{\alpha \in W} |w_\alpha - w_\alpha^\delta|^2 \overset{(2.100)}{\approx} \frac{1}{h^2} \int_{\mathbb{R}^2} |w_N(x) - w_N^\delta(x)|^2 \, dx$$

$$\overset{(C.6)}{=} \frac{1}{h^2} \int_{\mathbb{R}^2} |\widehat{w_N}(y) - \widehat{w_N^\delta}(y)|^2 \, dy \approx \frac{1}{h^2} \cdot \frac{1}{(2a)^2} \cdot \sum_{\beta \in W} \left| \widehat{w_N}\left(\frac{\beta}{2a}\right) - \widehat{w_N^\delta}\left(\frac{\beta}{2a}\right) \right|^2$$

$$\overset{(2.101)}{=} \frac{N^2}{(2a)^4} \sum_{\beta \in W} \sigma_\beta^2 \left| W_\beta - W_\beta^\delta \right|^2. \tag{3.114}$$

Typically, errors contain significant high-frequency components, which are neglected in the last finite sum. Thus, it is to be expected that (3.114) somewhat underestimates the error. From (3.114), $\lambda \ge 0$ will be determined such that

$$S(\lambda) := N^2 \sum_{\beta \in W} \left(\frac{\sigma_\beta}{4a^2}\right)^2 \left| U_\beta^\delta K_\beta - W_\beta^\delta \right|^2$$

$$\overset{(3.113)}{=} N^2 \sum_{\beta \in W} \left(\frac{\sigma_\beta}{4a^2}\right)^2 \left| W_\beta^\delta \right|^2 \left(\frac{\lambda}{\lambda + |K_\beta|^2}\right)^2 = \delta^2.$$

This nonlinear equation could be solved using Newton's method. It is preferable to set $\mu := 1/\lambda$ and determine a zero of the function $T :]0, \infty[\to \mathbb{R}$ defined by

$$T(\mu) := S(\lambda) = N^2 \sum_{\beta \in W} \left(\frac{\sigma_\beta}{4a^2}\right)^2 \left| W_\beta^\delta \right|^2 \left(\frac{1}{1 + \mu |K_\beta|^2}\right)^2, \tag{3.115}$$

---

[6]The values u_α^δ in general are complex numbers. To derive a real valued approximant of $u*$, we just take the real parts of u_α^δ.

since this function is monotonic decreasing and convex, as it can be checked by computing its first and second derivatives (compare [Hof99, p. 141]). Consequently, Newton's method applied to the equation $T(\mu) = \delta^2$ will converge monotonously, if

$$0 < \delta^2 < N^2 \sum_{\beta \in W} \left(\frac{\sigma_\beta}{4a^2}\right)^2 \left|W_\beta^\delta\right|^2 \approx \sum_{\alpha \in W} |w_\alpha|^2.$$

Application to Linear Inverse Gravimetry

Problem 1.10 concerning linearized inverse gravimetry was formulated in Sect. 1.3. The goal is to find the Mohorovičić discontinuity modelled by a function $\Delta u :$ $[-a, a]^2 \to \mathbb{R}$, which is implicitly determined as the solution of a convolutional equation (1.30). Continuing Δu by zero values to become a function $u : \mathbb{R}^2 \to \mathbb{R}$ and setting $w := f/u_0$, Eq. (1.30) can be written as a convolutional equation of the form (2.82), namely

$$w(x) = \int_{\mathbb{R}^2} k(x - y)u(y)\, dy, \quad x \in [-b, b]^2. \tag{3.116}$$

The function k is known, see (1.31), and can be Fourier transformed analytically:

$$\widehat{k}(y_1, y_2) = \frac{2\pi}{u_0} e^{-2\pi u_0 \sqrt{y_1^2 + y_2^2}}. \tag{2.84}$$

Example 3.34 (Linear Inverse Gravimetry) Let $a = b = 4$ and $N = 64$. An effect w was simulated for the step function u shown in Fig. 3.11 and sampled at $x_\alpha = (h\alpha_1, h\alpha_2)$, $\alpha \in W$, and $h = 2a/N$. To the samples $w_\alpha := w(x_\alpha)$, we added realizations of independent Gaussian random variables with zero mean and standard deviation σ, leading to perturbed data w_α^δ, $\alpha \in W$, and a value $\delta = N\sigma$ in (3.112). A regularized Fourier reconstruction as described by (3.111) and (3.113) was carried out, with the regularization parameter λ determined according to Morozov's discrepancy principle, see (3.114) and (3.115). Figure 3.12 shows the result we obtained for $\sigma = 10^{-2}$ and Fig. 3.13 shows the result we obtained for $\sigma = 10^{-5}$, illustrating the convergence of Fourier reconstruction. $\Diamond$

It should be mentioned that alternative approaches exist for Fourier reconstruction. A famous example is **Wiener filtering**, which is advocated in [Sam11]. For a readable introduction into Wiener filtering, see Sect. 13.3 of [PTVF92]. Wiener filtering relies on a stochastic modelling of data perturbations and requires more knowledge about the noise than we assumed in (3.112).

Exact function u

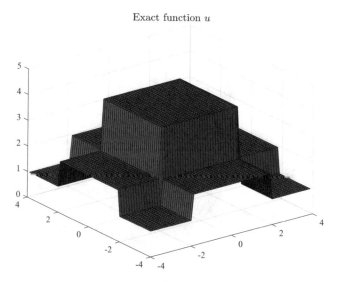

Fig. 3.11 Step function taking values 1, 2, and 4

Reconstructed function u^δ

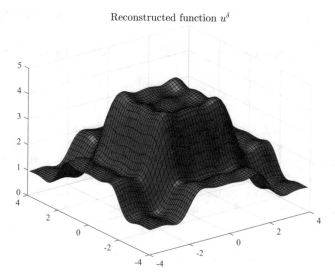

Fig. 3.12 Reconstructed step function, $\sigma = 10^{-2}$

Reconstructed function u^δ

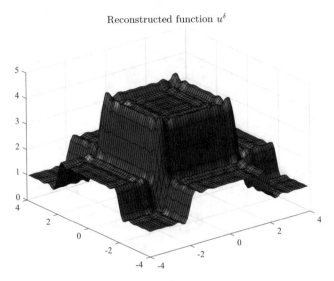

Fig. 3.13 Reconstructed step function, $\sigma = 10^{-5}$

3.9 Landweber Iteration and the Curve of Steepest Descent

Assuming that the matrix $A \in \mathbb{R}^{m,n}$ has full rank n, the least squares problem

$$\min_{x \in \mathbb{R}^n} F(x), \quad F(x) := \frac{1}{2}\|b - Ax\|_2^2, \tag{3.117}$$

has a unique solution $\hat{x}$, which can (theoretically) be found by solving the normal equations $A^\top Ax = A^\top b$, or using a QR decomposition or an SVD of A. When only an approximation $b^\delta \approx b$ is available and A is badly conditioned, the system $A^\top Ax = A^\top b^\delta$ must be replaced by a regularized variant, see Sects. 3.4 and 3.5. Alternatively, an iterative optimization algorithm can be used to solve (3.117) directly. If only perturbed data b^δ are available, such an algorithm would search for the minimizer $\bar{x}$ of

$$\min_{x \in \mathbb{R}^n} F^\delta(x), \quad F^\delta(x) := \frac{1}{2}\|b^\delta - Ax\|_2^2. \tag{3.118}$$

If A is badly conditioned, $\bar{x}$ may again be far away from $\hat{x}$. It will be seen that the iteration has to be stopped prematurely in order to obtain a regularized approximation of $\hat{x}$.

The simplest optimization method for (3.118) is the **method of steepest descent**, which is called **Landweber iteration** in the context of linear least squares problems. It is started with some initial guess $x^0 \in \mathbb{R}^n$ of $\bar{x}$, for example, $x^0 = 0$. Then, one iteratively computes

$$x^{k+1} := x^k - s_k \nabla F^\delta(x^k), \quad k = 0, 1, 2, \ldots,$$

where $s_k > 0$ is a **step size** to be chosen. If the step size is chosen to be constant, i.e., $s_k = s$ for all k, then this iteration, applied to (3.118), becomes

$$x^{k+1} = x^k - sA^\top(Ax^k - b^\delta) = (I - sA^\top A)x^k + sA^\top b^\delta. \tag{3.119}$$

For the choice $x^0 = 0$, one gets an explicit formula for the kth iterate:

$$x^k = s \sum_{j=0}^{k-1} (I - sA^\top A)^j A^\top b^\delta, \quad k = 1, 2, \ldots \tag{3.120}$$

This can be written as $x^k = R_k b^\delta$, where $(R_k)_{k \in \mathbb{N}}$ is the sequence of matrices

$$R_k = s \sum_{j=0}^{k-1} (I - sA^\top A)^j A^\top \quad \in \quad \mathbb{R}^{m,n}. \tag{3.121}$$

Using the reduced SVD $A = \hat{U} \hat{\Sigma} V^\top$ (see (A.2)), a short calculation shows that

$$R_k y = \sum_{i=1}^{n} \frac{[1 - (1 - s\sigma_i^2)^k]}{\sigma_i} \cdot (u_i^\top y) \cdot v_i, \quad y \in \mathbb{R}^m. \tag{3.122}$$

Here, u_i and v_i, $i = 1, \ldots, n$, are the columns of $\hat{U}$ and V, respectively, and σ_i, $i = 1, \ldots, n$, are the singular values of A. The vector $x^k = R_k b^\delta$ found after k steps of the Landweber iteration applied to (3.118) has to be compared with the vector one is actually interested in, namely

$$\hat{x} = A^+ b = \sum_{i=1}^{n} \frac{1}{\sigma_i} \cdot (u_i^\top b) \cdot v_i. \tag{3.123}$$

A comparison with (3.122) shows that R_k is an approximation of the matrix (operator) A^+, the factors $1/\sigma_i$ being modified by multiplication with attenuation factors $q(k, \sigma_i) = 1 - (1 - s\sigma_i^2)^k$. Under the condition

$$0 < s \le \frac{1}{\sigma_1^2} = \frac{1}{\|A\|_2^2}, \tag{3.124}$$

one gets $q(k, \sigma) \to 1$ for $k \to \infty$ and for $0 < \sigma \le \sigma_1 = \|A\|_2$. Consequently,

$$\|R_k y - A^+ y\|_2 \to 0 \quad \text{for} \quad k \to \infty \quad \text{and for all} \quad y \in \mathbb{R}^m,$$

if (3.124) holds. This shows that $(R_k)_{k \in \mathbb{N}}$ is converging to A^+, but it does not yet show the regularizing properties of the Landweber iteration. From (3.124), one deduces $0 < q(k, \sigma) \leq 1$ for all $k \in \mathbb{N}$ and for $0 < \sigma \leq \sigma_1$. From Bernoulli's inequality, one gets

$$1 - (1 - s\sigma^2)^k \leq \sqrt{1 - (1 - s\sigma^2)^k} \leq \sqrt{1 - (1 - ks\sigma^2)} = \sqrt{ks}\sigma,$$

and if this estimate is used in (3.122), one sees $\|R_k\|_2 \leq \sqrt{ks}$. Therefore,

$$\|R_k b^\delta - A^+ b\|_2 \leq \|R_k b^\delta - R_k b\|_2 + \|R_k b - A^+ b\|_2$$
$$\leq \sqrt{ks}\|b^\delta - b\|_2 + \|R_k b - A^+ b\|_2. \qquad (3.125)$$

This estimate evidently is the analog of (3.24). The second term on the right-hand side converges to 0 for $k \to \infty$. The total error $R_k b^\delta - A^+ b$ only converges to 0, if $\|b^\delta - b\|_2 \leq \delta \to 0$ and if $k\delta^2 \to 0$ for $\delta \to 0$ and $k \to \infty$. The iteration index k plays the role of a regularization parameter: the larger it is, the better R_k approximates A^+, but the more data perturbations $b^\delta - b$ are amplified. The following theorem, which is the analog of Theorem 3.26, suggests a choice for k.

Theorem 3.35 (Stopping Rule for the Landweber Iteration) *Let $A \in \mathbb{R}^{m,n}$, $m \geq n$, with $rank(A) = n$. Let $b, b^\delta \in \mathbb{R}^m$. Denote by*

$\hat{x} = A^+ b$ *the unique minimizer of $\|b - Ax\|_2 = $ min! (exact data) and by*
$x^k = R_k b^\delta$ *the kth iterate produced by (3.120) (perturbed data; $R_0 := 0$).*

The step size restriction (3.124) applies. Assume that for some known $\delta > 0$

$$\|b - b^\delta\|_2 + \|b - A\hat{x}\|_2 \leq \delta. \qquad (3.126)$$

Assume that $\|b^\delta\|_2 > \tau\delta$ for a fixed parameter $\tau > 1$. Then:

(1) There exists a smallest integer $k \in \mathbb{N}$ (depending on b^δ), such that

$$\|b^\delta - A R_k b^\delta\|_2 \leq \tau\delta < \|b^\delta - A R_{k-1} b^\delta\|_2. \qquad (3.127)$$

(2) With k as in (1), an estimate

$$\|\hat{x} - x^k\|_2 \leq \frac{\tau + 1}{\sigma_n}\delta \qquad (3.128)$$

holds, where σ_n is the minimal singular value of A.

Remarks In case $\|b^\delta\|_2 \leq \tau\delta$, $x^0 = 0$ will be declared a regularized solution. The estimate (3.128) holds in this case. The comparison of (3.122) and (3.123) shows that regularization of A^+ by the Landweber iteration relies on an approximation of the function $\sigma \mapsto 1/\sigma, \sigma > 0$, by the polynomial $[1 - (1 - s\sigma)^k]/\sigma$. Variants of the Landweber iteration can be obtained by choosing other polynomial approximations.

Proof The proof of part (1) is modelled after [Kir11, p. 51 ff]. Some modifications are needed, since we do not assume that the system $Ax = b$ has a solution. As already seen, the Landweber iteration produces iterates $x^k = R_k b^\delta$, $k = 1, 2, \ldots$, where R_k is the operator defined by

$$R_k y = \sum_{i=1}^{n} \frac{[1 - (1 - s\sigma_i^2)^k]}{\sigma_i} \cdot (u_i^\top y) \cdot v_i, \quad y \in \mathbb{R}^m, \tag{3.122}$$

which can be derived from a SVD $A = U \Sigma V^\top$ of A. The column vectors $u_i \in \mathbb{R}^m$ of U build an orthonormal basis of $\mathbb{R}^m$, and the column vectors $v_i \in \mathbb{R}^n$ of V build an orthonormal basis of $\mathbb{R}^n$. The values $\sigma_1 \geq \cdots \geq \sigma_n > 0$ are the singular values of A. Since $Av_i = \sigma_i u_i, i = 1, \ldots, n$, one gets

$$AR_k y = \sum_{i=1}^{n} [1 - (1 - s\sigma_i^2)^k](u_i^\top y)u_i \quad \text{for all } y \in \mathbb{R}^m. \tag{3.129}$$

Since $y = \sum_{i=1}^{m} (u_i^\top y)u_i$ for all $y \in \mathbb{R}^m$, one further gets

$$\|AR_k y - y\|_2^2 = \sum_{i=1}^{n} (1 - s\sigma_i^2)^{2k}(u_i^\top y)^2 + \sum_{i=n+1}^{m} (u_i^\top y)^2, \tag{3.130}$$

where the orthonormality of $\{u_1, \ldots, u_m\} \subset \mathbb{R}^m$ was used. By the step size restriction (3.124), all factors $(1 - s\sigma_i^2)^{2k}$ are bounded by 1, thus (3.130) shows

$$\|AR_k - I_m\|_2 \leq 1. \tag{3.131}$$

Set $\hat{b} := A\hat{x} \in \mathcal{R}_A \subset \mathbb{R}^m$, such that $\hat{b} = \sum_{i=1}^{n}(u_i^\top \hat{b})u_i$ ($\hat{b}$ has no components in directions $u_{n+1}, \ldots, u_m$). From

$$AR_k b^\delta - b^\delta = (AR_k - I_m)(b^\delta - \hat{b}) + AR_k\hat{b} - \hat{b}$$

and from (3.131), one gets

$$\|AR_k b^\delta - b^\delta\|_2 \leq 1 \cdot \|b^\delta - \hat{b}\|_2 + \|AR_k\hat{b} - \hat{b}\|_2. \tag{3.132}$$

The first term on the right-hand side is bounded by δ by assumption (3.126). For the second term we get

$$\|AR_k\hat{b} - \hat{b}\|_2^2 = \sum_{i=1}^{n}(1 - s\sigma_i^2)^{2k}(u_i^\top \hat{b})^2, \tag{3.133}$$

similarly as in (3.130), but with the last summands missing because of the special choice of $\hat{b}$. From (3.124) one concludes that $\|AR_k\hat{b} - \hat{b}\|_2 \to 0$ for $k \to \infty$, such that $\|AR_kb^\delta - b^\delta\|_2 \le \tau\delta$ for any $\tau > 1$, if only k is chosen large enough. This proves part (1). Concerning part (2), set $r^k = b^\delta - Ax^k = b^\delta - AR_kb^\delta$. From

$$A^\top r^k = A^\top b^\delta - A^\top Ax^k = \underbrace{A^\top b}_{= A^\top A\hat{x}} - A^\top(b - b^\delta) - A^\top Ax^k$$

$$= A^\top A(\hat{x} - x^k) - A^\top(b - b^\delta),$$

we get

$$\hat{x} - x^k = (A^\top A)^{-1}A^\top r^k + (A^\top A)^{-1}A^\top(b - b^\delta).$$

From $\|r^k\|_2 \le \tau\delta$, from $\|b - b^\delta\|_2 \le \delta$, and from $\|(A^\top A)^{-1}A^\top\|_2 \le \sigma_n^{-1}$ (using an SVD of A), the estimate (3.128) follows. $\square$

It is instructive to reconsider the Landweber iteration from a different perspective. The method of steepest descent for finding the minimum of $F^\delta(x) = \frac{1}{2}\|b^\delta - Ax\|_2^2$, when started at $x^0 = 0$, has a continuous analogon, namely the initial value problem

$$x'(t) = -\nabla F^\delta(x(t)) = A^\top b^\delta - A^\top Ax(t), \quad x(0) = 0. \tag{3.134}$$

The solution of (3.134) is called **curve of steepest descent**. Solving (3.134) numerically by the explicit Euler method with constant step size s produces approximations

$$x^k \approx x(t_k), \quad t_k = k \cdot s, \quad k \in \mathbb{N}_0,$$

which are exactly the same as the ones defined by the Landweber iteration (3.119). But the curve of steepest descent can also be found analytically. To see this, use the SVD $A = U\Sigma V^\top$ and the transformation $y(t) := V^\top x(t)$, which decouples (3.134) into n separate initial value problems

$$y_i'(t) + \sigma_i^2 y_i(t) = \sigma_i(u_i^\top b^\delta), \quad y_i(0) = 0, \quad i = 1, \ldots, n,$$

for the components of y, which can be solved easily. Transforming the solutions back gives the explicit solution formula

$$x(t) = \sum_{i=1}^n \frac{1 - \exp(-\sigma_i^2 t)}{\sigma_i} \cdot (u_i^\top b^\delta) \cdot v_i. \tag{3.135}$$

Evidently, $x(t) \to \bar{x} = A^+ b^\delta$ for $t \to \infty$, compare (3.123). For finite values $t = T$, the vector $x(T)$ defined by (3.135) is a regularized approximation of $\hat{x} = A^+ b$, as formally proven, e.g., in [Hof99, p. 153]. Now if some eigenvalues σ_i^2 of $A^\top A$ are much larger than others,[7] then some components of $x(t)$ converge much faster than others, as can immediately be seen from (3.135). In this case the differential equation in (3.134) is called **stiff**. It is well known in numerical analysis that solving stiff equations by explicit numerical methods requires small step sizes s for the approximate values $x^k \approx x(k \cdot s)$ to converge to $\bar{x}$ for $k \to \infty$. A step size restriction consequently reappears for the Landweber iteration in the form $s \le 1/\|A\|_2^2$ and might lead to a very slow convergence of this method. *Implicit* methods are more adequate for the numerical integration of stiff differential equations. The implicit Euler method with step sizes $s_k = t_{k+1} - t_k$, $k \in \mathbb{N}_0$, determines approximations $x^k \approx x(t_k)$, $k \in \mathbb{N}_0$, from the equations

$$x^{k+1} = x^k - s_k \nabla F^\delta(x^{k+1}) = x^k + s_k(A^\top b^\delta - A^\top A x^{k+1}), \quad x^0 = 0.$$

These identities can be reformulated to become

$$(I + s_k A^\top A)x^{k+1} = x^k + s_k A^\top b^\delta, \quad x^0 = 0, \tag{3.136}$$

or as well in the form

$$\left(A^\top A + \frac{1}{s_k}I\right)(x^{k+1} - x^k) = A^\top b^\delta - A^\top A x^k, \quad x^0 = 0. \tag{3.137}$$

There is a close relation between (3.137) and Newton's method for the minimization of $F^\delta(x)$. Since the Hessian of F^δ is given by $\nabla^2 F^\delta(x) = A^\top A$, Newton's method determines iteration updates by solving

$$\nabla^2 F^\delta(x^k)(x^{k+1} - x^k) = -\nabla F^\delta(x^k)$$

$$\Longleftrightarrow \quad A^\top A(x^{k+1} - x^k) = A^\top b^\delta - A^\top A x^k. \tag{3.138}$$

Of course, when started with $x^0 = 0$, Newton's method means nothing else than a solution of the normal equations and will converge in a single step. But this will be different in the nonlinear case to be considered in the next chapter. Comparing (3.138) to (3.137), one sees that solving the initial value problem (3.134) by the implicit Euler method exactly corresponds to a *regularized version of the Newton's method*. This kind of regularization is known as **trust region method** in optimization. The equivalence of trust region methods, regularization, and numerical solution of the initial value problem (3.134) is described in [Sch12]. The iteration (3.137) can be analyzed easily for constant step sizes $s_k = s$. From

---

[7]This is just the case when A is badly conditioned and thus the usual case for inverse problems.

(3.136) one gets the explicit formula

$$x^k = \sum_{j=1}^{k} s(I + sA^\top A)^{-j} A^\top b^\delta, \quad k \in \mathbb{N}.$$ (3.139)

This can be written as $x^k = R_k b^\delta$, where $(R_k)_{k \in \mathbb{N}}$ is the sequence of matrices

$$R_k = \sum_{j=1}^{k} s(I + sA^\top A)^{-j} A^\top \quad \in \quad \mathbb{R}^{m,n}.$$ (3.140)

Using an SVD $A = U \Sigma V^\top$, one finds that

$$R_k y = \sum_{i=1}^{n} \frac{1}{\sigma_i} \underbrace{\left(1 - \frac{1}{(1 + s\sigma_i^2)^k} \right)}_{=: q(k, \sigma_i)} \cdot (u_i^\top y) \cdot v_i, \quad y \in \mathbb{R}^m.$$ (3.141)

Now, $q(k, \sigma) \xrightarrow{k \to \infty} 1$ and thus $x^k \to \bar{x} = A^+ b^\delta$ independently of $s > 0$ und $\sigma > 0$. *The step size restriction required for the Landweber iteration is no longer needed.* From Bernoulli's inequality one further gets

$$\left(1 - \frac{s\sigma^2}{1 + s\sigma^2} \right)^k \geq 1 - \frac{ks\sigma^2}{1 + s\sigma^2}$$

for all $s, \sigma > 0$ and $k \in \mathbb{N}$, leading to

$$|q(k, \sigma)| \leq \sqrt{q(k, \sigma)} = \sqrt{1 - \left(1 - \frac{s\sigma^2}{1 + s\sigma^2} \right)^k} \leq \sqrt{\frac{ks\sigma^2}{1 + s\sigma^2}} \leq \sqrt{ks}\sigma.$$

From this estimate one can deduce that an estimate like (3.125) also holds for the iterates x^k determined by (3.137). Again, the iteration index k plays the role of a regularization parameter.

Theorem 3.36 (Stopping Rule for the Implicit Euler Method) *Let $A \in \mathbb{R}^{m,n}$ with $m \geq n$ and $\mathrm{rank}(A) = n$. Let $b, b^\delta \in \mathbb{R}^m$. Denote by*

$$\hat{x} = A^+ b \qquad \text{*the unique minimizer of* } \|b - Ax\|_2 = \min! \text{ *(exact data) and by*}$$
$$x^k = R_k b^\delta \qquad \text{*the kth iterate produced by (3.141) (perturbed data; } R_0 := 0\text{).}$$

Assume that for some known $\delta > 0$

$$\|b - b^\delta\|_2 + \|b - A\hat{x}\|_2 \leq \delta.$$ (3.142)

Assume that $\|b^\delta\|_2 > \tau\delta$ for a fixed parameter $\tau > 1$. Then:

(1) There exists a smallest integer $k \in \mathbb{N}$ (depending on b^δ), such that

$$\|b^\delta - AR_k b^\delta\|_2 \le \tau\delta < \|b^\delta - AR_{k-1}b^\delta\|_2. \tag{3.143}$$

(2) With k as in (1), an estimate

$$\|\hat{x} - x^k\|_2 \le \frac{\tau+1}{\sigma_n}\delta \tag{3.144}$$

holds, where σ_n is the smallest singular value of A.

Proof The kth iterate can be written in form $x^k = R_k b^\delta$, where $R_k \in \mathbb{R}^{n,m}$ is the matrix given by

$$R_k y = \sum_{i=1}^{n} \frac{1}{\sigma_i}\left(1 - \frac{1}{(1+s\sigma_i^2)^k}\right) \cdot (u_i^\top y) \cdot v_i, \quad y \in \mathbb{R}^m, \tag{3.145}$$

compare (3.141). This representation can be derived from a SVD $A = U\Sigma V^\top$ of A, as in (A.1). The column vectors $u_i \in \mathbb{R}^m$ of U build an orthonormal basis of $\mathbb{R}^m$, and the column vectors $v_i \in \mathbb{R}^n$ of V build an orthonormal basis of $\mathbb{R}^n$. The values $\sigma_1 \ge \cdots \ge \sigma_n > 0$ are the singular values of A. Since $Av_i = \sigma_i u_i, i = 1,\ldots,n$, one gets

$$AR_k y = \sum_{i=1}^{n}\left(1 - \frac{1}{(1+s\sigma_i^2)^k}\right)(u_i^\top y)u_i \quad \text{for all } y \in \mathbb{R}^m. \tag{3.146}$$

Since $y = \sum_{i=1}^{m}(u_i^\top y)u_i$ for all $y \in \mathbb{R}^m$, one further gets

$$\|AR_k y - y\|_2^2 = \sum_{i=1}^{n}\frac{1}{(1+s\sigma_i^2)^{2k}}(u_i^\top y)^2 + \sum_{i=n+1}^{m}(u_i^\top y)^2, \tag{3.147}$$

where the orthonormality of $\{u_1,\ldots,u_m\} \subset \mathbb{R}^m$ was used. Hence,

$$\|AR_k - I_m\|_2 \le 1. \tag{3.148}$$

Set $\hat{b} := A\hat{x} \in \mathcal{R}_A \subset \mathbb{R}^m$, such that $\hat{b} = \sum_{i=1}^{n}(u_i^\top \hat{b})u_i$ (since $\hat{b}$ has no components in directions $u_{n+1},\ldots,u_m$). From

$$AR_k b^\delta - b^\delta = (AR_k - I_m)(b^\delta - \hat{b}) + AR_k\hat{b} - \hat{b}$$

and from (3.148), one gets

$$\|AR_k b^\delta - b^\delta\|_2 \le 1 \cdot \|b^\delta - \hat{b}\|_2 + \|AR_k \hat{b} - \hat{b}\|_2. \tag{3.149}$$

The first term on the right-hand side is bounded by δ by assumption (3.142). For the second term we get

$$\|AR_k \hat{b} - \hat{b}\|_2^2 = \sum_{i=1}^{n} \frac{(u_i^\top \hat{b})^2}{(1 + s\sigma_i^2)^{2k}}, \tag{3.150}$$

similarly as in (3.147), but with the last summands missing because of the special choice of $\hat{b}$. Evidently, $\|AR_k \hat{b} - \hat{b}\|_2 \to 0$ for $k \to \infty$, such that $\|AR_k b^\delta - b^\delta\|_2 \le \tau\delta$ for any $\tau > 1$, if only k is chosen large enough. This proves part (1). The proof of part (2) is exactly the same as for Theorem 3.35. □

In practice, an automatic step size control has to be provided for the implicit Euler method (3.137). Let $\tilde{x}(t)$ be the solution of the differential equation $x'(t) = -\nabla F^\delta(x(t))$ *with initial value* $\tilde{x}(t_k) = x^k$. Then $\|x^{k+1} - \tilde{x}(t_k + s_k)\|_\infty$ is the magnitude of the error when taking a single step from t_k to t_{k+1}, neglecting the error already contained in x^k. It is the goal of a step size control to choose s_k as large as possible while keeping the error magnitude under control by requiring that

$$\frac{\|x^{k+1} - \tilde{x}(t_k + s_k)\|_\infty}{s_k} \le \varepsilon \tag{3.151}$$

for some chosen value ε (the left-hand side of (3.151) is commonly known as local discretization error). For the (implicit) Euler method, it is known that

$$\frac{x^{k+1} - \tilde{x}(t_k + s_k)}{s_k} \overset{\bullet}{=} \tau_k \cdot s_k, \tag{3.152}$$

for some constant vector τ_k, independent of s_k. The dotted equality sign in (3.152) means equality up to terms of order s_k^2 or higher. Thus, (3.151) can be achieved by choosing s_k small enough, but we cannot say offhand *how* small, since τ_k is not known. One therefore at first computes x^{k+1} according to (3.137) for some tentative value $s_k = s$. Furthermore, one computes an approximation $\tilde{x}^{k+1}$ by taking *two* subsequent steps of size $s/2$. Using (3.152) it can be shown that the corresponding error is given by

$$\frac{\tilde{x}^{k+1} - \tilde{x}(t_k + s)}{s} \overset{\bullet}{=} \tau_k \cdot \frac{s}{2}. \tag{3.153}$$

Comparing (3.152) (for $s_k = s$) and (3.153), one gets

$$\Delta := \frac{\|x^{k+1} - \tilde{x}^{k+1}\|_\infty}{s} \overset{\bullet}{=} \|\tau_k\|_\infty \cdot \frac{s}{2},$$

meaning that Δ can serve as an estimator for the magnitude of the error in (3.153). Moreover, this error scales with s, such that taking a step of rescaled size $s \cdot \frac{\varepsilon}{\Delta}$ should produce an error of size ε, the maximum allowed according to (3.151). Following common advice, we will choose a somewhat smaller new step size $0.9 \cdot s \cdot \frac{\varepsilon}{\Delta}$ and also take care to avoid extreme step size changes. We arrive at

---

Implicit Euler method with step size control for least squares problems. Let all conditions of Theorem 3.36 hold. Choose $\varepsilon \in]0, 1[$, $s > 0$ and $x^0 \in \mathbb{R}^m$ (e.g. $x^0 = 0$). Set $k = 0$.

Step 1: Compute $r^k := b^\delta - Ax^k$ and $c^k := A^\top r^k$. If $\|r^k\|_2 \le \tau\delta$, then stop. Otherwise, go to Step 2.

Step 2: If $A^\top A + \frac{1}{s}I$ is so badly conditioned that a numerical Cholesky factorization fails, then set $s := \frac{s}{2}$ and go to Step 2. Otherwise go to Step 3.

Step 3: Get the solution p of $(A^\top A + \frac{1}{s}I)x = c^k$ [via Cholesky factorization], set $\eta^s := x^k + p$, and goto Step 4.

Step 4: Get the solution p of $(A^\top A + \frac{2}{s}I)x = c^k$ and set $\eta^{s/2} := x^k + p$. Compute $\tilde{r} = b^\delta - A\eta^{s/2}$ and $\tilde{c} = A^\top\tilde{r}$. Get the solution q of $(A^\top A + \frac{2}{s}I)x = \tilde{c}$ and set $\tilde{\eta}^s = \eta^{s/2} + q$.

Step 5: Compute $\Delta := \|\eta^s - \tilde{\eta}^s\|_\infty/s$. If $\Delta < 0.1 \cdot \varepsilon$, set $f := 10$, else set $f := \varepsilon/\Delta$. If $f < 0.1$, set $f = 0.1$. Set $s := s \cdot f \cdot 0.9$. If $\Delta \le \varepsilon$, set $x^{k+1} := \tilde{\eta}^s$, set $k = k + 1$, and goto Step 1. Otherwise, go to Step 3.

Example 3.37 (Linear Waveform Inversion) We reconsider Example 3.28 of linearized waveform inversion, setting up A and b from collocation equations with $m = n = 100$. Perturbed data b^δ are simulated by adding realizations of independent normally distributed random variables with zero mean and standard deviation $\sigma = 10^{-5}$ to the components of b. Thus, we set $\delta := \sqrt{m+1} \cdot \sigma$. Figure 3.14 shows the result we achieved for the choice $\varepsilon = 10^{-2}$ and $\tau = 1$ (right) and also shows how well the minimization of $F(x) = \frac{1}{2}\|b^\delta - Ax\|_2$ proceeded with the iteration index k. The implicit Euler method stopped with $k = 10$. The result—of comparable quality as the original Tikhonov regularization for $L = I_{n+1}$—is not very satisfactory. $\Diamond$

Let us have one more look at formula (3.137), which we may rewrite in the form

$$\left(A^\top A + \frac{1}{s_k}I\right)u^k = A^\top r^k, \quad x^0 = 0, \tag{3.154}$$

if we set

$$u^k := x^{k+1} - x^k \quad \text{and} \quad r^k := b^\delta - Ax^k.$$

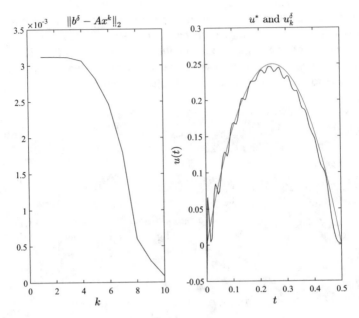

Fig. 3.14 Implicit Euler method, $\varepsilon = 10^{-2}$ and $\tau = 1$

A solution u^k of (3.154) can equivalently be characterized as the minimizer of

$$\|Au - r^k\|_2^2 + \frac{1}{s_k}\|u\|_2^2, \quad u \in \mathbb{R}^n,$$

which means that the above algorithm implicitly uses the Euclidean norm $\|u\|_2$ as regularity measure for computing iteration updates. More generally, an update u^k could be computed as the minimizer of

$$\|Au - r^k\|_2^2 + \frac{1}{s_k}\|Lu\|_2^2, \quad u \in \mathbb{R}^n,$$

but this will not be done by the implicit Euler method. However, we can force the implicit Euler method into behaving as if $\|L \bullet \|_2$ was used as a regularity measure by transforming the general Tikhonov regularization into a standard one. According to Theorem 3.32, we will do the following. First, compute

$$\tilde{A} \quad \text{and} \quad \tilde{b}^\delta \quad \text{as in} \quad (3.100) \quad [b \text{ replaced by } b^\delta]. \tag{3.155}$$

Second, apply the implicit Euler method to the minimization of

$$\|\tilde{A}z - \tilde{b}^\delta\|_2, \quad z \in \mathbb{R}^p. \tag{3.156}$$

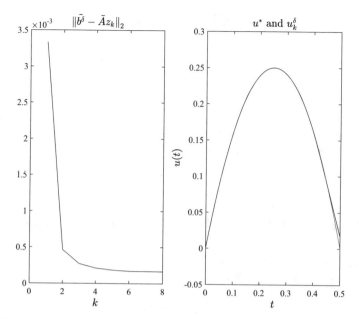

Fig. 3.15 Implicit Euler method with coordinate transform, $\varepsilon = 10^{-2}$ and $\tau = 1$

Finally, transform the obtained solution z^δ back to

$$x^\delta = (V_1 - V_2 U^{-1} Q_1^\top A V_1) R^{-\top} z^\delta + V_2 U^{-1} Q_1^\top b^\delta, \quad \text{as in (3.101).} \qquad (3.157)$$

This will be tested in the following example.

Example 3.38 (Linear Waveform Inversion) We solve the problem from Example 3.37 again, keeping all previous data, but including a coordinate transform. This means we follow the steps (3.155), (3.156), and (3.157), where we choose $L = L_2 \in \mathbb{R}^{p,n+1}$, $p = n - 1$, as in (3.39) ("second derivatives"). Figure 3.15 shows the result achieved, which is much better than in the previous example. ◊

The above example shows that using the implicit Euler iteration, one can only seemingly avoid the choice of a regularization term. The latter implicitly reappears in form of choosing a suitable coordinate transform. The same observation will be made for the other iterative methods to be presented below.

3.10   The Conjugate Gradient Method

The conjugate gradient method is an iterative algorithm to solve

$$\min_{x \in \mathbb{R}^n} F^\delta(x), \quad F^\delta(x) := \frac{1}{2} \|b^\delta - Ax\|_2^2, \tag{3.118}$$

where $A \in \mathbb{R}^{m,n}$, $m \geq n$, is a matrix of full rank n. Stopped prematurely, it will produce a regularized approximation of the unique minimizer $\hat{x}$ of $\|b - Ax\|_2$, as did the algorithms presented in Sect. 3.9.

Conjugate Gradient Method for Linear Systems of Equations

The conjugate gradient method will at first be described as an iterative method to solve a linear system of equations

$$Ax = b, \quad A \in \mathbb{R}^{n,n} \text{ symmetric and positive definite,} \quad b \in \mathbb{R}^n, \tag{3.158}$$

having the unique solution $x^* \in \mathbb{R}^n$. In the following paragraph, this iteration will be applied to the normal equations of (3.118), i.e., we will replace the matrix A in (3.158) by $A^\top A$ (which is symmetric and positive definite for any matrix $A \in \mathbb{R}^{m,n}$ of full rank) and the vector b by $A^\top b^\delta$.

Since A is symmetric and positive definite, the mapping

$$\| \bullet \|_A : \mathbb{R}^n \to \mathbb{R}, \quad x \mapsto \|x\|_A := \sqrt{x^* Ax} \tag{3.159}$$

defines a norm on $\mathbb{R}^n$, the so-called **A-norm**. The A-norm is induced by a scalar product on $\mathbb{R}^n$ given by

$$\langle x|y \rangle_A := x^\top Ay = \langle x|Ay \rangle = \langle Ax|y \rangle, \quad x, y \in \mathbb{R}^n.$$

We will further need the so-called **Krylov spaces**

$$\mathcal{K}_k := \langle b, Ab, \ldots, A^{k-1}b \rangle, \quad k = 1, \ldots, n.$$

There is a unique best approximation $x_k \in \mathcal{K}_k$ of x^*, i.e., a unique x_k such that

$$\|x_k - x^*\|_A \leq \|x - x^*\|_A \quad \text{for all} \quad x \in \mathcal{K}_k \tag{3.160}$$

This solution is characterized by

$$\langle x^* - x_k | x \rangle_A = 0 \quad \text{for all} \quad x \in \mathcal{K}_k, \tag{3.161}$$

Fig. 3.16 *A*-orthogonal
projection onto Krylov space

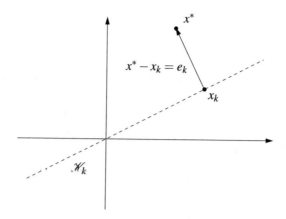

see Theorem B.13, meaning that x_k is the *A*-orthogonal projection of x^* onto $\mathscr{K}_k$, as
illustrated in Fig. 3.16. Two vectors $x, y \in \mathbb{R}^n$ are called *A*-orthogonal or **conjugate**
(with respect to *A*), iff $\langle x|y\rangle_A = 0$. In this case we write $x \perp_A y$, whereas $x \perp y$
designates orthogonality with respect to the Euclidean scalar product.

Assume we knew k *nonzero*, mutually conjugate vectors $p_1, \ldots, p_k \in \mathscr{K}_k$, i.e.
$\langle p_i|p_j\rangle_A = 0$ for $i \neq j$. Such vectors are necessarily linearly independent (implying
$\dim(\mathscr{K}_k) = k$), since $\alpha_1 p_1 + \ldots + \alpha_k p_k = 0$ implies

$$0 = \langle p_j|\sum_{i=1}^{k}\alpha_i p_i\rangle_A = \sum_{i=1}^{k}\alpha_i\langle p_j|p_i\rangle_A = \alpha_j\|p_j\|_A^2 \,,$$

and therefore $\alpha_j = 0$ for every j. Consequently, $\{p_1, \ldots, p_k\}$ is a basis of $\mathscr{K}_k$:

$$\mathscr{K}_k = \langle b, Ab, \ldots, A^{k-1}b\rangle = \langle p_1, \ldots, p_k\rangle. \qquad (3.162)$$

Then, the unknown solution x_k of (3.161) can be rated as a linear combination of
conjugate vectors: $x_k = \alpha_1 p_1 + \ldots + \alpha_k p_k$. From the definition of $\langle \bullet|\bullet\rangle_A$ we get an
equivalent formulation of (3.161), namely

$$\langle x^* - x_k|p_j\rangle_A = 0 \quad \Longleftrightarrow \quad \langle Ax^* - Ax_k|p_j\rangle = 0, \quad j = 1, \ldots, k, \qquad (3.163)$$

where the Euclidean scalar product is used on the right-hand side. Knowing $Ax^* = b$ and substituting the ansatz $x_k = \sum_i \alpha_i p_i$ one gets

$$\langle b|p_j\rangle = \langle Ax_k|p_j\rangle = \alpha_j\langle p_j|p_j\rangle_A, \quad j = 1, \ldots, k, \qquad (3.164)$$

which can be used to compute the coefficients α_j. Setting $r_k := b - Ax_k = A(x^* - x_k)$, the solution x_k of (3.160) can be written in the form

$$x_k = \sum_{j=1}^{k} \alpha_j \cdot p_j \,, \quad \alpha_j = \frac{\langle b | p_j \rangle}{\langle A p_j | p_j \rangle} = \frac{\langle x^* | p_j \rangle_A}{\langle p_j | p_j \rangle_A}. \tag{3.165}$$

It remains to find conjugate vectors. This could be achieved by Gram-Schmidt orthogonalization, but there is a cheaper way based on the following

Lemma 3.39 *Let $A \in \mathbb{R}^{n,n}$ be symmetric and positive definite, $b \in \mathbb{R}^n$, and $\mathscr{K}_k :=$ $\langle b, Ab, \ldots, A^{k-1}b \rangle$ for $k = 1, \ldots, n$. Let $x_k \in \mathscr{K}_k$ be the unique solution of (3.160) for $k = 1, \ldots, n$. Let $r_0 := b$ and $r_k := b - A x_k$, $k = 1, \ldots, n$. If $r_j \neq 0$ for $j = 0, \ldots, k$, then*

$$\mathscr{K}_{k+1} = \langle r_0, \ldots, r_k \rangle \tag{3.166}$$

and

$$\langle r_i | r_j \rangle = 0 \quad \text{for} \quad i \neq j, \quad i, j = 0, \ldots, k, \tag{3.167}$$

implying $\dim(\mathscr{K}_{k+1}) = k + 1$.

Proof The proof is by induction on k. For $k = 0$ one has by definition $\mathscr{K}_1 = \langle b \rangle = \langle r_0 \rangle$, which shows (3.166), whereas (3.167) is void. Assume the statement to be true for $k - 1$, i.e. $\mathscr{K}_k = \langle r_0, \ldots, r_{k-1} \rangle$ and $\langle r_i | r_j \rangle = 0$ for $i \neq j, i, j = 0, \ldots, k-1$. Although we have not yet come up with its explicit construction, the existence of an A-orthogonal basis $\{p_1, \ldots, p_k\}$ of $\mathscr{K}_k$ is guaranteed, so x_k can be written in the form (3.165). On the one hand, $r_k := b - A x_k = A x^* - A x_k \in \mathscr{K}_{k+1}$ by definition of $\mathscr{K}_{k+1}$ and since $x_k \in \mathscr{K}_k$. On the other hand, $r_k \perp \mathscr{K}_k = \langle p_1, \ldots, p_k \rangle$ by (3.163). Therefore, either $r_k = 0$ or (3.166) and (3.167) hold. $\qquad\square$

Conjugate directions p_k will be determined iteratively, starting with $p_1 = b$. Assume $p_1, \ldots, p_k$ to be already determined and $x_k \in \mathscr{K}_k$ to be computed by (3.165). Let $r_k = b - A x_k$. If $r_k = 0$, then $x_k = x^*$ and we are done—a solution of (3.158) is found. Otherwise $\mathscr{K}_{k+1} \ni r_k \perp \mathscr{K}_k = \langle p_1, \ldots, p_k \rangle$, so that $\mathscr{K}_{k+1} = \langle p_1, \ldots, p_k, r_k \rangle$. Now p_{k+1} can be determined by a Gram-Schmidt orthogonalization step, projecting r_k onto $\mathscr{K}_{k+1}$:

$$p_{k+1} = r_k - \sum_{j=1}^{k} \frac{\langle r_k | p_j \rangle_A}{\langle p_j | p_j \rangle_A} p_j = r_k + \beta_{k+1} p_k, \quad \beta_{k+1} := -\frac{\langle r_k | p_k \rangle_A}{\langle p_k | p_k \rangle_A}, \tag{3.168}$$

where (3.166) and (3.167) were used. The conjugate gradient iteration is defined by formulas (3.165) and (3.168), but the computation of α_k and β_{k+1} can still be simplified. Knowing $x_{k-1} \in \mathscr{K}_{k-1} \perp_A p_k$, (3.168) and $p_{k-1} \in \mathscr{K}_{k-1} \perp r_{k-1}$, one finds

$$\langle x^* | p_k \rangle_A = \langle x^* - x_{k-1} | p_k \rangle_A = \langle r_{k-1} | p_k \rangle = \langle r_{k-1} | r_{k-1} \rangle$$

and derives

$$\alpha_k = \frac{\langle r_{k-1} | r_{k-1} \rangle}{\langle p_k | p_k \rangle_A}. \tag{3.169}$$

Further on $r_k = b - Ax_k = r_{k-1} - \alpha_k A p_k$ and since $r_k \perp r_{k-1}$, one gets

$$-\alpha_k \langle r_k | p_k \rangle_A = \langle r_k | - \alpha_k A p_k \rangle = \langle r_k | r_k - r_{k-1} \rangle = \langle r_k | r_k \rangle.$$

This can be substituted together with (3.169) into (3.168) to give

$$\beta_{k+1} = \frac{\langle r_k | r_k \rangle}{\langle r_{k-1} | r_{k-1} \rangle}.$$

This gives the CG iteration of Hestenes and Stiefel:

The Conjugate Gradient (CG) Iteration.

$p_1 = r_0 = b$ and $x_0 = 0$

for $k = 1, 2, 3, \dots$

$\quad v := A p_k$

$\quad \alpha_k = \dfrac{\langle r_{k-1} | r_{k-1} \rangle}{\langle p_k | v \rangle}$

$\quad x_k = x_{k-1} + \alpha_k p_k$

$\quad r_k = r_{k-1} - \alpha_k v$

$\quad \beta_{k+1} = \dfrac{\langle r_k | r_k \rangle}{\langle r_{k-1} | r_{k-1} \rangle}$

$\quad p_{k+1} = r_k + \beta_{k+1} p_k$

The iteration surely has to be stopped as soon as $r_k = 0$, since this means $x_k = x^*$. To avoid near zero division, it should rather be stopped as soon as $\|r_k\|_2$ gets too small, see also Theorem 3.41 below.

For a convergence analysis of the CG iteration, we follow [TB97, p. 298]. Since $x_0 = 0$, for any $y \in \mathscr{K}_k = \langle b, Ab, \dots, A^{k-1} b \rangle$ we have

$$x^* - y = x^* - x_0 - \sum_{j=0}^{k-1} \gamma_j A^j b = x^* - x_0 - \sum_{j=0}^{k-1} \gamma_j A^j A (x^* - x_0)$$

$$=: I \cdot (x^* - x_0) - \sum_{j=1}^{k} \gamma_{j-1} A^j (x^* - x_0).$$

The above is the error when approximating the solution x^* of (3.158) by some $y \in \mathcal{K}_k$. This error can be written in the form

$$y \in \mathcal{K}_k \quad \Longrightarrow \quad x^* - y = p(A)(x^* - x_0) , \tag{3.170}$$

where p is a normalized polynomial of degree $\leq k$:

$$p \in P_k^0 := \{p \text{ polynomial of degree } \leq k; \ p(0) = 1\}.$$

Knowing that x_k is the best approximation of x^* from $\mathcal{K}_k$ with respect to the norm $\|\cdot\|_A$, one derives the optimality relation

$$\|x^* - x_k\|_A = \min_{p \in P_k^0} \|p(A)(x^* - x_0)\|_A, \tag{3.171}$$

relating the CG iteration to polynomial best approximation. Since A is symmetric and positive definite, it has n positive eigenvalues $\lambda_1 \geq \ldots \geq \lambda_n > 0$ with corresponding eigenvalues $v_1, \ldots, v_n$ forming an orthonormal basis of $\mathbb{R}^n$. Consequently, there are scalars $\gamma_1, \ldots, \gamma_n$ such that

$$x^* - x_0 = \sum_{j=1}^n \gamma_j v_j \quad \Longrightarrow \quad p(A)(x^* - x_0) = \sum_{j=1}^n \gamma_j p(\lambda_j) v_j .$$

But then

$$\|x^* - x_0\|_A^2 = \langle x^* - x_0 | A(x^* - x_0)\rangle = \langle \sum \gamma_j v_j | \sum \gamma_j \lambda_j v_j \rangle = \sum |\gamma_j|^2 \lambda_j$$

and for all $p \in P_k^0$:

$$\|p(A)(x^* - x_0)\|_A^2 = \langle \sum \gamma_j p(\lambda_j) v_j | \sum \gamma_j p(\lambda_j) \lambda_j v_j \rangle \leq \max_{\lambda \in \sigma(A)} |p(\lambda)|^2 \sum |\gamma_j|^2 \lambda_j ,$$

where $\sigma(A) = \{\lambda_1, \ldots, \lambda_n\}$ is called the **spectrum** of A. This proves the following

Theorem 3.40 (Convergence of CG Iteration)

$$\|x^* - x_k\|_A \leq \inf_{p \in P_k^0} \max_{\lambda \in \sigma(A)} |p(\lambda)| \cdot \|x^* - x_0\|_A . \tag{3.172}$$

From Theorem 3.40 one can conclude, that a clustering of eigenvalues is beneficial for the CG iteration. If A has only $k \leq n$ *distinct* eigenvalues, convergence will occur after k steps, since a polynomial $p \in P_k^0$ exists having these k eigenvalues as its zeros (since A is positive definite, 0 cannot be an eigenvalue). By a special choice of p in (3.172) one can show that

$$\|x^* - x_k\|_A \leq 2 \left(\frac{\sqrt{\kappa_2(A)} - 1}{\sqrt{\kappa_2(A)} + 1} \right)^k \|x^* - x_0\|_A, \tag{3.173}$$

where $\kappa_2(A)$ is the spectral condition number of A, see [TB97, Theorem 38.5].

Conjugate Gradient Method for Linear Least Squares Problems

To minimize $\|b^\delta - Ax\|_2$, where $A \in \mathbb{R}^{m,n}$ has full rank n, the CG iteration can directly be applied to the corresponding normal equations $A^\top Ax = A^\top b^\delta$. Denote the unique solution of these equations by $\bar{x}$, whereas $\hat{x}$ denotes the unique solution of $A^\top Ax = A^\top b$ (unperturbed data). One easily derives the following algorithm from the original CG iteration, which is known as **CGNE (CG iteration applied to normal equations)** algorithm.

CGNE: CG Iteration Applied to $A^\top Ax = A^\top b^\delta$.

$x_0 = 0, r_0 = b^\delta$ and $p_1 = A^\top r_0$

for $k = 1, 2, 3, \dots$

$$\alpha_k = \frac{\langle A^\top r_{k-1} | A^\top r_{k-1} \rangle}{\langle Ap_k | Ap_k \rangle}$$

$$x_k = x_{k-1} + \alpha_k p_k$$

$$r_k = r_{k-1} - \alpha_k Ap_k$$

$$\beta_{k+1} = \frac{\langle A^\top r_k | A^\top r_k \rangle}{\langle A^\top r_{k-1} | A^\top r_{k-1} \rangle}$$

$$p_{k+1} = A^\top r_k + \beta_{k+1} p_k$$

Here, $r_k := b^\delta - Ax_k$, as in the original CG iteration. However, the residual corresponding to the normal equations rather is

$$\rho_k := A^\top b^\delta - A^\top Ax_k = A^\top r_k. \tag{3.174}$$

For example, Lemma 3.39 in the context of CGNE applies to the vectors ρ_k, *not* to r_k! In analogy to (3.160), unless stopped by the occurrence of some $\rho_j = A^\top r_j = 0$, CGNE determines x_k as the minimizer of

$$
\begin{aligned}
\|x - \bar{x}\|_{A^\top A}^2 \quad &= \quad (x - \bar{x})^\top A^\top A(x - \bar{x}) \\
&= \quad x^\top A^\top Ax - 2x^\top A^\top A\bar{x} + (\bar{x})^\top A^\top A\bar{x} \\
\overset{A^\top A\bar{x} = A^\top b^\delta}{=}\quad & x^\top A^\top Ax - 2x^\top A^\top b^\delta + \left[(\bar{x})^\top A^\top A\bar{x} \right]
\end{aligned}
$$

in the Krylov space

$$\mathscr{K}_k = \langle A^\top b^\delta, (A^\top A) A^\top b^\delta, \ldots, (A^\top A)^{k-1} A^\top b^\delta \rangle. \tag{3.175}$$

The term in brackets is constant, so that omitting it does not change the minimizer. Also, the minimizer is not changed if the constant term $(b^\delta)^\top b^\delta$ is added. Therefore, the kth step of the CGNE iteration determines x_k as the minimizer of

$$x^\top A^\top A x - 2x^\top A^\top b^\delta + (b^\delta)^\top b^\delta = \|b^\delta - Ax\|_2^2, \quad x \in \mathscr{K}_k. \tag{3.176}$$

Since b^δ is only a perturbed version of b, it generally makes no sense to exactly solve $A^\top A x = A^\top b^\delta$. Rather, one will stop the iteration as soon as an iterate x_k is determined such that $\|b^\delta - Ax_k\|_2 \le \delta$, where δ is a bound on the error $b - b^\delta$ in the data. The following theorem tells us that this is a sensible criterion making CGNE a regularization method for ill-posed inverse problems.

Theorem 3.41 (Stopping Rule and Convergence Property of CGNE) *Let $A \in \mathbb{R}^{m,n}$, $m \ge n$, with $\operatorname{rank}(A) = n$. Let $b, b^\delta \in \mathbb{R}^m$. Denote by*

$\hat{x} = A^+ b$ *the unique minimizer of $\|b - Ax\|_2 = \min!$ (exact data) and by*
x_k *the kth iterate produced by CGNE (perturbed data b^δ).*

Assume that for some known $\delta > 0$

$$\|b - b^\delta\|_2 + \|b - A\hat{x}\|_2 \le \delta < \|b^\delta\|_2. \tag{3.177}$$

Then the following holds:

(1) There exists a smallest integer $k \in \mathbb{N}$ (depending on b^δ), such that

$$\|b^\delta - Ax_k\|_2 \le \delta < \|b^\delta - Ax_{k-1}\|_2. \tag{3.178}$$

(2) With k as in (1),

$$\|\hat{x} - x_k\|_2 \le 2\frac{\delta}{\sigma_n}, \tag{3.179}$$

with σ_n, the smallest singular value of A.

Remarks In case $\|b^\delta\|_2 \le \delta$, $x_0 := 0$ will be declared a solution. In this case, $\|b^\delta - Ax_0\|_2 \le \delta$ holds. *The bound given in (3.179) is worse than the comparable one from Theorem 3.6 for the unregularized solution $A^+ b^\delta$ and does not help to understand how CGNE can have a regularizing effect when solving a finite-dimensional least squares problem. To see this, refer to the technical proof presented in Appendix D.*

Proof By (3.177), one knows $\|b^{\delta} - Ax_0\|_2 > \delta$, so the iteration cannot stop with $k = 0$. CGNE will produce the minimizer $\bar{x}$ of $\|b^{\delta} - Ax\|_2$ after maximally n steps. Since

$$\|b^{\delta} - A\bar{x}\|_2 \leq \|b^{\delta} - A\hat{x}\|_2 \overset{(3.177)}{\leq} \delta,$$

this means that (3.178) will certainly be fulfilled after n steps of CGNE. This proves part (1). Assume now that k is determined such that (1) holds. With $r_k = b^{\delta} - Ax_k$, one gets

$$A^{\top} r_k = A^{\top} b^{\delta} - A^{\top} A x_k = \underbrace{A^{\top} b}_{= A^{\top} A \hat{x}} + A^{\top}(b^{\delta} - b) - A^{\top} A x_k$$

$$= A^{\top} A(\hat{x} - x_k) + A^{\top}(b^{\delta} - b),$$

showing that

$$\hat{x} - x_k = (A^{\top} A)^{-1} A^{\top} r_k - (A^{\top} A)^{-1} A^{\top}(b^{\delta} - b).$$

By (3.178), $\|r_k\|_2 \leq \delta$ and by (3.177), $\|b^{\delta} - b\|_2 \leq \delta$. Since $\|(A^{\top} A)^{-1} A^{\top}\|_2 \leq 1/\sigma_n$, as can be seen using an SVD of A, the estimate (3.179) follows. □

From Theorem 3.40, it is known that a clustering of eigenvalues of $A^{\top} A$ will be beneficial for the progress of the CGNE method. If the matrix A (and thus also $A^{\top} A$) is ill-conditioned, as it is expected to be for inverse problems, eigenvalues will cluster near 0, which is an advantage for the CG iteration. For further improvement of convergence, one can try to shape the rest of the spectrum by replacing the system $A^{\top} A x = A^{\top} b^{\delta}$ by an equivalent system

$$X^{-1} A^{\top} A X^{-\top} X^{\top} x = X^{-1} A^{\top} b^{\delta}, \quad X \in \mathbb{R}^{n,n} \text{ to be chosen.}$$

This is known as **preconditioning**. We will not deal with preconditioning, refer to Lecture 40 of [TB97] for a short introduction and pointers to the literature.

Example 3.42 (Linear Waveform Inversion) Example 3.37 is repeated with the same data, but using CGNE to solve the normal equations of the least squares problem. Figure 3.17 shows the result achieved, which is of comparable quality as the one obtained for the implicit Euler method presented in the last section. CGNE stopped after 36 iterations in this example. Note that one iteration step of CGNE is much cheaper than one iteration step of the implicit Euler method with step size control considered in Sect. 3.9, which requires at least the solution of three regularized systems of normal equations. To achieve regularization with respect to the semi-norm $x \mapsto \|Lx\|_2$, with $L = L_2 \in \mathbb{R}^{p,n+1}$, $p = n - 1$, as in (3.39) ("second derivatives"), we followed the steps (3.155), (3.156), and (3.157), as in Sect. 3.9 (but used CGNE to minimize (3.156)). Figure 3.18 shows the result

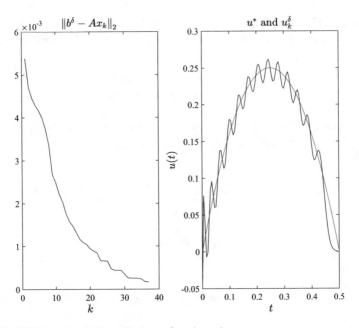

Fig. 3.17 CGNE iteration for linearized waveform inversion

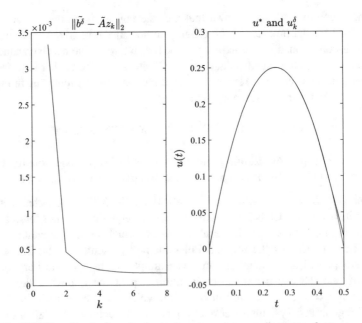

Fig. 3.18 CGNE iteration for linearized waveform inversion, coordinate transform

obtained in this case, where additionally the parameter m was increased from 100 to 300. This demonstrates the efficiency of CGNE to solve ill-conditioned linear problems. We also tested CGNE (with the transformations (3.155), (3.156), and (3.157) for $L = L_2$ from (3.39)) for the reconstruction of the step function from Example 3.29 and obtained results of comparable accuracy as in that example.   ◊

Chapter 4
Regularization of Nonlinear Inverse Problems

Nonlinear inverse problems are much more difficult to solve than linear ones and the corresponding theory is far less developed. Each particular problem may demand a specific regularization. Nevertheless, one can formulate and analyze basic versions of nonlinear Tikhonov regularization and nonlinear iterative methods, which will be done in Sects. 4.1 and 4.4, respectively. Various important numerical aspects of nonlinear Tikhonov regularization are discussed in Sects. 4.2 and 4.3. Finally, in Sect. 4.5 we solve our model problem of nonlinear inverse gravimetry and in Sect. 4.6, we solve our model problem of nonlinear waveform inversion.

4.1 Tikhonov Regularization of Nonlinear Problems

The following theoretical results are mostly taken from [EKN89]. We consider a nonlinear system of equations

$$F(x) = y, \quad F : D \subseteq \mathbb{R}^n \to \mathbb{R}^m, \quad x \in D, \quad y \in \mathbb{R}^m, \tag{4.1}$$

where $D \subseteq \mathbb{R}^n$ is *closed* and *convex* and where F is a continuous mapping. Problem (4.1) results from discretizing a nonlinear parameter identification problem, as demonstrated in Sects. 2.6 and 2.7 for examples from inverse gravimetry and waveform inversion. In these examples, D was a multi-dimensional interval ("a box"). *We assume that (4.1) has at least one solution $\hat{x}$.* This is a real restriction: even if the original parameter identification problem is always required to have a (unique) solution, the situation is different for its finite-dimensional counterpart, when discretization errors come into play. On the other hand, data errors are always present. This means that we only dispose of some approximation $y^\delta \approx y$ with

$$\|y - y^\delta\|_2 \leq \delta, \tag{4.2}$$

© The Author(s), under exclusive license to Springer Nature Switzerland AG 2020
M. Richter, *Inverse Problems*, Lecture Notes in Geosystems
Mathematics and Computing, https://doi.org/10.1007/978-3-030-59317-9_4

where $\delta > 0$ is assumed to be known. We will not assume that $F(x) = y^\delta$ has a
solution and therefore replace (4.1) by the **nonlinear least squares problem**[1]

$$\text{minimize} \quad \|F(x) - y^\delta\|_2, \quad x \in D. \tag{4.3}$$

But even (4.3) will only have a solution under appropriate conditions concerning D
and F. If a solution exists, it is not necessarily unique. Finally, if a unique solution
exists, it may be far away from $\hat{x}$ even if $\delta > 0$ is small—this will be the case
if (4.3) is badly conditioned, i.e., if a solution of (4.3) depends very sensitively
on y^δ. Motivated by the linear case, instead of trying to solve the minimization
problem (4.3), one might ask for some $\tilde{x} \in D$ which only fulfills $\|F(\tilde{x}) - y^\delta\|_2 \leq \delta$
but additionally has some other desirable quality which guarantees its being close to
$\hat{x}$. One possibility to achieve this is by considering the following nonlinear analogon
of Tikhonov regularization:

$$\text{minimize} \quad Z_\lambda(x) := \|F(x) - y^\delta\|_2^2 + \lambda \|Lx\|_2^2, \quad x \in D, \tag{4.4}$$

where $L \in \mathbb{R}^{p,n}$ and $\lambda \geq 0$ have to be chosen appropriately. Often, Lx will
correspond to some discretized derivative of the function represented by the vector
x. In the following analysis, only the case $L = I_n \in \mathbb{R}^{n,n}$ (identity matrix) will be
investigated, but (4.4) will also be slightly generalized: choose $x^* \in D$ and

$$\text{minimize} \quad T_\lambda(x) := \|F(x) - y^\delta\|_2^2 + \lambda \|x - x^*\|_2^2, \quad x \in D. \tag{4.5}$$

This means that one is looking for some x close to x^*, which at the same time fulfills
well the equality $F(x) = y^\delta$. The choice of x^* is crucial: only in case $x^* \approx \hat{x}$ one
can expect a meaningful result, but this choice requires an approximate knowledge
of the solution $\hat{x}$ of $F(x) = y$. Such knowledge may be very difficult to get, but is
also required from a practical perspective by the restricted convergence properties
of efficient numerical solvers for (4.5) (and for (4.4) as well)—see the remarks after
Theorem 4.3. It will now be shown that the regularized variant (4.5) of (4.3) always
has a solution.

Theorem 4.1 (Existence of a Minimizer) *Let $\lambda > 0$, let $D \subseteq \mathbb{R}^n$ be closed and
let $F : D \to \mathbb{R}^m$ be continuous. Then the function T_λ defined in (4.5) always has a
minimizer $x_\lambda \in D$.*

Proof Since $T_\lambda(x) \geq 0$ for all $x \in D$, the infimum $\mu := \inf\{T_\lambda(x); \ x \in D\}$ exists.
Thus for every $n \in \mathbb{N}$ there is an element $x_n \in D$ such that $T_\lambda(x_n) \leq \mu + 1/n$.
This means the sequences $(F(x_n))_{n \in \mathbb{N}}$ and $(x_n)_{n \in \mathbb{N}}$ are bounded. By the theorem

---

[1] If there is no $\hat{x} \in D$ with $F(\hat{x}) = y$ due to discretization errors, then hopefully there will at least
exist some $\bar{y} \approx y$ such that a solution $\hat{x} \in D$ of $F(x) = \bar{y}$ exists. If we know $\delta > 0$ such that
$\|\bar{y} - y^\delta\|_2 \leq \delta$, we can tacitly add discretization to measurement errors, interpreting $\bar{y}$ as "true"
right hand side.

of Bolzano–Weierstraß there is a convergent subsequence $(x_{n_k})_{k \in \mathbb{N}}$ of $(x_n)_{n \in \mathbb{N}}$ such that $x_{n_k} \to \bar{x} \in D$ (D is closed) and $F(x_{n_k}) \to \bar{y} = F(\bar{x})$ (F is continuous). From the continuity of T_λ one concludes $T_\lambda(\bar{x}) = \lim_{k \to \infty} T_\lambda(x_{n_k}) = \mu$, meaning that $x_\lambda = \bar{x}$ is a minimizer. $\qquad\square$

The minimizer x_λ of T_λ is stable in the following sense. Let $(y^{\delta_n})_{n \in \mathbb{N}} \subset Y$ be a sequence with $y^{\delta_n} \to y^\delta$ and let $(x_n)_{n \in \mathbb{N}} \subset D$ be a corresponding sequence of minimizers, i.e., x_n minimizes

$$T_{\lambda,n} : D \to \mathbb{R}, \quad x \mapsto T_{\lambda,n}(x) := \|F(x) - y^{\delta_n}\|_2^2 + \lambda \|x - x^*\|_2^2.$$

Then conclude from

$$T_{\lambda,n}(x_n) \leq \left(\|F(x_\lambda) - y^\delta\|_2 + \|y^\delta - y^{\delta_n}\|_2\right)^2 + \lambda \|x_\lambda - x^*\|_2^2 \to T_\lambda(x_\lambda),$$

that $(T_{\lambda,n}(x_n))_{n \in \mathbb{N}}$ is a bounded sequence. As in the proof of Theorem 4.1, this means that $(x_n)_{n \in \mathbb{N}}$ has a convergent subsequence. One also concludes that every convergent subsequence of $(x_n)_{n \in \mathbb{N}}$ converges to a minimizer of T_λ. If x_λ is unique, then one even gets $x_n \to x_\lambda$, meaning that in this case the minimization of T_λ is a properly posed problem in the sense of Hadamard (Definition 1.5).

It was assumed that $F(x) = y$ has a solution, so that $\{x \in D; \ F(x) = y\}$ is not empty. Since F is continuous, this set is closed and using the same arguments as in the proof of Theorem 4.1 one concludes that an element

$$\hat{x} \in D \text{ with } F(\hat{x}) = y \text{ and } \|\hat{x} - x^*\|_2 = \min\{\|x - x^*\|_2; \ x \in D, \ F(x) = y\} \tag{4.6}$$

exists. Any such (possibly not unique) $\hat{x}$ is called x^*-**minimum-norm solution** of the equation $F(x) = y$. The following theorem gives conditions for λ under which (4.5) defines a regularization for the computation of an x^*-minimum-norm solution of $F(x) = y$.

Theorem 4.2 (Regularization of x^*-Minimum-Norm Solutions) *Let D be closed and let F be continuous. Let $(\delta_n)_{n \in \mathbb{N}}$ be a sequence of positive numbers converging to zero. For every δ_n choose a parameter $\lambda_n = \lambda_n(\delta_n) > 0$ such that*

$$\lambda_n \to 0 \quad \text{and} \quad \frac{\delta_n^2}{\lambda_n} \to 0 \quad \text{for } n \to \infty. \tag{4.7}$$

For every δ_n let $y^{\delta_n} \in \mathbb{R}^m$ be such that $\|y - y^{\delta_n}\|_2 \leq \delta_n$ and let x_n be a minimizer of

$$T_{\lambda_n}(x) := \|F(x) - y^{\delta_n}\|_2^2 + \lambda_n \|x - x^*\|_2^2.$$

Then the sequence $(x_n)_{n \in \mathbb{N}}$ contains a subsequence converging to a x^*-minimum-norm solution $\hat{x}$ of $F(x) = y$. If there is a unique x^*-minimum-norm solution, then $x_n \to \hat{x}$ for $n \to \infty$.

Proof Being a minimizer of T_{λ_n}, the vector x_n fulfills the inequality

$$T_{\lambda_n}(x_n) \leq T_{\lambda_n}(\hat{x}) \leq \delta_n^2 + \lambda_n \|\hat{x} - x^*\|_2^2 \tag{4.8}$$

(here we made use of $\|F(\hat{x}) - y^{\delta_n}\|_2 = \|y - y^{\delta_n}\|_2 \leq \delta_n$, where $\hat{x}$ is any fixed x^*-minimum-norm solution). The right hand side of (4.8) converges to 0, therefore $\|F(x_n) - y\|_2 \leq \|F(x_n) - y^{\delta_n}\|_2 + \|y - y^{\delta_n}\|_2 \to 0$ and this means $F(x_n) \to y$. Division of (4.8) by λ_n shows

$$\frac{1}{\lambda_n} \|F(x_n) - y^{\delta_n}\|_2^2 + \|x_n - x^*\|_2^2 \leq \frac{\delta_n^2}{\lambda_n} + \|\hat{x} - x^*\|_2^2 \to \|\hat{x} - x^*\|_2^2.$$

Consequently, $(x_n)_{n \in \mathbb{N}}$ is bounded with $\limsup_{n \to \infty} \|x_n - x^*\|_2 \leq \|\hat{x} - x^*\|_2$. By the theorem of Bolzano–Weierstraß there exists a convergent subsequence $(x_{n_k})_{k \in \mathbb{N}}$ with $x_{n_k} \to \bar{x} \in D$. Conclude $F(\bar{x}) = y$ from the continuity of F. In addition,

$$\|\bar{x} - x^*\|_2 = \lim_{k \to \infty} \|x_{n_k} - x^*\|_2 \leq \limsup_{n \to \infty} \|x_n - x^*\|_2 \leq \|\hat{x} - x^*\|_2,$$

meaning that $\bar{x}$ is itself a x^*-minimum-norm solution. It was just shown that any sequence $(x_n)_{n \in \mathbb{N}}$ of minimizers has at least one limit point, which necessarily is a x^*-minimum-norm solution. So when there is exactly one x^*-minimum-norm solution, then the bounded sequence $(x_n)_{n \in \mathbb{N}}$ has exactly one limit point and therefore is convergent. $\square$

The following theorem makes a statement about the convergence rate of the regularized solution in case the regularization parameter is chosen according to a variant of the discrepancy principle. To state this theorem, let us first recall a well known formula for the linearization error of a vector valued smooth function. Let $F : U \subseteq \mathbb{R}^n \to \mathbb{R}^m$ (U open) be two times continuously differentiable. Let $x_0, h \in \mathbb{R}^n$ be such that $x_0, x_0 + h$, and the whole straight line connecting them is contained in U. Then

$$F(x_0 + h) = F(x_0) + F'(x_0)h + r(x_0, h),$$

where $r(x_0, h)$ is vector valued with components

$$r_i(x_0, h) = \sum_{j,k=1}^{n} \left(\int_0^1 \frac{\partial^2 F_i}{\partial x_j \partial x_k}(x_0 + th)(1 - t)\, dt \right) h_j h_k, \quad i = 1, \dots, m. \tag{4.9}$$

Here, h_j, h_k are the components of h.

Theorem 4.3 (Rate of Convergence of Regularization) *Let $\lambda > 0$, let $D \subseteq U \subseteq \mathbb{R}^n$ with U open and D closed and convex, let $F : U \to \mathbb{R}^m$ be twice continuously differentiable, and let x_0 be a x^*-minimum-norm solution of equation $F(x) = y$. Let $y^\delta \in \mathbb{R}^m$ with $\|y - y^\delta\|_2 \leq \delta$ for some $\delta > 0$. Let x_λ be a minimizer of (4.5). Assume that there exists some $w \in \mathbb{R}^m$ such that*

$$x_0 - x^* = F'(x_0)^\top w. \tag{4.10}$$

Set $h := x_\lambda - x_0$, define $r(x_0, h)$ as in (4.9) and assume that for w as in (4.10) one has

$$2|w^\top r(x_0, h)| \leq \varrho \|h\|_2^2, \quad \varrho < 1. \tag{4.11}$$

If for some constants $C_1, C_2 > 0$ or $1 \leq \tau_1 \leq \tau_2$ we have

$$C_1 \delta \leq \lambda \leq C_2 \delta \quad or \quad \tau_1 \delta \leq \|F(x_\lambda) - y^\delta\|_2 \leq \tau_2 \delta, \tag{4.12}$$

respectively, then

$$\|x_\lambda - x_0\|_2 \leq C \sqrt{\delta} \tag{4.13}$$

with some constant C. Only a single x^*-minimum-norm solution x_0 can fulfill conditions (4.10) and (4.11).

Proof According to Theorem 4.1 a minimizer x_λ of T_λ exists. For any such minimizer one has

$$T_\lambda(x_\lambda) = \|F(x_\lambda) - y^\delta\|_2^2 + \lambda\|x_\lambda - x^*\|_2^2 \leq T_\lambda(x_0) \leq \delta^2 + \lambda\|x_0 - x^*\|_2^2,$$

because of $F(x_0) = y$ and $\|y - y^\delta\|_2 \leq \delta$. Consequently,

$$\|F(x_\lambda) - y^\delta\|_2^2 + \lambda\|x_\lambda - x_0\|_2^2$$
$$= \|F(x_\lambda) - y^\delta\|_2^2 + \lambda\left(\|x_\lambda - x^*\|_2^2 + \|x_\lambda - x_0\|_2^2 - \|x_\lambda - x^*\|_2^2\right)$$
$$\leq \delta^2 + \lambda\left(\|x_0 - x^*\|_2^2 + \|x_\lambda - x_0\|_2^2 - \|x_\lambda - x^*\|_2^2\right)$$
$$= \delta^2 + 2\lambda\left(x_0 - x^*\right)^\top (x_0 - x_\lambda) = \delta^2 + 2\lambda w^\top \left(F'(x_0)(x_0 - x_\lambda)\right),$$

where (4.10) was used to derive the last identity. Using $F(x_0) = y$ one gets

$$F'(x_0)(x_0 - x_\lambda) = \left(y - y^\delta\right) + \left(y^\delta - F(x_\lambda)\right) + \left(F(x_\lambda) - F(x_0) - F'(x_0)(x_\lambda - x_0)\right).$$

Moreover, $F(x_\lambda) - F(x_0) - F'(x_0)(x_\lambda - x_0) = r(x_0, h)$ and using (4.11) one derives from the above inequality

$$\|F(x_\lambda) - y^\delta\|_2^2 + \lambda\|x_\lambda - x_0\|_2^2 \leq \delta^2 + 2\lambda\delta\|w\|_2 +$$
$$2\lambda\|w\|_2\|F(x_\lambda) - y^\delta\|_2 + \lambda\varrho\|x_\lambda - x_0\|_2^2.$$

This shows the inequality

$$\|F(x_\lambda) - y^\delta\|_2^2 + \lambda(1 - \varrho)\|x_\lambda - x_0\|_2^2$$
$$\leq \delta^2 + 2\lambda\delta\|w\|_2 + 2\lambda\|w\|_2\|F(x_\lambda) - y^\delta\|_2, \qquad (4.14)$$

which can also be written in the form

$$\left(\|F(x_\lambda) - y^\delta\|_2 - \lambda\|w\|_2\right)^2 + \lambda(1 - \varrho)\|x_\lambda - x_0\|_2^2 \leq (\delta + \lambda\|w\|_2)^2. \qquad (4.15)$$

Because of $\varrho < 1$, this inequality is true a fortiori, if the first summand on the left hand side is omitted, leading to

$$\|x_\lambda - x_0\|_2 \leq \frac{\delta + \lambda\|w\|_2}{\sqrt{\lambda} \cdot \sqrt{1 - \varrho}} \leq \frac{\delta + C_2\delta\|w\|_2}{\sqrt{C_1}\sqrt{\delta} \cdot \sqrt{1 - \varrho}} = C\sqrt{\delta},$$

whenever $C_1\delta \leq \lambda \leq C_2\delta$, proving (4.13). Otherwise, if $\tau_1\delta \leq \|F(x_\lambda) - y^\delta\|_2 \leq \tau_2\delta$, then from (4.14) one gets

$$\tau_1^2\delta^2 + \lambda(1 - \varrho)\|x_\lambda - x_0\|_2^2 \leq \delta^2 + 2\lambda\delta\|w\|_2 + 2\lambda\|w\|_2\tau_2\delta.$$

Using $(1 - \tau_1^2) \leq 0$ this shows

$$\lambda(1 - \varrho)\|x_\lambda - x_0\|_2^2 \leq 2\lambda\|w\|_2(1 + \tau_2)\delta$$

and so leads again to the estimate (4.13). Since $C\sqrt{\delta}$ gets arbitrarily small for $\delta \to 0$, a x^*-minimum-norm solution fulfilling the conditions of the theorem necessarily is unique (although multiple x^*-minimum-norm solutions may exist).[2]   □

Remarks Condition (4.10) is called **source condition**. It will certainly hold if the functional matrix $F'(x_0)$ has full rank n, which we reasonably expect in the setting of parameter identification problems. The source condition corresponds to some "abstract smoothness" condition for the solution x_0 and thus represents a kind of a priori knowledge about x_0, see for example Sect. 1.3 in [Kir11] or Sect. 3.2 in [EHN96]. Condition (4.11) is fulfilled if either the second derivatives of F are small enough (meaning the function F is only "weakly nonlinear") or if x^* is close to x_0, meaning that x^* must be chosen close to a solution of $F(x) = y$ (one must have a

---

[2]If x_0 and x_1 were two x^*-minimum-norm solutions both fulfilling (4.10) and (4.11), then $\|x_\lambda - x_0\|_2 \leq C_1\sqrt{\delta}$ and $\|x_\lambda - x_1\|_2 \leq C_2\sqrt{\delta}$, by the above arguments. Therefore, $\|x_0 - x_1\|_2 \leq (C_1 + C_2)\sqrt{\delta}$. Now let $\delta \to 0$.

good a priori knowledge of a desired solution). Both requirements are restrictive, but the latter usually is indispensable also from a practical point of view, since in general T_λ is non-convex and possesses stationary points, which are not minimizers. But efficient numerical methods to minimize T_λ are iterative and can only be guaranteed to converge to a stationary point next to where they are started. They will thus fail to detect a global minimizer unless started close enough to one. A final remark concerns the choice of the regularization parameter. It may well happen that T_λ has no unique minimizer. In that case the set

$$M_\lambda := \{\hat{x} \in D; \ T_\lambda(\hat{x}) \leq T_\lambda(x) \text{ for all } x \in D\}$$

contains more than one element. Nevertheless, T_λ takes the same value for all elements $x_\lambda \in M_\lambda$, thus $T_\lambda(x_\lambda)$ is in fact a function of $\lambda \in]0, \infty[$. Using arguments similar to the ones in the proof of Theorem 4.1, it is not difficult to show that this function is continuous. However, the term

$$J(x_\lambda) := \|F(x_\lambda) - y^\delta\|_2 = T_\lambda(x_\lambda) - \lambda\|x_\lambda - x^*\|_2,$$

in general is *not* a function of λ, possibly taking different values for different elements $x_\lambda \in M_\lambda$. Using the same technique as in the proof of Theorem 3.21 it is still possible to show

$$J(x_{\lambda_1}) \leq J(x_{\lambda_2}) \quad \text{for} \quad 0 < \lambda_1 < \lambda_2, \quad x_{\lambda_1} \in M_{\lambda_1}, \quad x_{\lambda_2} \in M_{\lambda_2}. \tag{4.16}$$

Since $J(x_\lambda)$ is not a continuous function of λ (unless T_λ has unique minimizers), one cannot guarantee that $J(x_\lambda)$ takes any specific value and the discrepancy principle must be formulated with some care, as in the second alternative considered in (4.12). Still, additional conditions are required to guarantee the existence of some minimizer x_λ such that $\tau_1\delta \leq \|F(x_\lambda) - y^\delta\|_2 \leq \tau_2\delta$. This problem is considered in [Ram02].

4.2 Nonlinear Least Squares Problems

In this section practical methods are considered to solve nonlinear least squares problems of the form

$$\text{minimize} \quad Z(x) := \frac{1}{2}\|F(x) - y\|_2^2, \quad x \in D, \tag{4.17}$$

where $D \subseteq U \subseteq \mathbb{R}^n$ with U open and D closed and convex and where $F : U \to \mathbb{R}^m$ is assumed to be twice continuously differentiable. Nonlinear Tikhonov regularization as in (4.4) and (4.5) leads to optimization problems of this kind with

$$Z(x) = \left\| \begin{pmatrix} F(x) \\ \sqrt{\lambda}Lx \end{pmatrix} - \begin{pmatrix} y^\delta \\ 0 \end{pmatrix} \right\|_2^2 \quad \text{or} \quad Z(x) = \left\| \begin{pmatrix} F(x) \\ \sqrt{\lambda}x \end{pmatrix} - \begin{pmatrix} y^\delta \\ \sqrt{\lambda}x^* \end{pmatrix} \right\|_2^2,$$

respectively. In these cases an exact solution of (4.17) is requested, which means the solution of a global optimization problem. The original, unregularized problem (4.3) also has the form (4.17). As an alternative to Tikhonov regularization, we can directly tackle the original, unregularized problem, but stop an iterative solver prematurely, *before* it has found a minimizer. This method, which produces a regularized solution as in the linear case, will be investigated in Sect. 4.4.

Problem (4.17) can be quite difficult to solve. In the present section, we make the following four restrictions.

First, we only consider the unconstrained case $D = \mathbb{R}^n$. For the solution of our nonlinear model problems, we actually have to deal with so-called **box constraints**, i.e., we have to consider the case where $D \subset \mathbb{R}^n$ is a multi-dimensional interval (a box). Some remarks about the consideration of box constraints are to follow at the end of the present section.

Second, the computation of derivatives of Z, as needed by the following algorithms, can mean a prohibitive arithmetical effort, when not done efficiently. This will only be discussed in Sect. 4.3.

Third, all efficient standard solvers applied to (4.17) are only appropriate to detect *local,* not *global* minimizers. Since we need global minima, special measures have to be taken which are specific for the model problem to be solved. The corresponding discussion is postponed to Sects. 4.5 and 4.6.

Fourth, we assume that an adequate regularization parameter λ can be found according to the discrepancy principle. In our model problems, it was (nearly) always possible to tune λ in accordance with the discrepancy principle.

One idea to solve (4.17) for $D = \mathbb{R}^n$ is to use **Newton's method**. Gradient and Hessian of Z are easily computed:

$$\nabla Z(x) = F'(x)^\top (F(x) - y),$$
$$\nabla^2 Z(x) = F'(x)^\top F'(x) + \sum_{i=1}^m (F_i(x) - y_i)\nabla^2 F_i(x). \tag{4.18}$$

Here, F_i means the ith component of the vector field F, $\nabla^2 F_i$ is the corresponding Hessian and $F'(x)$ is the Jacobian of F evaluated at x. Newton's method tries to find some $\hat{x}$ with $\nabla Z(\hat{x}) = 0$. Beginning at a **start value** $x^0 \in D$, the original Newton method successively computes approximations x^i of $\hat{x}$, defined by the rule

$$x^{i+1} := x^i + s \quad \text{where} \quad \nabla^2 Z(x^i)s = -\nabla Z(x^i), \quad i = 0, 1, 2, \ldots$$

This means that s is a zero of the linearization

$$\nabla Z(x^i + s) \stackrel{\bullet}{=} \nabla Z(x^i) + \nabla^2 Z(x^i)s.$$

One obtains a so-called **Quasi-Newton method** if the Hessian $\nabla^2 Z(x)$ is approximated by some other matrix. The choice

$$\nabla^2 Z(x) \approx F'(x)^\top F'(x)$$

suggests itself, since we may reasonably assume $F(x) = y$ to be a "nearly consistent" system of equations, unsolvable only because of perturbed data y. Therefore, as long as x is close enough to a solution $\hat{x}$ of (4.17), the contributions of the Hessians $\nabla^2 F_i(x)$ to $\nabla^2 Z(x)$ in (4.18) are damped by small values $|F_i(x) - y_i|$. Using the approximation $\nabla^2 Z(x) \approx F'(x)^\top F'(x)$, one gets the **Gauß–Newton method**. Starting from an approximation x^i of $\hat{x}$ it computes a next approximation x^{i+1} according to the following rules:

Gauß–Newton Step

(a) Compute $b := F(x^i) - y$ and $J := F'(x^i)$.
(b) Solve $J^\top b + J^\top J s = 0$ for s.
(c) Set $x^{i+1} := x^i + s$.

The same method results if F is linearized at x^i:

$$F(x^i + s) - y \overset{\bullet}{=} F(x^i) - y + F'(x^i)s = b + Js$$

and if then the *linear* least squares problem

$$\text{minimize} \quad \|b + Js\|_2, \quad s \in \mathbb{R}^n,$$

is solved. The vector s is called a **search direction**. In case $\operatorname{rank}(J) = n$, the matrix $J^\top J$ is positive definite and so is $(J^\top J)^{-1}$. Consequently

$$-\nabla Z(x^i)^\top s = -b^\top J s = b^\top J (J^\top J)^{-1} J^\top b > 0$$

(unless $b = 0$, which means that $F(x^i) = y$). Consequently, s points into a direction of decreasing values of Z, as does $-\nabla Z(x^i)$, the negative gradient.[3] But this does not yet mean $Z(x^{i+1}) < Z(x^i)$, since the step might go into a downhill direction, but take us too far across a valley and uphill again on the opposite side. To guarantee a descent, part (c) of the above rule needs to be modified by a so-called **step size control**:

---

[3] This is generally true even if $\operatorname{rank}(J) = n$ does not hold, provided only that $\nabla Z(x^i) \neq 0$, see [Bjö96, p. 343].

$$\textit{Do not set} \quad x^{i+1} = x^i + s, \quad \textit{but rather} \quad x^{i+1} = x^i + \mu s,$$

where $\mu \in]0, 1]$ is chosen such that $Z(x^{i+1}) < Z(x^i)$. Such a value μ always exists and there are a number of well known algorithms to find one, see [LY08, Sect. 8.5].

A different idea is to keep (c), but to replace (b) in the above rule by defining s as the solution of the optimization problem

$$\text{minimize} \quad \|b + Js\|_2 \quad \text{under the constraint} \quad \|s\|_2 \le \Delta,$$

where Δ is a parameter to be chosen. The corresponding algorithm is known as **Levenberg–Marquardt method** and is also called a **trust region method**. The constraint defines a ball of radius Δ, centered at x^i, where the linearization is trusted to be a good approximation of F. More generally, one can consider the optimization problem

$$\text{minimize} \quad \|b + Js\|_2 \quad \text{under the constraint} \quad \|Ds\|_2 \le \Delta \qquad (4.19)$$

with some positive definite (diagonal) matrix D. This modification means to distort the ball where the linearization $F(x^i + s) - y \overset{\bullet}{=} b + Js$ is trusted to allow for different change rates of F in different directions. The parameters D and Δ have to be changed as the iteration proceeds—details are described in [Mor78]. Using the method of Lagrange multipliers, (4.19) is seen to be equivalent to the linear system

$$(J^\top J + \lambda D^\top D)s = -J^\top b \qquad (4.20)$$

where λ is the Lagrange parameter corresponding to Δ. From (4.20) it can be seen that a search direction s will be found which is different from the one in the Gauß–Newton method. The parameter λ has to be found iteratively as described in Sect. 3.4. This means a considerable arithmetical effort, if the dimension n is high. To cut down the costs, variants of (4.19) have been developed, which look for a solution s of (4.19) only in a carefully chosen, two-dimensional subspace of $\mathbb{R}^n$. This is described in [BCL99].

It is possible to handle box constraints, i.e., to solve (4.17) with

$$D = \{x \in \mathbb{R}^n; \ \ell_i \le x_i \le r_i, \ i = 1, \dots, n\},$$

where $-\infty \le \ell_i < r_i \le \infty$ for $i = 1, \dots, n$. For example one may continue $F : D \to \mathbb{R}^m$ by reflection and periodic continuation to become a function defined on $\mathbb{R}^n$. Then the unconstrained problem is solved. Any minimizer found can be related back to a corresponding minimizer in D. These ideas are also developed in [BCL99].

To set up the system (4.20), the Jacobian J has to be known. If an iterative algorithm like CGNE is used to solve (4.20), then at least an efficient method to compute Js and $J^\top Js$ for any given direction s has to be provided. This will be done in the next section.

4.3 Computation of Derivatives

Applying the optimization methods of the previous section to problem (4.17) in its regularized form we have to compute Jacobians

$$J = \begin{pmatrix} F'(x) \\ \sqrt{\lambda}L \end{pmatrix} \quad \text{or} \quad J = \begin{pmatrix} F'(x) \\ \sqrt{\lambda}I \end{pmatrix}$$

or at least—in case iterative methods like CGNE are used to solve (4.20)—have to compute products of the form Js or $J^\top Js$, with J as above. Only the part $F'(x)$ is non-trivial and will be considered below.

Sometimes, the function F is given by an explicit formula and the computation of $F'(x)$ is straightforward.

Example 4.4 (Derivatives for Inverse Gravimetry) Inverse Gravimetry in its discretized form means to solve the nonlinear least squares problem (2.117) with $F : C \to \mathbb{R}^M$ defined by

$$F_\beta(c) = \sum_{\alpha \in G_n} \omega_\alpha k(\hat{x}_\beta, x_\alpha, c_\alpha), \quad \beta \in B,$$

see (2.115). In this case, derivatives are easily computed as

$$\frac{\partial F_\beta(c)}{\partial c_\alpha} = -\omega_\alpha c_\alpha \left(\|\hat{x}_\beta - x_\alpha\|_2^2 + c_\alpha^2 \right)^{-3/2}. \tag{4.21}$$

Thus we find

$$J = \left(\frac{\partial F_\beta(c)}{\partial c_\alpha} \right)_{\beta \in B, \alpha \in G_n} \in \mathbb{R}^{M,N}$$

by direct computation. ◇

Often, however, function values $F(x)$ are implicitly determined as solutions of systems of equations. Derivatives can then be found by implicit differentiation. If $F(x)$ is defined by a *linear* system of equations, derivatives are given as solutions of linear systems of equations again. As a prototypical example, assume that

$$F(x) = [A(x)]^{-1} \cdot b, \quad \text{i.e.,} \quad A(x) \cdot F(x) = b,$$

where $A = A(x) \in \mathbb{R}^{n,n}$ depends on x. The jth column of J, i.e., the column vector $\partial F(x)/\partial x_j$, is then given by the equation

$$\frac{\partial A}{\partial x_j} \cdot F(x) + A \cdot \frac{\partial F(x)}{\partial x_j} = 0 \quad \Longleftrightarrow \quad A \cdot \frac{\partial F(x)}{\partial x_j} = -\frac{\partial A}{\partial x_j} \cdot F(x).$$

A computation of J column by column requires the solution of n systems of equations, albeit with identical coefficient matrix A. To compute a product Js for some $s \in \mathbb{R}^n$, formally define the matrix

$$M := \left(\frac{\partial A}{\partial x_1} \cdot F(x), \ldots, \frac{\partial A}{\partial x_n} \cdot F(x) \right).$$

Consequently

$$J = F'(x) = -A^{-1} \cdot M \quad \Longrightarrow \quad Js = -A^{-1} \cdot M \cdot s =: v$$

and this vector v can be computed as follows:

1. Compute $w := -M \cdot s$.
2. Solve the system $Av = w$.

Often, the matrices $\partial A/\partial x_j$ are sparse, so the computation of Ms does not mean a big arithmetical effort. This is the case for waveform inversion.

Example 4.5 (Derivatives for Waveform Inversion) According to (2.148), we have

$$F(\mathbf{x}) = M \cdot A^{-1} \cdot \mathbf{g}, \quad A = A(G\mathbf{x}). \tag{4.22}$$

As in Sect. 2.7, let $\mathbf{k} = (\kappa_0, \ldots, \kappa_m)^\top = G\mathbf{x}$ and let $\mathbf{u} = \mathbf{u}(\mathbf{k}) := A^{-1}\mathbf{g}$, where $A = A(\mathbf{k}) = A(G\mathbf{x}) \in \mathbb{R}^{(m+1)n,(m+1)n}$ depends on $\mathbf{k}$ and thus on $\mathbf{x}$. In this case, $\mathbf{u}$ is determined by the system

$$A(\mathbf{k}) \cdot \mathbf{u}(\mathbf{k}) = \mathbf{g}.$$

Implicitly differentiating the last equality, we get

$$\frac{\partial \mathbf{u}}{\partial \kappa_j} = -A^{-1} \cdot \frac{\partial A}{\partial \kappa_j} \cdot \mathbf{u}, \quad j = 0, \ldots, m,$$

for the columns of the Jacobian matrix $\partial \mathbf{u}/\partial \mathbf{k}$. Formally setting

$$C := \left(\frac{\partial A}{\partial \kappa_0} \cdot \mathbf{u}, \ldots, \frac{\partial A}{\partial \kappa_m} \cdot \mathbf{u} \right), \tag{4.23}$$

we may write

$$\frac{\partial \mathbf{u}}{\partial \mathbf{k}} = -A^{-1} \cdot C \in \mathbb{R}^{n(m+1), m+1}.$$

Applying the chain rule, we get

$$\frac{\partial \mathbf{u}}{\partial \mathbf{x}} = -A^{-1} \cdot C \cdot G \in \mathbb{R}^{n(m+1), k+1},$$

and, applying by the chain rule once more, we end up with

$$F'(\mathbf{x}) = -M \cdot A^{-1} \cdot C \cdot G \in \mathbb{R}^{n, k+1}. \tag{4.24}$$

Of course, we will not compute the matrix inverse A^{-1} but solve a linear system $Ax = CGv$, whenever asked to compute $x = A^{-1}CGv$. As can be seen from (2.140) and from the definitions of T_1, T_2, and T_3 preceding this equation, the matrices $\partial A / \partial \kappa_j$ are very easy to compute, "almost constant" (only a few components depend on $\mathbf{x}$), and sparse. $\diamondsuit$

Another way to compute gradient information is known as the **adjoint method**. We will outline this method following Chap. 2 of [Cha09], but restrict ourselves to the finite-dimensional case. The method is not new—Chavent already used it in 1974, see [Cha74]. The task is to compute the gradient ∇Z of the objective function Z defined in (4.17), or the Jacobian $F'(x)$ of the corresponding function $F : D \subset \mathbb{R}^n \to \mathbb{R}^m$. It will be assumed that a function evaluation $v = F(x)$ can be decomposed into two steps. *First*, given $x \in D$, solve an equation

$$e(x, z) = 0 \quad \text{for} \quad z \in \mathbb{R}^p \tag{4.25}$$

and *second*, compute

$$v = M(z) \quad [= F(x)]. \tag{4.26}$$

Here, $e(x, z) = 0$ is called **state equation**, defined by a mapping $e : D \times \mathbb{R}^p \to \mathbb{R}^p$, and $M : \mathbb{R}^p \to \mathbb{R}^m$ is an **observation operator**.

Example 4.6 (Function Evaluations for Waveform Inversion) Function evaluations $v = F(\mathbf{x})$ as in (4.22) can be decomposed according to (4.25) and (4.26). Let $x = \mathbf{x} \in D = [\kappa_-, \kappa_+]^{k+1} \subset \mathbb{R}^{k+1}$ and let $z = \mathbf{u} \in \mathbb{R}^{(m+1)n}$. The state equation takes the form

$$e(\mathbf{x}, \mathbf{u}) = A \cdot \mathbf{u} - \mathbf{g} = 0, \tag{4.27}$$

where $A = A(\mathbf{x}) \in \mathbb{R}^{(m+1)n, (m+1)n}$ is a matrix (of full rank) depending on $\mathbf{x}$. The observation operator is just a matrix multiplication

$$M : \mathbb{R}^{(m+1)n} \to \mathbb{R}^n, \quad \mathbf{u} \mapsto M \cdot \mathbf{u}. \tag{4.28}$$

Here, M has a double meaning as a function and as a matrix.   ◇

Assumption 4.7 *Concerning e and M in (4.25) and (4.26), it will always be required that*

- *Equation (4.25) has a unique solution $z \in \mathbb{R}^p$ for every $x \in D$,*
- *$e : D \times \mathbb{R}^p \to \mathbb{R}^p$ is continuously differentiable in all points of $\overset{\circ}{D} \times \mathbb{R}^p$, where $\overset{\circ}{D}$ is the set of all interior points of D,*
- *the matrix $\partial e(x, z)/\partial z \in \mathbb{R}^{p,p}$ is invertible for all $x \in \overset{\circ}{D}$, and that*
- *the mapping M is continuously differentiable.*

The first requirement means that (4.25) defines a mapping $D \mapsto \mathbb{R}^p$, $x \mapsto z = z(x)$. From the next two requirements it follows (by virtue of the implicit function theorem) that this mapping is continuously differentiable with Jacobian

$$z'(x) = -\left[\frac{\partial e(x, z)}{\partial z}\right]^{-1} \frac{\partial e(x, z)}{\partial x} \in \mathbb{R}^{p,n}$$

for all $x \in \overset{\circ}{D}$. It is easy to check that all requirements are fulfilled in Example 4.6. Now assume further that a function

$$G : \mathbb{R}^n \times \mathbb{R}^m \to \mathbb{R}, \quad G \text{ continuously differentiable,}$$

is given and that we wish to compute the derivative (gradient) ∇G of $x \mapsto G(x, F(x))$ with F from (4.17). For the choice

$$G(x, v) := \frac{1}{2}\|v - y\|_2^2, \tag{4.29}$$

(the function values are actually independent of x in this case), we get

$$G(x, F(x)) = Z(x) \text{ for } x \in D, \text{ with } Z \text{ from (4.17).}$$

In this case $\nabla G = \nabla Z$, the gradient of Z as in (4.18). If, on the other hand, one chooses

$$G(x, v) = v^\top w, \quad w \in \mathbb{R}^m \quad \text{(fixed)}, \tag{4.30}$$

then

$$G(x, F(x)) = F(x)^\top w$$

and $\nabla G = F'(x)^\top w$. In the special case $w = e^j$, with e^j the jth unit vector from $\mathbb{R}^m$, this would mean to compute the jth column of $F'(x)^\top$, which is the jth row of $F'(x)$.

Theorem 4.8 (Computation of Derivatives by the Adjoint Method) *Let (4.25) and (4.26) define a decomposition of the mapping $F : D \to \mathbb{R}^m$ and let Assumption 4.7 hold. Further let $G : \mathbb{R}^n \times \mathbb{R}^m \to \mathbb{R}$ be continuously differentiable and define*

$$L : \mathbb{R}^n \times \mathbb{R}^p \times \mathbb{R}^p \to \mathbb{R}, \quad L(x, z, \lambda) := G(x, M(z)) + e(x, z)^{\top} \lambda. \tag{4.31}$$

Then the mapping $D \to \mathbb{R}$, $x \mapsto G(x, F(x))$, is continuously differentiable on $\overset{\circ}{D}$, and its gradient ∇G is given by

$$\nabla G = \frac{\partial L}{\partial x}(x, z, \lambda) = \frac{\partial G}{\partial x}(x, M(z)) + \frac{\partial e}{\partial x}(x, z)^{\top} \lambda, \tag{4.32}$$

*where z is the unique solution of the state equation (4.25) and where $\lambda \in \mathbb{R}^p$ is the unique solution of the so-called **adjoint state equation***

$$\frac{\partial L}{\partial z}(x, z, \lambda) = 0. \tag{4.33}$$

Proof From Assumption 4.7 it follows, as already noted above, that $e(x, z) = 0$ defines a mapping $D \to \mathbb{R}^p$, $x \mapsto z = z(x)$, which is continuously differentiable on $\overset{\circ}{D}$. One therefore has $F(x) = M(z(x))$ and $G(x, F(x)) = G(x, M(z(x)))$, the latter being a continuously differentiable function on $\overset{\circ}{D}$, as follows from the corresponding assumptions on M and G. Furthermore, for $z = z(x)$, one has $e(x, z) = 0$ and the definition of L from (4.31) reduces to

$$L(x, z(x), \lambda) = G(x, F(x)) \quad \text{for all} \quad x \in D, \ \lambda \in \mathbb{R}^p.$$

Differentiating this identity with respect to x, one gets

$$\nabla G = \frac{\partial L}{\partial x}(x, z, \lambda) + z'(x)^{\top} \cdot \frac{\partial L}{\partial z}(x, z, \lambda), \quad x \in \overset{\circ}{D}.$$

This shows that (4.32) is true, if (4.33) holds. It remains to show that for any $x \in \overset{\circ}{D}$ and $z = z(x)$, there is a unique $\lambda \in \mathbb{R}^p$ solving (4.33). Differentiating, one obtains

$$\frac{\partial L}{\partial z}(x, z, \lambda) \overset{(4.31)}{=} \frac{d}{dz}\left[G(x, M(z)) + e(x, z)^{\top} \lambda \right]$$

$$= M'(z)^{\top} \cdot \frac{\partial G}{\partial v}(x, M(z)) + \frac{\partial e}{\partial z}(x, z)^{\top} \lambda.$$

Here, $\partial G / \partial v$ means differentiation of G with respect to its second argument. By Assumption 4.7, the matrix $\partial e(x, z)/\partial z$ is invertible and thus (4.33) can indeed be uniquely solved for λ. □

Remark The function L is the Lagrangian to be used for finding an extremum of $G(x, M(z))$ under the constraint $e(x, z) = 0$, which is the same as finding an extremum of $G(x, F(x))$.

Remark It was shown above, using the definition (4.30) for G, how the adjoint method can be used to compute the Jacobian $F'(x)$ row by row. A much more straightforward way to compute this matrix column by column is by direct differentiation

$$\frac{\partial F}{\partial x_j} = M'(z) \frac{\partial z}{\partial x_j}. \tag{4.34}$$

Further, implicit differentiation of $e(x, z) = 0$ leads to

$$\frac{\partial e}{\partial x_j}(x, z) + \frac{\partial e}{\partial z}(x, z) \cdot \frac{\partial z}{\partial x_j} = 0. \tag{4.35}$$

This is a linear system of equations, which can be solved for the vector $\partial z / \partial x_j$, to be used afterwards in (4.34) to compute $\partial F / \partial x_j$, the jth column of $F'(x)$—this is how we proceeded in Example 4.5. To get all columns of $F'(x)$, one has to solve n systems of equations of the form (4.35), which is the main effort with this approach. This effort must also be spent if one is only interested in getting the gradient $\nabla Z = F'(x)^\top (F(x) - y)$. On the other hand, ∇Z can be computed by the adjoint method for the choice (4.29) of G by solving only a single adjoint state equation (4.33).

Remark Equation (4.33) can equivalently be stated in "variational form," setting all possible directional derivatives to zero:

$$\left[\frac{\partial L}{\partial z}(x, z, \lambda) \right]^\top \delta z = 0 \quad \text{for all} \quad \delta z \in \mathbb{R}^p. \tag{4.36}$$

Directional derivatives are also known under the name of "first variations," whence the name "variational form."

Example 4.9 We repeat the computation of derivatives already accomplished in Example 4.5. Precisely, let us compute the ℓth row of $F'(\mathbf{x})$, i.e., the ℓth column of $F'(\mathbf{x})^\top$, with $\ell \in \{1, \ldots, n\}$. As seen in Example 4.6, the state equation takes the form (4.27) and the observation operator is the one defined in (4.28). The components of $\mathbf{u}$ are written in their double indexed form as in (2.138), the components of $\mathbf{g}$, given by (2.139), are written in the form g_i^j (in accordance with the components of $\mathbf{u}$). The components of matrix A will be indexed in an unusual way:

$$A = (a_{(p,q),(i,j)}) \quad i, p = 0, \ldots, m, \quad j, q = 1, \ldots, n,$$

i.e., we use *double indices* to index rows and columns of A, but assume that these double indices are ordered in accordance with (2.138). The vector $\lambda \in \mathbb{R}^{(m+1)n}$ of Lagrange multipliers will be indexed in the form

$$\lambda = (\lambda_{(p,q)}), \quad p = 0, \ldots, m, \quad q = 1, \ldots, n,$$

again in accordance with the indexing scheme used for $\mathbf{u}$. Finally, let $e_\ell \in \mathbb{R}^n$ be the ℓth unit vector. The Lagrange function now takes the form

$$L(\mathbf{x}, \mathbf{u}, \lambda) = (M\mathbf{u})^\top e_\ell + e(\mathbf{x}, \mathbf{u})^\top \lambda$$

$$= u_0^\ell + \sum_{p=0}^{m} \sum_{q=1}^{n} \left[\sum_{i=0}^{m} \sum_{j=1}^{n} a_{(p,q),(i,j)} u_i^j - g_i^j \right] \lambda_{(p,q)}.$$

From this, we get the adjoint state equation (4.33) in its variational form (4.36):

$$\frac{\partial L}{\partial \mathbf{u}} \cdot \delta \mathbf{u} = \delta u_0^\ell + \sum_{p=0}^{m} \sum_{q=1}^{n} \left[\sum_{i=0}^{m} \sum_{j=1}^{n} a_{(p,q),(i,j)} \delta u_i^j \right] \lambda_{(p,q)} = 0.$$

Reordering, this becomes

$$\frac{\partial L}{\partial \mathbf{u}} \cdot \delta \mathbf{u} = \delta u_0^1 \cdot \sum_{p,q} a_{(p,q),(0,1)} \lambda_{(p,q)} + \ldots + \delta u_0^{\ell-1} \cdot \sum_{p,q} a_{(p,q),(0,\ell-1)} \lambda_{(p,q)} +$$

$$+ \delta u_0^\ell \cdot \left(1 + \sum_{p,q} a_{(p,q),(0,\ell)} \lambda_{(p,q)} \right) +$$

$$+ \delta u_0^{\ell+1} \cdot \sum_{p,q} a_{(p,q),(0,\ell+1)} \lambda_{(p,q)} + \ldots + \delta u_0^n \cdot \sum_{p,q} a_{(p,q),(0,n)} \lambda_{(p,q)} +$$

$$\sum_{i=1}^{m} \sum_{j=1}^{n} \delta u_i^j \cdot \left[\sum_{p,q} a_{(p,q),(i,j)} \lambda_{(p,q)} \right] = 0,$$

which must hold for all choices of $\delta \mathbf{u}$, i.e., the multiplier of any term δu_i^j must vanish. This condition can be written concisely as

$$A^\top \lambda = -M^\top e_\ell. \tag{4.37}$$

Since the quadratic matrix A has full rank, λ is determined uniquely by (4.37). The function $M(\mathbf{u}) = M\mathbf{u}$ does not depend explicitly on $\mathbf{x}$, so we get

$$\frac{\partial L}{\partial \mathbf{x}}(\mathbf{x}, \mathbf{u}, \lambda) \overset{(4.32)}{=} \frac{\partial e}{\partial \mathbf{x}}(\mathbf{x}, \mathbf{u})^\top \lambda. \tag{4.38}$$

Next we need to determine $\partial e / \partial \mathbf{x}$. From $e(\mathbf{x}, \mathbf{u}) = A(\mathbf{x})\mathbf{u} - \mathbf{g}$ we conclude

$$\frac{\partial e}{\partial \mathbf{x}}(\mathbf{x}, \mathbf{u}) = \left(\frac{\partial A}{\partial x_1} \cdot \mathbf{u}, \dots, \frac{\partial A}{\partial x_{k+1}} \cdot \mathbf{u} \right) =: B.$$

On the other hand, as in Example 4.5, based on $\mathbf{k} = G\mathbf{x}$, we get the equality

$$B = C \cdot G \quad \text{with} \quad \left(\frac{\partial A}{\partial \kappa_0} \cdot \mathbf{u}, \dots, \frac{\partial A}{\partial \kappa_m} \cdot \mathbf{u} \right) =: C.$$

Putting all together, we arrive at

$$F'(\mathbf{x})^\top e_\ell = \nabla G \stackrel{(4.32)}{=} \frac{\partial L}{\partial \mathbf{x}}(\mathbf{x}, \mathbf{u}, \lambda) \stackrel{(4.37),(4.38)}{=} -G^\top C^\top A^{-\top} M^\top e_\ell.$$

We have found the result (4.24) from Example 4.5 again.   ◊

4.4   Iterative Regularization

Iterative regularization of a nonlinear system of equations $F(x) = y$ means to stop an iterative solver prematurely, before it converges. This was investigated for linear systems $Ax = b$ in Sects. 3.9 and 3.10, but the same idea works in the nonlinear case. For example one can generalize the Landweber iteration to be used for nonlinear equations, see Chap. 11 of [EHN96]. In the following, we rather consider algorithms based on Newton's method. The results presented below are taken from [Han97a] and [Han97b].

We wish to solve the nonlinear system of equations

$$F(x) = y, \quad F : D \subseteq \mathbb{R}^n \to \mathbb{R}^m, \quad x \in D, \quad y \in \mathbb{R}^m, \tag{4.39}$$

where $D \subseteq \mathbb{R}^n$ is *closed* and *convex* and where $F \in C^2(D)$. Below, we will simply take $D = \mathbb{R}^n$, not considering any constraints, which would be possible, but would complicate matters. We assume that a solution $\hat{x}$ of (4.39) exists, but that we only dispose of some approximation $y^\delta \approx y$ with

$$\| y - y^\delta \|_2 \leq \delta, \tag{4.40}$$

for $\delta > 0$ known. With $\hat{x}$ a solution of (4.39) and x_n some approximation of it, one has

$$F(\hat{x}) - F(x_n) = F'(x_n)(\hat{x} - x_n) + R(\hat{x}; x_n), \tag{4.41}$$

where $R(\hat{x}; x_n)$ is the Taylor remainder. Adding y^δ on both sides of (4.41) and using $F(\hat{x}) = y$, one gets

$$F'(x_n)(\hat{x} - x_n) = \underbrace{y^\delta - F(x_n)}_{=: \tilde{y}_n} + y - y^\delta - R(\hat{x}; x_n) =: y_n \qquad (4.42)$$

If y_n was known, then an ideal update $z = \hat{x} - x_n$ of x_n (which directly leads to $\hat{x}$) could be computed as a solution of the linear system of equations

$$A_n r = y_n, \quad A_n := F'(x_n). \qquad (4.43)$$

However, only the approximation $\tilde{y}_n$ of y_n is known with

$$\|y_n - \tilde{y}_n\|_2 \le \delta + \|R(\hat{x}; x_n)\|_2. \qquad (4.44)$$

One could try to solve $A_n x = \tilde{y}_n$ instead of $A_n x = y_n$. This would give a non-ideal update $\tilde{z}$ and a next, hopefully improved, approximation $x_{n+1} = x_n + \tilde{z}$ of $\hat{x}$. According to the discrepancy principle, the iterative procedure should be stopped as soon as $\|y^\delta - F(x_n)\|_2 \le \tau\delta$, where $\tau \ge 1$ is a chosen parameter. However, if the solution $\hat{x}$ of the nonlinear problem $F(x) = y$ is badly conditioned, then chances are high that so is the matrix A_n, and if this is the case, then a solution $\tilde{z}$ of $A_n x = \tilde{y}_n$ may be far away from a solution z of $A_n x = y_n$. Therefore, it is advisable to use a regularization method for solving $A_n x = \tilde{y}_n$. Hanke [Han97b] proposes to use the powerful CG method applied to the normal equations

$$A_n^\top A_n x = A_n^\top \tilde{y}_n,$$

which is known as CGNE method. The CGNE method as an iterative equation solver was investigated in Sect. 3.10. In the present case it is used as an *inner* iteration within the *outer* Newton iteration and will produce a sequence $z_0, z_1, z_2, \ldots$ of approximations to the solution z of $A_n x = y_n$. It should be stopped according to the discrepancy principle as soon as

$$\|\tilde{y}_n - A_n z_k\|_2 \le \|y_n - \tilde{y}_n\|_2 \qquad (4.45)$$

holds for an index $k \in \mathbb{N}_0$, see Theorem 3.41. However, not even the upper bound (4.44) for the right hand side of (4.45) is known, unless we make assumptions on the Taylor remainder term in (4.41). Following [Han97b], it will be required that for some ball $B_r(\hat{x}) \subset D = \mathbb{R}^n$ around the solution $\hat{x}$ of (4.39), there exists a constant $C > 0$ such that

$$\|F(x) - F(\tilde{x}) - F'(\tilde{x})(x - \tilde{x})\|_2 \le C\|x - \tilde{x}\|_2 \|F(x) - F(\tilde{x})\|_2 \qquad (4.46)$$

for all $x, \tilde{x} \in B_r(\hat{x})$. The left hand side of (4.46) is the magnitude of the linearization error $R(x; \tilde{x})$ for F, as in (4.41). Since $F \in C^2(D)$, it is clear that the linearization error can be bounded in the form

$$\|F(x) - F(\tilde{x}) - F'(\tilde{x})(x - \tilde{x})\|_2 \leq C \|x - \tilde{x}\|_2^2,$$

but (4.46) additionally requires the linearization error to be controlled by the size of the nonlinear residual $F(x) - F(\tilde{x})$. This can be interpreted as a restrictive assumption on F, which is not allowed to be "too nonlinear" near $\hat{x}$. Using (4.46), the requirement (4.45) can be replaced by

$$\|\tilde{y}_n - A_n z_k\|_2 \leq \delta + C \|\hat{x} - x_n\|_2 \|y - F(x_n)\|_2,$$

which still is not directly usable, since $\hat{x}$ and y are unknown. But as long as the outer Newton iteration is not terminated by the discrepancy principle, one has $\delta < \|y^\delta - F(x_n)\|_2$ and a sufficiently large fraction of $\|y^\delta - F(x_n)\|_2$ can serve as an upper bound for $\delta + C \|\hat{x} - x_n\|_2 \|y - F(x_n)\|_2$, if x_n is close enough to $\hat{x}$ (how close it has to be depends on the constant C). We take the following algorithm from [Han97b], which is based on CGNE from Sect. 3.10.

Inexact Newton-CG Method for Minimizing $\|F(x) - y^\delta\|_2$.

choose parameters $\tau \geq 1$ and $0 < \rho < 1$

choose starting point $x_0 \in \mathbb{R}^n$ and set $n = 0$

while $\|y^\delta - F(x_n)\|_2 > \tau \delta$ % outer (Newton) iteration

   $b = y^\delta - F(x_n), A = F'(x_n)$

   $z_0 = 0, w_0 = b, r_0 = b, k = 0$

   repeat   % inner (CGNE) iteration: $A^\top A z = A^\top b$

      $d_k = A^\top w_k$

      $\alpha_k = \|A^\top r_k\|_2^2 / \|A d_k\|_2^2$

      $z_{k+1} = z_k + \alpha_k d_k$

      $r_{k+1} = r_k - \alpha_k A d_k$

      $\beta_k = \|A^\top r_{k+1}\|_2^2 / \|A^\top r_k\|_2^2$

      $w_{k+1} = r_{k+1} + \beta_k w_k$

      $k = k + 1$

   until $\|r_k\|_2 < \rho \|b\|_2$ % end inner iteration

(continued)

$$x_{n+1} = x_n + z_k$$

$$n = n + 1$$

end % end outer iteration

The method is called an **inexact Newton method**, because the linear "update equation" $Az = b$ is not solved exactly, but only approximately (in order to obtain a regularized solution). In [Han97b] it is proven that under the conditions

$$\tau \rho^2 > 2 \quad \text{and} \quad x_0 \text{ close enough to } \hat{x} \tag{4.47}$$

the above algorithm works and terminates after a finite number of iterations. It produces a regularized solution x^δ, i.e., x^δ converges to $\hat{x}$ for $\delta \to 0$. In practice, one would like to choose τ close to 1 in order to obtain a good data fit, but (4.47) requires $\tau > 2/\rho^2 > 2$. For example, the combination $\rho = 0.9$ and $\tau = 2.5$ would be in accordance with (4.47). A further general restriction is the requirement (4.46), as already mentioned.

4.5 Solution of Model Problem for Nonlinear Inverse Gravimetry

Problem 1.9 concerning nonlinear inverse gravimetry was discretized in Sect. 2.6. For convenience, we repeat the main points. One has to solve a nonlinear Fredholm equation of the first kind

$$w(x_1, x_2) = \int_{-a}^{a} \int_{-a}^{a} k(x_1, x_2, t_1, t_2, u(t_1, t_2)) \, dt_2 \, dt_1, \tag{4.48}$$

where the kernel function is defined by

$$k(x_1, x_2, y_1, y_2, z) := \frac{1}{\sqrt{(x_1 - y_1)^2 + (x_2 - y_2)^2 + z^2}} \tag{4.49}$$

for $(x_1, x_2), (y_1, y_2) \in \mathbb{R}^2$ and $z > 0$. A unique solution exists, which is a continuous function $\hat{u} : [-a, a]^2 \to [b_1, b_2]$, $0 < b_1 < b_2$. As a discrete version of (4.48) we derived the nonlinear system of equations

$$F(c) = y, \quad c \in C. \tag{4.50}$$

Here,

- $y \in \mathbb{R}^M$ is a vector with components[4]

$$y_\beta = w(\hat{x}_\beta), \quad \hat{x}_\beta \in [-b, b]^2, \quad \beta \in B,$$

  for an index set B comprising M elements,
- $C = [b_1, b_2]^N$ is an N-dimensional interval (a box),
- $c \in C \subset \mathbb{R}^N$ is a vector with components c_α, to be interpreted as samples of the sought-after function $u \in C([-a, a]^2)$:

$$c_\alpha = u(x_\alpha) \quad \text{with} \quad x_\alpha = h\alpha, \quad h = \frac{a}{n}, \quad \alpha \in G_n,$$

  where

$$G_n = \left\{ (\alpha_1, \alpha_2) \in \mathbb{Z}^2; \; -n \le \alpha_j \le n \right\}, \quad |G_n| = N = (2n+1)^2,$$

- $F(c) \in \mathbb{R}^M$ is a vector with components

$$F_\beta(c) = \sum_{\alpha \in G_n} \omega_\alpha \left(\|\hat{x}_\beta - x_\alpha\|_2^2 + c_\alpha^2 \right)^{-1/2}, \quad \beta \in B, \tag{4.51}$$

  where the sum on the right hand side means an approximate numerical evaluation of (4.48) by the trapezoidal rule for

  - $(x_1, x_2) = \hat{x}_\beta$, and
  - constant factors ω_α defined in (2.110).

System (4.50) is to be replaced by the minimization problem

$$\text{minimize} \quad \frac{1}{2} \|y - F(c)\|_2^2, \quad c \in C, \tag{4.52}$$

as discussed in Sect. 2.6. Let us define a vector $\hat{c} \in \mathbb{R}^N$ by

$$\hat{c}_\alpha := \hat{u}(x_\alpha), \quad \alpha \in G_n, \tag{4.53}$$

---

[4]To arrange multi-indexed components into a vector, a certain index ordering has to be fixed. In case of rectangular index grids, we will always use a rowwise ordering, i.e., we will use the first index component as "inner index." For example, the indices $\alpha \in G_n$ for the grid G_n defined below, will always be ordered to form the sequence

$$(-n, -n), \ldots, (n, -n), \quad \ldots, \quad (-n, n), \ldots, (n, n).$$

with $\hat{u}$ the exact solution of (4.48) for given w. *Deviating from the notation in (4.50), let us now redefine*

$$y := F(\hat{c}), \qquad (4.54)$$

i.e., y will no longer be the discretized true effect, but a (discretized) effect simulated by $\hat{c}$. This way, the system (4.50) is artificially made to have a solution (namely $\hat{c}$). Perturbed data values

$$y_\beta^\delta \approx y_\beta, \quad \beta \in B, \qquad (4.55)$$

thus include measurement *and* discretization errors. If we require $\|y^\delta - y\|_2 \leq \delta$, then δ has to be an upper bound for both errors together, unless it can be argued that discretization errors are negligible as compared to measurement errors. If a discretization method is convergent (as it should be), the latter will be achievable by choosing a discretization fine enough. To reconstruct an approximation $\tilde{c}$ of $\hat{c}$ from inexact data y^δ, two variants of Tikhonov regularization will be tested as well as iterative regularization.

Reconstruction by Smoothing Regularization

The first reconstruction method consists in minimizing

$$Z_\lambda(c) := \|F(c) - y^\delta\|_2^2 + \lambda \|Lc\|_2^2, \quad c \in C, \qquad (4.56)$$

which is of the form (4.4) and which we prefer to (4.5), since we have no a priori information about the location of the solution $\hat{c}$. The matrix L is chosen as follows. Define $A_n, B_n \in \mathbb{R}^{2n+1,2n+1}$ by

$$A_n := \begin{pmatrix} 2 & & & & \\ -1 & 4 & -1 & & \\ & \ddots & \ddots & \ddots & \\ & & -1 & 4 & -1 \\ & & & & 2 \end{pmatrix} \quad \text{and} \quad B_n := \begin{pmatrix} 1 & -1 & & & \\ -1 & 2 & -1 & & \\ & \ddots & \ddots & \ddots & \\ & & -1 & 2 & -1 \\ & & & -1 & 1 \end{pmatrix},$$

set $I = I_{2n+1} \in \mathbb{R}^{2n+1,2n+1}$ and set

$$L := \begin{pmatrix} B_n \\ -I & A_n & -I \\ & \ddots & \ddots & \ddots \\ & & -I & A_n & -I \\ & & & B_n \end{pmatrix} \in \mathbb{R}^{(2n+1)^2,(2n+1)^2}. \tag{4.57}$$

For a vector $c = (c_\alpha)_\alpha$, $\alpha \in G_n$, ordered rowwise

$$c_{(-n,-n)}, \ldots, c_{(n,-n)}, \quad \ldots, \quad c_{(-n,n)}, \ldots, c_{(n,n)},$$

Lc corresponds to

- a discrete version of the (negative) Laplace operator $-\Delta$ applied to u at grid points x_α in the interior of the box C,
- a discrete version of $-u_{xx}$ at the grid points x_α on the lower and upper boundary of the box C (excluding the corner points),
- a discrete version of $-u_{yy}$ at the grid points x_α on the left and right boundary of the box C (excluding the corner points),
- a discrete version of $+u_x$ or $-u_x$ at the four corner points x_α of the box C.

The term $\|Lc\|_2^2$ thus inversely measures the smoothness of a reconstruction u.

Example 4.10 (Reconstruction of a Smooth Function) Let $a = 4$, let $b = 4$, and let $n = 16$, such that $N = 1089$. Let

$$\hat{u}(x) = \left[1 + \frac{1}{10}\cos\left(\frac{\pi x_1}{a}\right)\right]\left[1 + \frac{1}{10}\sin\left(\frac{\pi x_2}{a}\right)\right], \quad x = (x_1, x_2) \in [-a, a]^2$$

be the function which we want to reconstruct and let its samples be given by

$$\hat{c}_\alpha = \hat{u}(h\alpha), \quad \alpha \in G_n, \quad h = a/n.$$

The exact function $\hat{u}$ is shown in Fig. 4.1. For $m \in \mathbb{N}$, let

$$\hat{h} := b/m, \quad B := \{(\beta_1, \beta_2) \in \mathbb{Z}^2; \ -m \le \beta_j < m\}, \quad \hat{x}_\beta := \hat{h}\beta, \ \beta \in B.$$

Set $m = 64$, such that $M = 16,384$, and let

$$y_\beta := F_\beta(\hat{c}), \quad \beta \in B.$$

These values are samples of an artificial effect w *simulated* by replacing function u in (4.48) by the bilinear spline interpolant of $(x_\alpha, \hat{c}_\alpha)$, $\alpha \in G_n$. In contrast to y_β, let

$$w_\beta := \int_{-a}^{a}\int_{-a}^{a} k(\hat{x}_\beta, y, \hat{u}(y)) \, dy_2 \, dy_1, \quad y = (y_1, y_2),$$

Exact function $\hat{u}$

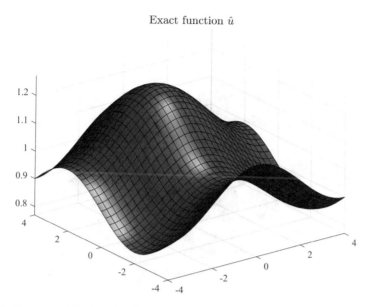

Fig. 4.1 Exact smooth function $\hat{u}$ to be reconstructed

be the samples of the true effect caused by $\hat{u}$. We found

$$\frac{1}{M} \sum_{\beta \in B} |y_\beta - w_\beta|^2 \approx 6 \cdot 10^{-5} \qquad (4.58)$$

for the mean square discretization error. Values y_β^δ were defined by adding to y_β realizations of independent random variables with normal distribution, having mean value 0 and standard deviation $\sigma = 10^{-1}$. The corresponding mean square error

$$\frac{1}{M} \sum_{\beta \in B} |y_\beta^\delta - y_\beta|^2 \approx \sigma^2 = 10^{-2}$$

is dominating the discretization error (4.58). The relative data error resulting from the perturbation of y is

$$\|y - y^\delta\|_2 / \|y\|_2 \approx 5 \cdot 10^{-3}.$$

In Fig. 4.2, we show a reconstruction u^δ of $\hat{u}$ as the bilinear spline interpolant of $(x_\alpha, c_\alpha^\delta)$, $\alpha \in G_n$, where the values c_α^δ were found by numerically determining the minimizer of (4.56), with $\lambda = 10^2$. This choice of λ is in accordance with Morozov's principle, since it leads to

Reconstructed function u^δ

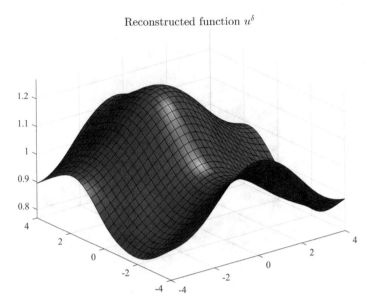

Fig. 4.2 Reconstructed approximation u^δ of $\hat{u}$ from noisy data, $\sigma = 10^{-1}$, $b = 4$

$$\|F(c^\delta) - y^\delta\|_2 \approx 12.7825 \approx 12.7993 \approx \|y - y^\delta\|_2.$$

The numerical minimization was performed using Matlab's function lsqnonlin, which implements an algorithm for solving nonlinear least squares problems with box constraints developed in [BCL99]. An outline of some ideas behind this algorithm was given in Sect. 4.2. The derivatives needed by lsnonlin were provided by direct computation as in Example 4.4. We chose the box constraints determined by $b_1 = 0.5$ and $b_2 = 5.5$ and set the constant vector $c^0 = (1, 1, \ldots, 1)$ as an initial approximation (start value) to be used by lsqnonlin. In this case, lsqnonlin stopped after 6 iterations. In Fig. 4.3, the singular values of the Jacobian $DF(c^\delta) = F'(c^\delta)$ are shown. This tells us how badly conditioned this problem is.

We repeated the above experiment, keeping all parameter values except for setting $b = 2$. This means that we restricted ourselves to a much smaller data window. As expected, the reconstruction deteriorates notably, as shown in Fig. 4.4.

Reconstruction by Variation Diminishing Regularization

For the reconstruction of non-smooth functions u, using the regularization as in (4.56) with L defined by (4.57) would be misleading, since in this case the term

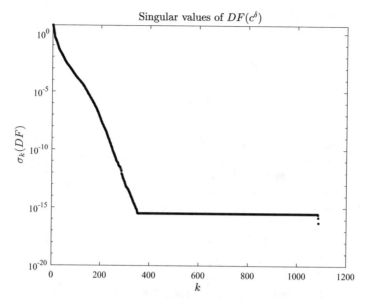

Fig. 4.3 Singular values of Jacobian at optimal point

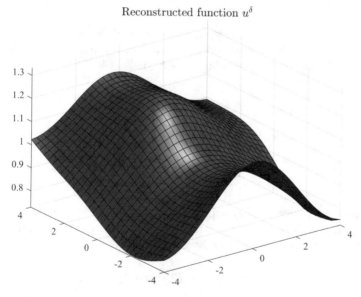

Fig. 4.4 Reconstructed approximation u^δ of $\hat{u}$ from noisy data, $\sigma = 10^{-1}, b = 2$

$\|Lc\|_2$ generally will not be small for the samples c_α of the sought-after function u. We still can resort to generalized Tikhonov regularization of the form (3.66). As recommended in [ROF92] for image reconstruction from noisy data, we will use a regularization term G, which is a discrete analogon of the total variation

$$\int_{-a}^{a} \int_{-a}^{a} \|\nabla u(x, y)\|_2 \, dx \, dy \tag{4.59}$$

of the function $u : [-a, a]^2 \to \mathbb{R}$. The benefit of using total variation for regularization will be explained in detail in Sect. 4.6. A discrete version of (4.59) will be computed by numerical integration via the rectangular rule, based on the samples $c_\alpha = u(x_\alpha)$, $\alpha \in G_n$, already introduced above. Let $N' := (2n)^2$, $G'_n := \{\alpha \in G_n; \ \alpha_1 < n \text{ and } \alpha_2 < n\}$ and set

$$h \cdot u_x(x_\alpha) \approx d_{x,\alpha} := c_{(\alpha_1+1,\alpha_2)} - c_{(\alpha_1,\alpha_2)}, \quad \alpha \in G'_n,$$

$$h \cdot u_y(x_\alpha) \approx d_{y,\alpha} := c_{(\alpha_1,\alpha_2+1)} - c_{(\alpha_1,\alpha_2)}, \quad \alpha \in G'_n.$$

From these finite differences, the integral (4.59) can be approximated by

$$G(c) = \sum_{\alpha \in G'_n} h \sqrt{d_{x,\alpha}^2 + d_{y,\alpha}^2},$$

but this function G is not differentiable, which impedes the use of Newton methods to solve (3.66). We therefore introduce a new "small" parameter $\varepsilon > 0$, define (assuming the usual rowwise ordering of double indices α)

$$V(c) := \left(\sqrt{h} \cdot \left[d_{x,\alpha}^2 + d_{y,\alpha}^2 + h^2 \varepsilon \right]^{\frac{1}{4}} \right)_{\alpha \in G'_n} \in \mathbb{R}^{N'}, \tag{4.60}$$

and use

$$G(c) := \|V(c)\|_2^2, \tag{4.61}$$

as a smoothed (differentiable) approximation of (4.59). We will use the same approximation in Sect. 4.6, compare formulas (4.74) and (4.75). For a regularized solution of (4.52), we now consider the minimization of

$$Y_\lambda(c) := \|F(c) - y^\delta\|_2^2 + \lambda G(c), \quad c \in C, \tag{4.62}$$

instead of (4.56). For the perturbed data y^β, we still assume (4.54) and (4.55) to hold.

Example 4.11 (Reconstruction of a Step Function) Let $a = 4$, let $b = 2$, and let $n = 16$. Let $\hat{u}$ be the two-dimensional step function

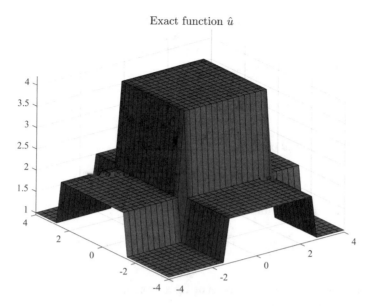

Exact function $\hat{u}$

Fig. 4.5 Exact step function $\hat{u}$ to be reconstructed

$$\hat{u} : [-a, a]^2 \to \mathbb{R}, \quad (x_1, x_2) \mapsto v(x_1) \cdot v(x_2),$$

where

$$v : [-a, a] \to \mathbb{R}, \quad x \mapsto \begin{cases} 2, & \text{if } -\frac{a}{3} \le x \le \frac{2a}{3}, \\ 1, & \text{else.} \end{cases}$$

This function is shown in Fig. 4.5. Let $h = a/n$ and $\hat{c}_\alpha = \hat{u}(h\alpha)$, $\alpha \in G_n$, as in Example 4.10. Also, let $m = 64$, $\hat{h} = b/m$, $B = \{(\beta_1, \beta_2) \in \mathbb{Z}^2; \ -m \le \beta_j < m\}$, and $\hat{x}_\beta = \hat{h}\beta$, $\beta \in B$, also as in Example 4.10. Further, and once more as in Example 4.10, let $y_\beta = F_\beta(\hat{c})$, $\beta \in B$, be the samples of an artificial effect w *simulated* by replacing function u in (4.48) by the bilinear spline interpolant of $(x_\alpha, \hat{c}_\alpha)$, $\alpha \in G_n$, and let $w_\beta = \int_{-a}^a \int_{-a}^a k(\hat{x}_\beta, y, \hat{u}(y)) \, dy_2 \, dy_1$, $\beta \in B$, be the samples of the true effect caused by $\hat{u}$. The "artificial" parameter in Eq. (4.60) was set to $\varepsilon := 10^{-10}$. The box constraints $b_1 = 0.5$ and $b_2 = 5.5$ from Example 4.10 were reused.

We first tried to reconstruct $\hat{u}$ from exact data values y_β, not contaminated by noise. For the computation, we relied on Matlab's function lsqnonlin, providing the exact Jacobian of F and V from (4.62) and (4.61). In the absence of noise, the regularization parameter λ in (4.62) cannot be determined by Morozov's principle. We chose a "small" value $\lambda = 10^{-7}$. The optimizer was started with function $u_0 \equiv 2$ as initial guess for $\hat{u}$. For this start value, lsqnonlin stopped after 540 iterations with the message "local minimum possible." The stopping criterion was a stagnation in step size, which however occurred when the optimizer was still far away from the global minimum point of (4.62). To get closer to the global minimum point, we resorted to multiscale optimization, as described in Sect. 2.6. Deviating notationally slightly from (2.118), for $k \in \mathbb{N}_0$ we set

$$n_k := 2^k, \quad h_k := \frac{2a}{n_k}, \quad G^{(k)} := \left\{ (\alpha_1, \alpha_2); \ 0 \le \alpha_j \le n_k, \ j = 1, 2 \right\}.$$

We used $(2^k + 1)^2$ grid points

$$x_\alpha^{(k)} := (-a, -a) + h_k \alpha, \quad \alpha \in G^{(k)},$$

and looked for approximations

$$u^{(k)}(x) = \sum_{\alpha \in G^{(k)}} c_\alpha^{(k)} \Phi([x + (a, a)]/h_k - \alpha) \tag{4.63}$$

"at level k." The unknown parameters ("optimization parameters") at level k were ordered into a column vector

$$c^{(k)} := (c_\alpha^{(k)}; \ \alpha \in G^{(k)}) \in \mathbb{R}^{N_k}, \quad N_k = (2^k + 1)^2.$$

We used $K = 5$ as finest level and looked for optimal parameter vectors $c^{(k)}$ at levels $k = 0, 1, \ldots, K = 5$. At each level, the optimization was performed using Matlab's lsqnonlin. At level 0, an initial guess $u_0 \equiv 2$ was used to start the optimization. The computed optimizer $u^{(k)}$ (found by lsqnonlin) was then interpolated to the next finer grid to serve as a start value for the optimization at level $k + 1$. At levels $k = 0, 1, \ldots, K - 1$, the number of iterations to be performed by lsqnonlin was restricted to a maximal value of 20. At the finest level, no such restriction was applied and lsqnonlin stopped after (the high number of) 4949 iterations with the (very good) reconstruction shown in Fig. 4.6. The peak to be seen in the north eastern corner of the graph corresponds to coefficient $c_{(n_K, n_K)}^{(K)}$. From (4.60) it becomes clear that regularization does not apply to this one coefficient. Its reconstruction can therefore not be trusted.

We next tried to reconstruct $\hat{u}$ from noisy data values y_β^δ, which were obtained from y_β by adding realizations of independent random variables with normal distribution, having mean value 0 and standard deviation $\sigma = 10^{-1}$. Multiscale optimization was performed as in the case of exact data. At levels $k = 0, 1, \ldots, K - 1$, the number of iterations to be performed by lsqnonlin was again restricted to a maximal value of 20. At the finest level, minimization was stopped after 1000 iterations. The regularization parameter was set to $\lambda = 0.5$ according to the discrepancy principle, since this led to

$$\|y - y^\delta\|_2 \approx 12.7993 \approx 12.8090 \approx \|F(c^\delta) - y^\delta\|_2.$$

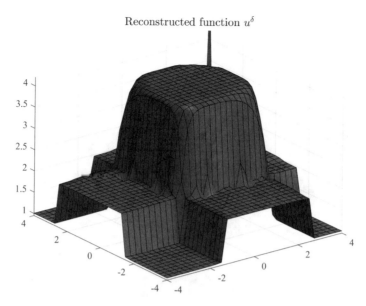

Fig. 4.6 Reconstructed approximation u of $\hat{u}$ from exact data, $b = 2$

Here, we write $c_\alpha^\delta = c_\alpha^{(K)}$ for the coefficients at the finest level. The bilinear spline interpolant u^δ of $(x_\alpha, c_\alpha^\delta)$, $\alpha \in G_n$, is shown in Fig. 4.7. Apparently, u^δ gives an idea of how the sought-after function $\hat{u}$ looks like, but is of mediocre quality only. The reconstruction improves significantly, if more information is available, e.g., if we can collect more data. Increasing the value of parameter b to $b = 4$ and keeping all other parameters exactly as just before led to the improved reconstruction shown in Fig. 4.8. $\Diamond$

Reconstruction by Iterative Regularization

Iterative regularization as described in Sect. 4.4 can be applied to solve the minimization problem (4.52) with F defined in (4.51). I have not been successful in proving that condition (4.46) holds for this function F, but used the inexact Newton-CG method nevertheless. The parameters $\tau = 1.01$ and $\rho = 0.99$ were always kept fixed in the following, against the requirement $\tau\rho^2 > 2$ from (4.47).

Example 4.12 (Iterative Regularization of Smooth Function) The numerical values from Example 4.10 were retained, especially we used $a = 4$ and $b = 2$ and the same exact function $\hat{u}$ to be reconstructed. The values $c_\alpha = \hat{u}(x_\alpha)$, $\alpha \in G_n$, define values $y_\beta = F_\beta(c)$, $\beta \in B$, as in Example 4.10. To these values y_β we added random numbers from a normal distribution with mean value 0 and standard

Reconstructed function u^δ

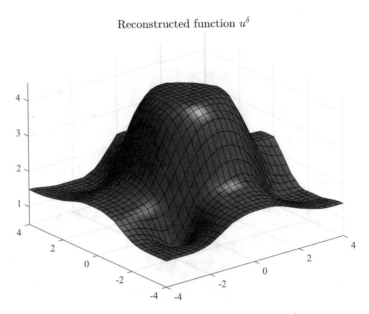

Fig. 4.7 Reconstructed approximation u^δ of $\hat{u}$ from noisy data, $\sigma = 10^{-1}$, $b = 2$.

Reconstructed function u^δ

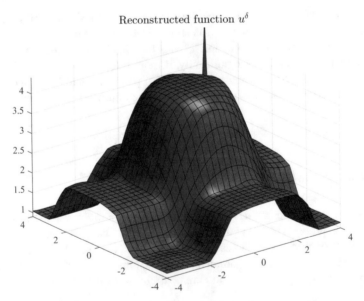

Fig. 4.8 Reconstructed approximation u^δ of $\hat{u}$ from noisy data, $\sigma = 10^{-1}$, $b = 4$

Reconstructed function u^δ

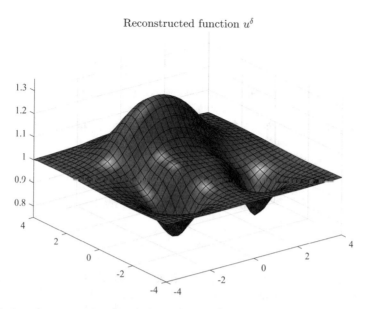

Fig. 4.9 Reconstruction of $\hat{u}$, straightforward application of inexact Newton-CG method

deviation $\sigma = 10^{-1}$, resulting in perturbed samples y_β^δ. The inexact Newton-CG method was straightforwardly applied to the system of equations $F(c) = y^\delta$. This gave a regularized solution c^δ, which defines a bilinear spline function u^δ shown in Fig. 4.9. Visibly, u^δ is only a poor approximation of $\hat{u}$ (the latter was shown in Fig. 4.1). $\Diamond$

Even if the inexact Newton-CG method formally produces a regularized reconstruction u^δ, which converges to $\hat{u}$ for $\delta \to 0$, this does not mean that the reconstruction is good for a finite value of δ. If only a single data set is given, containing a finite error, then the careful choice of a regularization term matters more than convergence of the method. Now for the inexact Newton-CG method regularization is achieved by prematurely stopping the inner CG iteration before it converges. It was observed in Sect. 3.10 that this implicitly corresponds to using a minimum-norm regularization criterion, related to approximating a solution of $Ax = b$ by minimizing $\|Ax - b\|_2^2 + \lambda \|x\|_2^2$. But there is no obvious physical justification of why this regularization term is a good one. One could as well compute x as a minimizer of $\|Ax - b\|_2^2 + \lambda \|Lx\|_2^2$. In case L is invertible, the coordinate transform

$$Lx = z \quad \Longleftrightarrow \quad x = L^{-1}z \tag{4.64}$$

leads to $\|Ax - b\|_2^2 + \lambda \|Lx\|_2^2 = \|Bz - b\|_2^2 + \lambda \|z\|_2^2$ with matrix $B = AL^{-1}$. This means that applying CGNE to the transformed problem and then transforming back the obtained solution makes CGNE mimic a regularization not with respect to $\|x\|_2$, but with respect to $\|Lx\|_2$.

Example 4.13 (Iterative Regularization of a Smooth Function with Coordinate Transform) All the settings of Example 4.12 are retained, but the inexact Newton-CG method will now be applied to a transformed problem, i.e., to problem $\tilde{F}(z) := F(L^{-1}z) = y^\delta$ where $Lc = z$ for an invertible matrix L. The obtained solution z will then be transformed back. It remains to choose L. The following choice is based on the discrete Laplace operator used in (4.56). Let

$$u_{i,j} := u(x_\alpha) = c_\alpha \text{ for } \alpha = (i, j) \in G_n,$$

where $u : [-a, a]^2 \to \mathbb{R}$. As usual, one approximates by finite differencing:

$$\Delta u(x_\alpha) \approx \frac{1}{h^2} \left[c_{i-1,j} + c_{i+1,j} - 4c_{i,j} + c_{i,j-1} + c_{i,j+1} \right], \quad \text{if } \alpha = (i, j),$$
$$(4.65)$$

which requires evaluation of u at five grid points, which are located to the west (index $(i - 1, j)$), to the east (index $(i + 1, j)$), to the south (index $(i, j - 1)$), and to the north (index $(i, j + 1)$) with respect to the central point with index (i, j). A perfectly regular vector c is supposed to fulfill

$$c_{i-1,j} + c_{i+1,j} - 4c_{i,j} + c_{i,j-1} + c_{i,j+1} = 0 \qquad (4.66)$$

at all interior grid points. At boundary grid points, (4.66) makes no sense: if, e.g., $i = -n$ and $-n < j < n$, then $c_{i-1,j}$ is not defined. To deal with boundary grid points, additional boundary conditions have to be introduced. One possibility, much in use for elliptic partial differential equations, is **Robin's boundary condition**. For a function $u : D \to \mathbb{R}$ defined on some region $D \subset \mathbb{R}^s$ it requires that

$$\nabla_n u(x) + \alpha u(x) = 0 \quad \text{for} \quad x \in \partial D, \qquad (4.67)$$

where n is the outer normal vector at $x \in \partial D$ and where $\nabla_n u(x) = (\nabla u(x))^\top n$ is the exterior normal derivative of u in $x \in \partial D$. If the parameter $\alpha \in \mathbb{R}$ is chosen equal to zero, (4.67) becomes Neumann's boundary condition. If $D = [-a, a]^2$ is a rectangle, and x is located on the left boundary, then n points westwards. At a grid point indexed $\alpha = (i, j)$ with $i = -n$ and $-n < j < n$, (4.67) can formally be discretized by finite differencing, leading to

$$\frac{c_{i-1,j} - c_{i+1,j}}{2h} + \alpha c_{i,j} = 0. \qquad (4.68)$$

Taking this equation as a *definition* of $c_{i-1,j}$ and inserting into (4.66) lead to

$$2c_{i+1,j} - (4 + 2\alpha h)c_{i,j} + c_{i,j-1} + c_{i,j+1} = 0. \qquad (4.69)$$

For reasons of symmetry, which will become apparent in a moment, this equation is scaled by $\frac{1}{2}$, leading to

$$c_{i+1,j} - (2 + \alpha h)c_{i,j} + \frac{1}{2}c_{i,j-1} + \frac{1}{2}c_{i,j+1} = 0. \tag{4.70}$$

The same approach can be used to get equations at the right, the lower, and the upper boundary. There are still four corner points to consider, e.g., the lower left corner point with index $(i, j) = (-n, -n)$. Here, neither $c_{i-1,j}$ nor $c_{i,j-1}$ are defined in (4.66). But now, instead of (4.68), we get two equations

$$\frac{c_{i-1,j} - c_{i+1,j}}{2h} + \alpha c_{i,j} = 0 \text{ and}$$

$$\frac{c_{i,j-1} - c_{i,j+1}}{2h} + \alpha c_{i,j} = 0,$$

corresponding to two normal vectors, one pointing to the west and the other pointing to the south. These two equations are used to define $c_{i-1,j}$ and $c_{i,j-1}$. Inserting into (4.66) leads to

$$2c_{i+1,j} - (4 + 4\alpha h)c_{i,j} + 2c_{i,j+1} = 0.$$

After scaling by $\frac{1}{4}$, this becomes

$$\frac{1}{2}c_{i+1,j} - (1 + \alpha h)c_{i,j} + \frac{1}{2}c_{i,j+1} = 0. \tag{4.71}$$

Equations (4.66), (4.70), and (4.71) (with obvious modifications for the other boundary and corner points) can consistently be written in matrix form $Lc = 0$, where c is derived from $\{c_\alpha, \ \alpha \in G_n\}$ by rowwise ordering and where L is defined as follows. Set

$$A_n := \begin{pmatrix} -(2+\alpha h) & 1 & & & \\ 1 & -4 & 1 & & \\ & & \ddots & \ddots & \ddots & \\ & & & 1 & -4 & 1 \\ & & & & 1 & -(2+\alpha h) \end{pmatrix} \quad \text{and} \quad B_n := \begin{pmatrix} \frac{1}{2} & 0 & & & \\ 0 & 1 & 0 & & \\ & & \ddots & \ddots & \ddots & \\ & & & 0 & 1 & 0 \\ & & & & 0 & \frac{1}{2} \end{pmatrix}$$

both being $(2n + 1) \times (2n + 1)$-matrices, and define

Reconstructed function u^δ

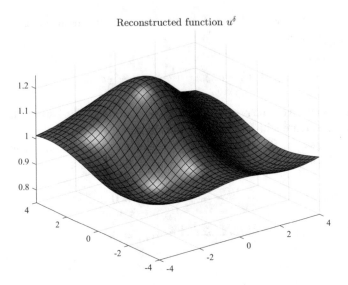

Fig. 4.10 Reconstruction of $\hat{u}$, inexact Newton-CG method applied to transformed problem

$$
\tilde{L} := \begin{pmatrix}
\frac{1}{2}A_n - \alpha h I_{2n+1} & B_n & & & \\
B_n & A_n & B_n & & \\
& \ddots & \ddots & \ddots & \\
& & B_n & A_n & B_n \\
& & & B_n & \frac{1}{2}A_n - \alpha h I_{2n+1}
\end{pmatrix} \in \mathbb{R}^{(2n+1)^2,(2n+1)^2},
$$

which is a symmetric matrix. Finally, L is defined by adding $\alpha h/2$ to the diagonal of $\tilde{L}$ at the positions of the four corner points. In case $\alpha = 0$, matrix L becomes singular, its kernel being spanned by the vector $e = (1, 1, \ldots, 1)^\top$. In case $\alpha \neq 0$, this matrix can be used to define a coordinate transform as in (4.64). Applying the inexact Newton-CG method to the transformed problem means to implicitly use $\|Lc\|_2$ as an inverse measure for the regularity of c and the function u it defines. In Fig. 4.10 the result is shown which was obtained for $\alpha = 1$. Obviously, a great improvement was achieved as compared to the result obtained in Example 4.12. $\lozenge$

From the above example one can learn that the inexact Newton-CG method has the potential to achieve good reconstructions. It does not require the full Jacobian $F'(z)$ to be computed, but only relies on the computation of matrix times vector products of the form $F'(z) \cdot v$ and $F'(z)^\top \cdot v$ in the inner iteration, so this method can be quite efficient. On the negative side we note the restrictive choices of τ and ρ and the restrictive condition (4.46) on the nonlinearity of F, which were required to prove convergence of the method. However, all these restrictions were ignored or not checked in the present example.

We repeat an observation already made for linear problems in Sects. 3.9 and 3.10. *Working with iterative methods one can only seemingly avoid the choice of a regularity measure.* In fact, an inverse regularity measure, which appears in Tikhonov regularization as a penalty term, also influences iterative methods via coordinate transforms. It has become apparent that the success of an iterative method depends on the proper choice of this transform. How to choose it well, or, equivalently, how to choose well a regularity measure, depends on whether one can successfully capture a priori information about the sought-after solution.

Further Reading

A summary of regularization methods applicable to the problem of inverse gravimetry is given by Freeden and Nashed in [FN18b]. This article complements [FN18a], which was already mentioned in Sect. 2.6. For a discussion of total variation regularization in connection with inverse gravimetry see, e.g., [BACO02]. Inverse gravimetry as a problem of shape calculus like in our model problem is solved by a level set method (by iterated shape approximation) in [ILQ11]. An introduction to level set methods can be found in [BO05]. For a monograph dealing extensively with iterative regularization methods for nonlinear inverse problems, see [KNS08].

4.6 Solution of Model Problem for Nonlinear Waveform Inversion

Problem 1.11 of nonlinear waveform inversion in its discretized form was presented in Sect. 2.7. We derived a nonlinear least squares problem

$$\text{minimize} \quad \frac{1}{2}\|\mathbf{y} - F(\mathbf{x})\|_2^2, \quad \mathbf{x} \in D = [\kappa_-, \kappa_+]^{k+1}. \tag{2.150}$$

Here, $\mathbf{y}$ corresponds to the observed seismogram and the function F has the form

$$F : D \to \mathbb{R}^n, \quad \mathbf{x} \mapsto M \cdot [A(G\mathbf{x})]^{-1} \mathbf{g}, \tag{2.148}$$

where the vector $\mathbf{g}$ and the matrices G, A, and M were defined in Sect. 2.7. The restriction $\|\kappa'\|_{L_2(0,z_0)} \leq M$ for the elasticity coefficient has not yet been translated into its discrete counterpart—this will implicitly be achieved by the following regularization. Instead of an exact seismogram, only perturbed data $\mathbf{y}^\delta \approx \mathbf{y}$ are available, so that (2.150) will have to be replaced by the minimization problem

$$\min_{\mathbf{x} \in D} \frac{1}{2}\|F(\mathbf{x}) - \mathbf{y}^\delta\|_2^2, \tag{4.72}$$

which in turn will be replaced by its regularized variant

$$\min_{\mathbf{x} \in D} \frac{1}{2} \| F(\mathbf{x}) - \mathbf{y}^\delta \|_2^2 + \frac{\alpha}{2}\, G(\mathbf{x}), \tag{4.73}$$

where

$$G(\mathbf{x}) = \sum_{i=1}^{k} |x_{i+1} - x_i| \tag{4.74}$$

is a discrete analog of the total variation of κ. Optimization problems of the form (4.73) were mentioned as a generalization of standard Tikhonov regularization in Sect. 3.4, see (3.66). They were already used to regularize the problem of inverse gravimetry, see (4.62). Total variation as a regularization term was introduced for "image denoising" by Rudin, Osher, and Fatemi in [ROF92], but was already recommended in the present case by Bamberger, Chavent, and Lailly in [BCL77]. In contrast to regularization terms measuring smoothness by taking first or second derivatives, total variation can be defined for step functions. Total variation as a regularization term is useful, if a function shall be reconstructed which is known not be oscillatory, but which may contain jumps. Also, it implicitly captures restrictions of the form $\|\kappa'\|_{L_2(0,Z_0)} \le M$. A disadvantage of (4.74) is its non-differentiability, which complicates the solution of (4.73). Either minimization algorithms have to be used which can deal with non-differentiable objective functions or one uses an approximation of G by a smooth function. We opted for the latter approach and set

$$G_\beta : D \to \mathbb{R}^k, \quad \mathbf{x} \mapsto \begin{pmatrix} \sqrt[4]{(x_2 - x_1)^2 + \beta} \\ \sqrt[4]{(x_3 - x_2)^2 + \beta} \\ \vdots \\ \sqrt[4]{(x_{k+1} - x_k)^2 + \beta} \end{pmatrix} \quad \text{for} \quad \beta > 0 \quad \text{``small,''}$$

$$\tag{4.75}$$

such that

$$G(\mathbf{x}) \approx \| G_\beta(\mathbf{x}) \|_2^2,$$

compared with (4.60) and (4.61) for inverse gravimetry. In the following examples, $\beta = 10^{-12}$ was used. The minimization problem (4.73) can now be approximated by a standard nonlinear least squares problem, namely

$$\min_{\mathbf{x} \in D} \frac{1}{2} \| F(\mathbf{x}) - \mathbf{y}^\delta \|_2^2 + \frac{\alpha}{2} \| G_\beta(\mathbf{x}) \|_2^2 \quad = \quad \min_{\mathbf{x} \in D} \frac{1}{2} \left\| \begin{pmatrix} F(\mathbf{x}) - \mathbf{y}^\delta \\ \sqrt{\alpha} G_\beta(\mathbf{x}) \end{pmatrix} \right\|_2^2. \tag{4.76}$$

For the practical solution of this minimization problem in all the following examples, Matlab's function `lsqnonlin` was used. The Jacobian $F'(\mathbf{x})$ can be computed analytically as explained in Example 4.5, the Jacobian $G'_\beta(\mathbf{x})$ is easy to compute by direct differentiation.

Specification of an Example Case for Waveform Inversion

We undertake to solve problem (4.72), or rather its regularized variant (4.76), in a situation similar to the one described in [BG09]. The subsurface is to be explored for depths ranging from 0 to $Z_0 = 900$ m. Mass density is assumed to be equal to a known constant $\rho_0 = 2.3 \cdot 10^3$ [kg/m³]. The sought-after exact elasticity coefficient is assumed to equal the piecewise linear interpolant of the samples

$$(z_i, \kappa(z_i)), \quad z_i = i \cdot \Delta z, \quad \Delta z = Z_0/k, \quad i = 0, \dots, k \in \mathbb{N}. \quad (4.77)$$

We will investigate the cases $k = 64$ and $k = 256$, leading to a "coarse model" $\kappa = \kappa_{64}$ and to a "fine model" $\kappa = \kappa_{256}$, respectively. The fine model coefficient is shown in Fig. 4.11. It corresponds to a resolution of $900/256 \approx 3.5$ [m]. The coarse model coefficient will be shown in figures below. It is defined by taking the first and then every fourth sample of κ_{256}. As a source signal (function g in (1.54)), we used the Ricker wavelet already introduced in Example 2.7. The central frequency was chosen to be $f_0 = 30$ [Hz] for the coarse model and $f_0 = 100$ [Hz] for the fine

Fig. 4.11   Exact elasticity coefficient $\kappa(z)$. Wave velocities range from 1626 to 5299 [m/s]

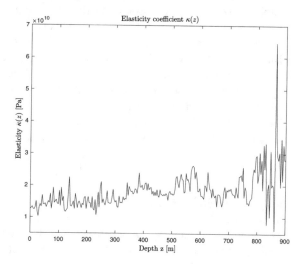

model.[5] An amplitude value of $a = 10^7$ [Pa] = 10 [MPa] for the Ricker wavelet was chosen in both cases. With average wave speeds above 2000 m/s, we can be sure that the total travel time of a wave from the surface down to a depth of 900 m and back to the surface will not be longer than 1 s, so we restricted the observation time of the seismogram to an interval $[0, T_0]$ with $T_0 = 1$ [s]. A **synthetic seismogram** is produced from κ_k, $k = 64$ or $k = 256$, as described in Sect. 2.7, evaluating function F from (2.148) for exact samples $\mathbf{x}$ of κ_k. In our example case:

- The vector $\mathbf{x}$ is composed of the $k + 1$ samples $\kappa(z_i)$, $i = 0, \ldots, k$, from (4.77). We will write $\mathbf{x} = \mathbf{x}_{64}$ in case $k = 64$ and $\mathbf{x} = \mathbf{x}_{256}$ in case $k = 256$.
- The discretization parameters (2.128) were chosen such that m is the smallest power of two meeting condition (2.144) and such that n is the smallest integer meeting the tightened CFL condition $\Delta t \leq 0.7 \cdot \Delta z / c_{\max}$, i.e.,

$$n = \lceil T_0 / (0.7 \cdot \Delta z / c_{\max}) \rceil, \quad \Delta t = T_0 / n. \qquad (4.78)$$

In our example case, for $k = 64$ this meant the choice $m = 512$ ($\Delta z \approx 1.76$ [m]) and $n = 4307$ ($\Delta t \approx 2.3 \cdot 10^{-4}$ [s]), whereas for $k = 256$ it meant $m = 2048$ and $n = 17{,}227$.

The vector $\mathbf{y} = F(\mathbf{x})$ is considered (a vector of samples of) an "exact seismogram."

As stated above, to reconstruct the unknown vector $\mathbf{x}$, we used Matlab's optimizer lsqnonlin applied to problem (4.76), starting from an initial guess $\mathbf{x}_0$ and using exact derivatives. If the initial guess is not very good, then even for *exact* data $\mathbf{y}^\delta = \mathbf{y}$, lsqnonlin *cannot find the global minimizer* of (4.76), but gets stuck in a local minimum. This can be explained by a well known phenomenon called **cycle skipping**. Consider some elasticity coefficient $\tilde{\kappa}$ (and its samples $\tilde{\mathbf{x}}$), derived from κ by a smooth ("large scale") perturbation. This changes the wave velocity from c to a smoothly perturbed velocity $\tilde{c}$ and thus leads to a time shift in the recorded seismogram. But a time shifted oscillatory signal may be in good partial agreement with the non-shifted signal. In this case, there is a local minimum of (2.150) (and, very likely, also of (4.76)) at $\tilde{\mathbf{x}}$. Consequently, a local minimization routine as lsqnonlin cannot take a step from $\tilde{\mathbf{x}}$ toward $\mathbf{x}$, since this would mean to leave a local minimum. To circumvent this problem, we used the following remedy. We cut the time interval $[0, T_0]$ into time slices of equal length

$$dT \approx 1/(2 f_0), \qquad (4.79)$$

with f_0 the central frequency of the Ricker pulse used as a source signal. Within one time slice, the Ricker wave cannot advance more than half its (dominant) cycle length—no cycle skipping can occur within such a time interval. Starting from an initial guess $\mathbf{x}_0$, we will successively use lsqnonlin to find vectors $\mathbf{x}_i \in \mathbb{R}^{k+1}$,

---

[5]An increased central frequency in the received signal is required, when a higher resolution is demanded for the reconstruction.

$i = 1, 2, \ldots$, such that $F(\mathbf{x}_i)$ matches best the exact seismogram *up to time* $i \cdot dT$. More precisely, we let

$$\mathbf{x}_i = \operatorname{argmin}_{\mathbf{x} \in D} \frac{1}{2} \| P_i (F(\mathbf{x}) - \mathbf{y}^\delta) \|_2^2 + \frac{\alpha}{2} \| G_\beta(\mathbf{x}) \|_2^2, \qquad (4.80)$$

where $P_i : \mathbb{R}^n \to \mathbb{R}^{N_i}$ means the projection of vectors $y \in \mathbb{R}^n$ to their first

$$N_i = \lceil i \cdot dT / \Delta t \rceil \quad (\Delta t \text{ from } (4.78))$$

components. When $\mathbf{x}_i$ is found, it will serve as an initial guess to find $\mathbf{x}_{i+1}$. Advancing from one time slice to the next will subsequently be called **iterated reconstruction**.

Example 4.14 (Coarse Model Iterated Reconstruction) We investigated the coarse model corresponding to $k = 64$. The central frequency of the Ricker wavelet was chosen to be $f_0 = 30$ [Hz] and the observation interval was cut into slices of length $dT = 0.01$ [s], according to (4.79). A solution of (4.76) was sought within the box $D = [\kappa_-, \kappa_+]^{k+1}$, defined by setting $\kappa_- := 0.5 \cdot 10^{10}$ [Pa] and $\kappa_+ := 6.5 \cdot 10^{10}$ [Pa]. For an initial guess $\mathbf{x}_0$, we set all components equal to $x_i = 2 \cdot 10^{10}$ [Pa].

We first tested iterated reconstruction for exact data $\mathbf{y} = F(\mathbf{x}_{64})$. We set $\alpha = 0$ in (4.76) in this case, i.e., we did not regularize at all and in fact solved (2.150). Starting from $\mathbf{x}_0$, a reconstruction was found that could not visibly be distinguished from κ_{64}. This showed that iterated reconstruction is effective to find a global minimum point.

We next experimented with a noisy seismogram $\mathbf{y}^\delta$ obtained from $\mathbf{y}$ via additive perturbation:

$$y_i^\delta = y_i + \eta_i, \quad i = 1, \ldots, n, \qquad (4.81)$$

where η_i were realizations of independent normal random variables with expectation 0 and standard deviation $\sigma = 2 \cdot 10^{-4}$. This led to a relative data error

$$\| \mathbf{y}^\delta - \mathbf{y} \|_2 / \| \mathbf{y} \|_2 \approx 0.137.$$

Exact and noisy seismogram are shown in Fig. 4.12 as black and red line, respectively. For the choice $\alpha = 2 \cdot 10^{-7}$ we found a minimizer $\mathbf{x}^\delta$ of (4.76), which fulfilled

$$\| F(\mathbf{x}^\delta) - \mathbf{y}^\delta \|_2 \approx 1.30 \cdot 10^{-2} \approx 1.31 \cdot 10^{-2} \approx \| \mathbf{y} - \mathbf{y}^\delta \|_2,$$

i.e., which was in accordance with the discrepancy principle—this value of α was found by trial and error. We derived a reconstruction κ_{64}^δ of κ as the piecewise linear spline interpolant of the components of $\mathbf{x}^\delta$. This reconstruction is shown in Fig. 4.13 (red line) alongside with the exact elasticity coefficient $\kappa = \kappa_{64}$ (black line). Increasing the noise level quickly led to a deterioration of the reconstruction.

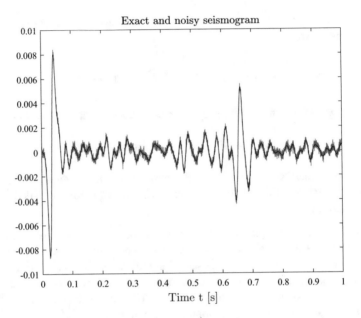

Fig. 4.12 Exact and noisy seismogram, $\sigma = 2 \cdot 10^{-4}$, $k = 64$

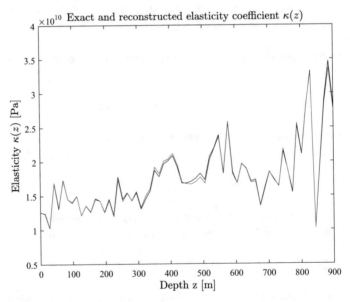

Fig. 4.13 Exact and reconstructed elasticity coefficient, $\sigma = 2 \cdot 10^{-4}$, $\alpha = 2 \cdot 10^{-7}$

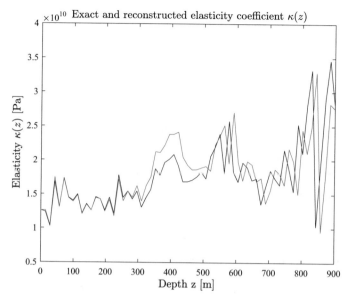

Fig. 4.14 Exact and reconstructed elasticity coefficient, $\sigma = 3 \cdot 10^{-4}$, $\alpha = 4 \cdot 10^{-7}$

For example, defining $\mathbf{y}^\delta$ as above, but with increased parameter $\sigma = 3 \cdot 10^{-4}$ (instead of $\sigma = 2 \cdot 10^{-4}$), we got $\|\mathbf{y}^\delta - \mathbf{y}\|_2 / \|\mathbf{y}\|_2 \approx 0.206$. We chose $\alpha = 4 \cdot 10^{-7}$, since this led to

$$\|F(\mathbf{x}^\delta) - \mathbf{y}^\delta\|_2 \approx 1.97 \cdot 10^{-2} \approx 1.96 \cdot 10^{-2} \approx \|\mathbf{y} - \mathbf{y}^\delta\|_2.$$

The corresponding reconstruction, as shown in Fig. 4.14, becomes poor for depths below 300 m.

At the end of this example we want to stress the fact, that all above results are "random" insofar as they depend on the random perturbations η_i in (4.81). We encountered examples, where the reconstruction went well even for $\sigma = 3 \cdot 10^{-4}$. $\Diamond$

With data disturbed by Gaussian noise, the idea suggests itself to preprocess data for noise reduction, *before* any reconstruction is attempted. There is an elaborate theory on noise removal in signals. We do not delve into this and just report on a single experiment which already shows that data preprocessing can be quite effective.

Example 4.15 (Coarse Model Iterated Reconstruction with Data Preprocessing) We used the setup of the previous Example 4.14. Noisy data $\mathbf{y}^\delta$ were simulated by adding to the components y_i of $\mathbf{y}$ realizations η_i of independent Gaussian random variables with zero mean and standard deviation $\sigma = 3 \cdot 10^{-4}$. For the present example, we used *exactly* the same perturbations η_i as in the previous example, i.e., we reused exactly the same data for which a reconstruction went wrong. Now, we define a **binomial filter**:

- let $\ell \in \mathbb{N}$ be an even integer,
- let

$$b_j := 2^{-\ell} \cdot \binom{\ell}{j}, \quad j = 0, \dots, \ell,$$

- and let $a_k := b_{k+\ell/2}$ for $k = -\ell/2, \dots, \ell/2$.

Then define

$$y_i^P := \sum_{k=-\ell/2}^{\ell/2} a_k \cdot y_{i+k}^{\delta} =: y_i + \eta_i^P, \quad i = 1, \dots, n, \tag{4.82}$$

where $y_{i+k}^{\delta} := 0$ for $i + k \le 0$ and $i + k \ge n + 1$. The first equality defines the preprocessed data, the second equality defines $\eta_i^P := y_i^P - y_i$ as noise in the preprocessed data. From $y_{i+k}^{\delta} = y_{i+k} + \eta_{i+k}$ we get

$$\sum_{k=-\ell/2}^{\ell/2} a_k \cdot y_{i+k}^{\delta} = \sum_{k=-\ell/2}^{\ell/2} a_k \cdot y_{i+k} + \sum_{k=-\ell/2}^{\ell/2} a_k \cdot \eta_{i+k},$$

where the first sum on the right is an approximation of y_i and the second sum is an approximation of η_i^P. From known properties of the Gaussian distribution we conclude that η_i^P can approximately be considered a realization of a Gaussian random variable with zero mean and variance

$$\sigma^2 \cdot A_\ell^2 \quad \text{with} \quad A_\ell^2 := \left(\sum_{k=-\ell/2}^{\ell/2} a_k^2 \right).$$

Therefore

$$\|\mathbf{y}^P - \mathbf{y}\|_2 \approx \sqrt{n \cdot A_\ell^2 \cdot \sigma^2} \approx A_\ell \cdot \|\mathbf{y}^{\delta} - \mathbf{y}\|_2. \tag{4.83}$$

We chose $\ell = 32$ and thus $A_\ell \approx 0.3152$. The original noisy and the preprocessed seismogram are shown in Fig. 4.15 (red lines), as compared to exact data (black lines). We then tried regularized reconstruction (4.76) based on preprocessed data $\mathbf{y}^P$ instead of $\mathbf{y}^{\delta}$. We chose a regularization parameter $\alpha = 1 \cdot 10^{-6}$, which led to

$$\|F(\mathbf{x}^{\delta}) - \mathbf{y}^P\|_2 \approx 5.77 \cdot 10^{-2} \approx 6.26 \cdot 10^{-2} \approx 0.3152 \cdot \|\mathbf{y}^{\delta} - \mathbf{y}\|_2.$$

For this parameter, we got the reconstruction shown in Fig. 4.16, which is of much better quality than the one shown in Fig. 4.14. In the present example, we could not find a parameter α in better accordance with the discrepancy principle. For example,

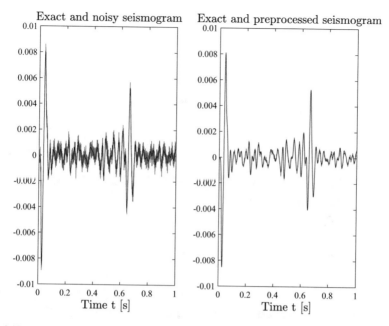

Fig. 4.15 Noisy and preprocessed seismograms, $\sigma = 3 \cdot 10^{-4}$, $k = 64$

for $\alpha = 1.1 \cdot 10^{-6}$ we got $\| F(\mathbf{x}^{\delta}) - \mathbf{y}^{p} \|_2 \approx 5.78 \cdot 10^{-2}$ and for $\alpha = 1.15 \cdot 10^{-6}$, we got $\| F(\mathbf{x}^{\delta}) - \mathbf{y}^{p} \|_2 \approx 6.37 \cdot 10^{-2}$ ◊

Example 4.16 (Fine Model Iterated Reconstruction) We investigated the fine model corresponding to $k = 256$. The central frequency of the Ricker wavelet was chosen to be $f_0 = 100$ [Hz] and the observation interval was cut into slices of length $dT = 0.005$ [s]. A solution of (4.76) was sought within the box $D = [\kappa_-, \kappa_+]^{k+1}$, defined by setting $\kappa_- := 0.5 \cdot 10^{10}$ [Pa] and $\kappa_+ := 6.5 \cdot 10^{10}$ [Pa]. For an initial guess $\mathbf{x}_0$, we set all components equal to $x_i = 2 \cdot 10^{10}$ [Pa].

Again, we first tested iterated reconstruction for exact data $\mathbf{y} = F(\mathbf{x}_{256})$. Setting $\alpha = 0$ in (4.76) and starting from $\mathbf{x}_0$, we again found a reconstruction that could not visibly be distinguished from the exact function κ_{256}. Thus, iterated reconstruction again is effective to find a global minimum point.

We next experimented with a noisy seismogram $\mathbf{y}^{\delta}$ obtained from $\mathbf{y}$ as in Example 4.14, where η_i were realizations of independent normal random variables with expectation 0 and standard deviation $\sigma = 1 \cdot 10^{-5}$. This led to

$$\| \mathbf{y}^{\delta} - \mathbf{y} \|_2 / \| \mathbf{y} \|_2 \approx 0.031.$$

Choosing a regularization parameter $\alpha = 5.0 \cdot 10^{-10}$, we got

$$\| F(\mathbf{x}^{\delta}) - \mathbf{y}^{\delta} \|_2 \approx 1.33 \cdot 10^{-3} \approx \| \mathbf{y}^{\delta} - \mathbf{y} \|_2$$

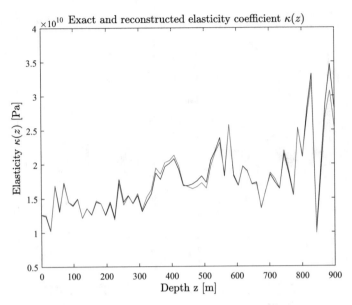

Fig. 4.16 Exact and reconstructed elasticity coefficient, $\sigma = 3 \cdot 10^{-4}, \alpha = 1 \cdot 10^{-6}$, preprocessed

in accordance with the discrepancy principle. In Fig. 4.17 we show the reconstruction κ_{256}^{δ} we achieved (red line) and which is hardly distinguishable from the exact function $\kappa = \kappa_{256}$ (black line).

Increasing the noise level by setting $\sigma = 2 \cdot 10^{-5}$ caused an increased relative data perturbation:

$$\|\mathbf{y}^{\delta} - \mathbf{y}\|_2 / \|\mathbf{y}\|_2 \approx 0.062.$$

For a reconstruction we preprocessed data, using the binomial filter as in Example 4.15, this time with $\ell = 10$, which led to $A_\ell = 0.4198$. The reconstruction κ_{256}^{δ} is shown in Fig. 4.18 (red line), as compared to the exact elasticity coefficient κ_{256} (black line). The regularization parameter $\alpha = 5 \cdot 10^{-9}$ was chosen such that

$$\|F(\mathbf{x}^{\delta}) - \mathbf{y}^p\|_2 \approx 1.11 \cdot 10^{-3} \approx 1.12 \cdot 10^{-3} \approx 0.4198 \cdot \|\mathbf{y}^{\delta} - \mathbf{y}\|_2,$$

i.e., in good accordance with the discrepancy principle.   ◊

Further Reading

Our main concern in this section was the practical solution of (2.150) or rather (4.72) and its regularized variant (4.73)—which is complicated because of the presence

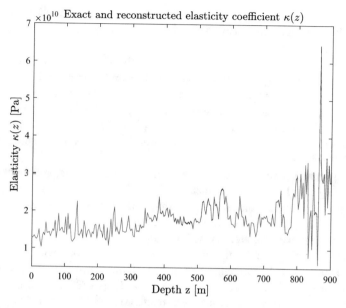

Fig. 4.17 Exact and reconstructed elasticity coefficient, $\sigma = 1 \cdot 10^{-5}, \alpha = 5 \cdot 10^{-10}$

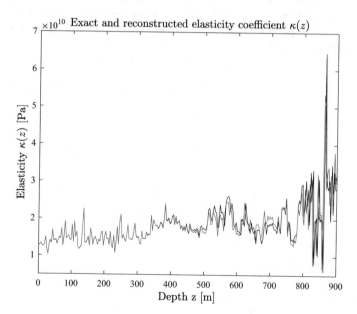

Fig. 4.18 Exact and reconstructed elasticity coefficient, $\sigma = 2 \cdot 10^{-5}, \alpha = 5 \cdot 10^{-9}$, preprocessed

of many local minima. This difficulty was observed by many authors ([JBC$^+$89] is an early reference) and many attempts have been made to overcome it. The one we proposed above, namely matching observed and simulated seismograms for increasing time slots, is also known under the name of time windowing or layer stripping [WR09]. This technique can be complemented by considering the wave equation in the frequency domain, matching lowest frequency components of observed and simulated data first (see [Pra99]) and then increasingly including higher frequencies in the match (see, e.g., [BSZC95], [SP04]). The practical difficulty with this approach is the lack of sufficient low frequency components in recorded seismograms.

In a different attempt, one can try to derive a "better" objective function (i.e., one exposing less local minima) than (4.72), which was based on measuring the misfit between simulated and observed seismogram in the L_2-norm. We recommend [MBM$^+$16] for a discussion of various approaches undertaken in these directions. The authors give many pointers to the literature.

Besides changing the misfit function, one can also work with reparameterizations. The so-called Difference Semblance Optimization (DSO, [SC91]) and the Migration-Based Travel-time approach (MBTT, [CCG01]) both consider the acoustic wave equation in the frequency domain. For MBTT, the velocity parameter is decomposed as the sum of a smooth background velocity and a high wavenumber reflectivity function. The latter is then parameterized via the smooth background velocity. One achieves an optimization problem for the background velocity alone, which behaves better (has less local minima) than the original optimization problem. MBTT is taken up in [BCF19] and [FCBC20].

Appendix A
Results from Linear Algebra

We assume the reader is familiar with the mathematical concepts of "linear dependence," "linear independence," "dimension," "subspace," and "basis" in the real and complex Euclidean spaces $\mathbb{R}^n$ and $\mathbb{C}^n$.

$$x \in \mathbb{R}^n \text{ or } \mathbb{C}^n \quad \Longleftrightarrow \quad x = \begin{pmatrix} x_1 \\ \vdots \\ x_n \end{pmatrix}, \quad \text{all } x_i \in \mathbb{R} \text{ or } \mathbb{C}, \text{ respectively.}$$

We will also write $x = (x_1, \ldots, x_n)^\top$. The superindex $\top$ means **transposed** and makes a row vector out of a column vector and vice versa. For the subspace of $\mathbb{R}^n$ (or $\mathbb{C}^n$) spanned by k vectors $b_1, \ldots, b_k \in \mathbb{R}^n$ (or $\mathbb{C}^n$) we use the notation

$$\langle b_1, \ldots, b_k \rangle := \text{span}\{b_1, \ldots, b_k\} := \{\lambda_1 b_1 + \ldots + \lambda_k b_k; \ \lambda_1, \ldots, \lambda_k \in \mathbb{R}(\text{or } \mathbb{C})\}.$$

A **matrix** is defined by its components

$$A \in \mathbb{R}^{m,n} \text{ or } \mathbb{C}^{m,n} \quad \Longleftrightarrow \quad A = \begin{pmatrix} a_{11} & a_{12} & \cdots & a_{1n} \\ a_{21} & a_{22} & \cdots & a_{2n} \\ \vdots & \vdots & & \vdots \\ a_{m1} & a_{m2} & \cdots & a_{mn} \end{pmatrix}, \quad \text{all } a_{ij} \in \mathbb{R} \text{ or } \mathbb{C}, \text{ resp.,}$$

or equally by its columns

$$A = \begin{pmatrix} \, \\ a_1 \,\Big|\, a_2 \, \Big| \cdots \Big|\, a_n \\ \, \end{pmatrix}$$

$$A \in \mathbb{R}^{m,n} \text{ or } \mathbb{C}^{m,n} \iff A = \begin{pmatrix} a_1 \,\Big|\, a_2 \,\Big| \cdots \Big|\, a_n \end{pmatrix}, \quad \text{all } a_j \in \mathbb{R}^m \text{ or } \mathbb{C}^m, \text{resp.}$$

The rules for multiplying matrices should be known as well as the fact that a matrix $A \in \mathbb{R}^{m,n}$ defines a linear mapping $f : \mathbb{R}^n \to \mathbb{R}^m$, $x \mapsto Ax$. Conversely, every linear mapping $f : \mathbb{R}^n \to \mathbb{R}^m$ can be represented in the form $f(x) = Ax$. The same equivalence holds for complex valued matrices $A \in \mathbb{C}^{m,n}$ and linear mappings $A : \mathbb{C}^n \to \mathbb{C}^m$. It is assumed the reader knows what is meant by the inverse of a matrix, by its determinant, and by its rank. The columns of a given matrix $A \in \mathbb{R}^{m,n}$ ($A \in \mathbb{C}^{m,n}$) span the linear space

$$\mathscr{R}_A := \{Ax = \sum_{j=1}^n x_j a_j;\ x_j \in \mathbb{R}\ (\text{or } \mathbb{C})\} \subseteq \mathbb{R}^m\ (\text{or } \mathbb{C}^m)$$

of dimension rank(A). The kernel or nullspace of A is defined as

$$\mathscr{N}_A := \{x \in \mathbb{R}^n\ (\text{or } \mathbb{C}^n);\ Ax = 0\} \subseteq \mathbb{R}^n\ (\text{or } \mathbb{C}^n)$$

and has dimension $n - \text{rank}(A)$.

The columns of the unity matrix $I_n \in \mathbb{C}^{n,n}$ are designated by $e_1, \ldots, e_n$ and are called **canonical unity vectors** (of $\mathbb{R}^n$ and $\mathbb{C}^n$). A matrix $A \in \mathbb{C}^{m,n}$ with components a_{ij} has a **transposed** $A^\top \in \mathbb{C}^{n,m}$ with components

$$(A^\top)_{ij} := a_{ji}, \quad i = 1, \ldots, n \text{ and } j = 1, \ldots, m,$$

and an **adjoint** or **Hermitian conjugate** $A^* \in \mathbb{C}^{n,m}$ with components

$$(A^*)_{ij} := \overline{a_{ji}}, \quad i = 1, \ldots, n \text{ and } j = 1, \ldots, m,$$

where $\bar{z}$ is the conjugate complex number of $z \in \mathbb{C}$. For a real number $z \in \mathbb{R}$ we have $\bar{z} = z$ and therefore $A^* = A^\top$ in case of a real valued matrix A. We have $(AB)^\top = B^\top A^\top$, $(AB)^* = B^* A^*$, and $(A^{-1})^* = (A^*)^{-1} =: A^{-*}$, whenever the inverse A^{-1} of A exists. In case $A = A^\top$ the matrix A is called **symmetric** and in case $A = A^*$ it is called **Hermitian** or **self-adjoint**. For $x, y \in \mathbb{C}^n$, we define the **Euclidean scalar product** (Euclidean inner product) by

$$\langle x|y \rangle := \overline{x^* y} = \sum_{i=1}^n x_i \overline{y_i}.$$

Here, the row vector $x^* = (\overline{x_1}, \ldots, \overline{x_n})$ is the adjoint of the column vector x and x^*y is a matrix product. The scalar product for real vectors is defined in the same way, but the overline has no effect in this case: $\langle x|y \rangle = \sum\limits_{i=1}^{n} x_i y_i$ for $x, y \in \mathbb{R}^n$. Vectors $x, y \in \mathbb{C}^n$ are called **orthogonal**, if $x^*y = 0$. In this case, we write $x \perp y$. Vectors $b_1, \ldots, b_k \in \mathbb{C}^n$ are called **orthonormal**, if $b_i^* b_j = 0$ for $i \neq j$ and $b_i^* b_i = 1$ for all i. If additionally $k = n$, then $\{b_1, \ldots, b_n\}$ is called an **orthonormal basis (ONB)** of $\mathbb{C}^n$. A matrix $V \in \mathbb{C}^{n,n}$ is called **unitary** (in case $A \in \mathbb{R}^{n,n}$ also: **orthogonal**), if its columns are an orthonormal basis of $\mathbb{C}^n$ (or $\mathbb{R}^n$). This is equivalent to the identities

$$V^*V = I_n \quad \Longleftrightarrow \quad V^{-1} = V^*.$$

A matrix $A \in \mathbb{R}^{n,n}$ or $A \in \mathbb{C}^{n,n}$ is said to have an **eigenvalue** $\lambda \in \mathbb{C}$ (possibly complex valued even if the matrix is real valued!) and corresponding **eigenvector** $v \in \mathbb{C}^n$, if

$$Av = \lambda v \quad \text{and} \quad v \neq 0.$$

If A is Hermitian, all eigenvalues are real valued and there exists an orthonormal basis $\{v_1, \ldots, v_n\} \subset \mathbb{C}^n$ of eigenvectors. Thus

$$Av_i = \lambda_i v_i, \ i = 1, \ldots, n \quad \Longleftrightarrow \quad AV = V\Lambda \quad \Longleftrightarrow \quad V^*AV = \Lambda,$$

where $V = (v_1| \cdots |v_n)$ (columns are eigenvectors) and $\Lambda = \mathrm{diag}(\lambda_1, \ldots, \lambda_n)$. In case A is real valued, V is real valued, too.

A matrix $A \in \mathbb{C}^{n,n}$ is called **positive definite**, if it is Hermitian *and* $x^*Ax > 0$ for all $x \in \mathbb{C}^n \setminus \{0\}$. It is called **positive semidefinite**, if it is Hermitian *and* $x^*Ax \geq 0$ for all $x \in \mathbb{C}^n$. A matrix $A \in \mathbb{C}^{n,n}$ is positive definite, if and only if it is Hermitian and all its eigenvalues are positive and it is positive semidefinite, if and only if it is Hermitian and has no negative eigenvalue. A is positive definite, if and only if there exists a nonsingular upper triangular matrix $R \in \mathbb{C}^{n,n}$ such that

$$A = R^*R.$$

This is called the **Cholesky factorization** of A. The matrix R can be chosen real valued if A is real valued.

A.1 The Singular Value Decomposition (SVD)

Let $m \geq n$ and $A \in \mathbb{C}^{m,n}$ with $\mathrm{rank}(A) = r$. Then $A^*A \in \mathbb{C}^{n,n}$ is positive semidefinite. Let $\sigma_1^2 \geq \ldots \geq \sigma_r^2 > 0$ and $\sigma_{r+1}^2 = \ldots = \sigma_n^2 = 0$ be its eigenvalues and let $v_1, \ldots, v_n$ be the corresponding orthonormal eigenvectors:

$$A^*Av_k = \sigma_k^2 v_k, \quad k = 1, \ldots, n.$$

Then $u_k := Av_k/\sigma_k \in \mathbb{C}^m, k = 1, \ldots, r$, are eigenvectors of AA^*, since $AA^*u_k = AA^*Av_k/\sigma_k = A\sigma_k v_k = \sigma_k^2 u_k$. The vectors u_k are also orthonormal:

$$u_i^* u_k = v_i^* A^* A v_k/(\sigma_i \sigma_k) = v_i^* v_k \sigma_k/\sigma_i = \delta_{i,k}.$$

Here, we make use of the Kronecker symbol defined by $\delta_{i,k} := 0$ for $i \neq k$ and $\delta_{i,i} := 1$. The set $\{u_1, \ldots, u_r\}$ is completed with $m - r$ orthonormal vectors $u_{r+1}, \ldots, u_m \in \mathbb{C}^m$ which span the $(m - r)$-dimensional nullspace $\mathcal{N}_{A^*}$:

$$A^*u_k = 0, \quad k = r + 1, \ldots, m,$$

and which are the remaining eigenvectors of AA^*. For $i \leq r < k$ we get $u_i^* u_k = v_i^* A^* u_k/\sigma_i = v_i^* 0/\sigma_i = 0$, so that $U := (u_1| \cdots |u_m) \in \mathbb{C}^{m,m}$ is a unitary matrix as is $V := (v_1| \cdots |v_n) \in \mathbb{C}^{n,n}$. From the definitions of u_k and v_k we get $Av_k = \sigma_k u_k$ for $k = 1, \ldots, r$ and $Av_k = 0$ for $k = r + 1, \ldots, n$. Together

$$AV = U\Sigma \quad \Longleftrightarrow \quad A = U\Sigma V^* \quad \text{with} \quad \Sigma_{i,j} = \sigma_i \delta_{i,j}. \tag{A.1}$$

One can drop the last $m - n$ columns of matrix U and the last $m - n$ rows of Σ to get

$$A = \hat{U}\hat{\Sigma}V^*, \quad \hat{U} := (u_1| \cdots |u_n) \in \mathbb{C}^{m,n}, \quad \hat{\Sigma} = \text{diag}(\sigma_1, \ldots, \sigma_n) \in \mathbb{R}^{n,n} \tag{A.2}$$

instead of (A.1). If $m < n$, a factorization (A.1) of A^* can be derived as above. Afterwards one takes the Hermitian conjugate of the result. This shows

Theorem A.1 (Singular Value Decomposition (SVD)) *Let $A \in \mathbb{C}^{m,n}$ have rank r. Then there exist unitary matrices $U \in \mathbb{C}^{m,m}$ and $V \in \mathbb{C}^{n,n}$ and a matrix $\Sigma \in \mathbb{R}^{m,n}$ with components $\Sigma_{i,j} = \sigma_i \delta_{i,j}$ and*

$$\sigma_1 \geq \ldots \geq \sigma_r > 0, \quad \sigma_{r+1} = \ldots = \sigma_{\min\{m,n\}} = 0$$

such that

$$A = U\Sigma V^*.$$

This factorization is called **singular value decomposition (SVD)** *and the numbers $\sigma_1 \geq \ldots \geq \sigma_{\min\{m,n\}} \geq 0$ are called* **singular values** *of A. For $m \geq n$, a factorization (A.2) exists, which is called* **reduced SVD**.

Using appropriate coordinate transforms $y = U\eta$ in $\mathbb{C}^m$ and $x = V\xi$ in $\mathbb{C}^n$ any linear map $\mathbb{C}^n \to \mathbb{C}^m, x \mapsto y = Ax$, thus splits into r one-dimensional maps $\eta_i = \sigma_i \xi_i$ and $m - r$ trivial maps $\eta_i = 0$. For $A \in \mathbb{R}^{m,n}$, U and V can be chosen real orthonormal matrices. For a numerical computation of the SVD, one must never explicitly form any of the matrices A^*A or AA^*, since this would lead to a numerically unstable algorithm. Instead, more involved algorithms have to be used like the method of Golub and Reinsch, see, e.g., [Dem97, Sect. 5.4].

A pair of matrices $A, B \in \mathbb{C}^{n,n}$ has **generalized eigenvalue** $\lambda \in \mathbb{C}$ and corresponding **generalized eigenvector** $v \in \mathbb{C}^n$, if

$$Av = \lambda Bv \quad \text{and} \quad v \neq 0.$$

Now let A be positive semidefinite and let B be positive definite. Using the factorization $B = R^*R$ and the transformation $Rv = w$ we can equivalently reformulate the generalized eigenvalue problem as an ordinary one:

$$R^{-*}AR^{-1}w = \lambda w, \quad w \neq 0,$$

where $R^{-*}AR^{-1}$ is a positive semidefinite matrix. Thus, there is an orthonormal basis $\{w_1, \ldots, w_n\}$ of eigenvectors corresponding to eigenvalues $\lambda_1, \ldots, \lambda_n \geq 0$ of $R^{-*}AR^{-1}$. Defining the orthogonal matrix $W := (w_1 | \cdots | w_n)$ and the nonsingular matrix $V := R^{-1}W$ we get

$$V^*BV = W^*R^{-*}R^*RR^{-1}W = W^*W = I_n$$

and we also get

$$V^*AV = W^*(R^{-*}AR^{-1})W = W^*W \text{diag}(\lambda_1, \ldots, \lambda_n) = \text{diag}(\lambda_1, \ldots, \lambda_n).$$

To summarize: If $A \in \mathbb{C}^{n,n}$ is positive semidefinite and $B \in \mathbb{C}^{n,n}$ is positive definite, then there is a nonsingular matrix $V \in \mathbb{C}^{n,n}$ such that

$$V^*AV = \text{diag}(\lambda_1, \ldots, \lambda_n), \ \lambda_1, \ldots, \lambda_n \geq 0, \quad \text{and} \quad V^*BV = I_n. \tag{A.3}$$

We define the **Euclidean norm** on $\mathbb{R}^n$ (or $\mathbb{C}^n$) by

$$\|x\| = \|x\|_2 := \sqrt{|x_1|^2 + \ldots + |x_n|^2} = \sqrt{x^*x}.$$

The **inequality of Cauchy–Schwarz** reads

$$|x^*y| \leq \|x\|_2 \|y\|_2.$$

By **Pythagoras' Theorem** we have

$$\|b_1 + \ldots + b_k\|_2^2 = \|b_1\|_2^2 + \ldots + \|b_k\|_2^2,$$

if $b_1, \ldots, b_k \in \mathbb{C}^n$ are mutually orthogonal. If $V \in \mathbb{C}^{n,n}$ is a unitary matrix, then

$$\|Vx\|_2^2 = x^*V^*Vx = x^*x = \|x\|_2^2 \quad \text{for all} \quad x \in \mathbb{C}^n.$$

We define the **spectral norm** of a matrix $A \in \mathbb{C}^{m,n}$ by

$$\|A\|_2 := \max \left\{ \frac{\|Ax\|_2}{\|x\|_2}; \ x \in \mathbb{C}^n \setminus \{0\} \right\} = \max \left\{ \|Ax\|_2; \ \|x\|_2 = 1 \right\}$$

(analogous definition in the real valued case). The spectral norm has the following properties (which must hold for every norm):

$$\|A\|_2 = 0 \iff A = 0, \quad \|\lambda A\|_2 = |\lambda| \|A\|_2, \quad \text{and } \|A + B\|_2 \le \|A\|_2 + \|B\|_2$$

for $A, B \in \mathbb{C}^{m,n}$ and $\lambda \in \mathbb{C}$. Additionally, it is **consistent** with the Euclidean norm—from which it is induced—and it also is consistent with itself. This means that

$$\|Ax\|_2 \le \|A\|_2 \|x\|_2 \quad \text{and} \quad \|AB\|_2 \le \|A\|_2 \cdot \|B\|_2$$

for all $x \in \mathbb{C}^n$, $A \in \mathbb{C}^{m,n}$, and $B \in \mathbb{C}^{n,k}$. If $V \in \mathbb{C}^{n,n}$ is unitary, then $\|V\|_2 = 1$. If $U \in \mathbb{C}^{m,m}$ also is unitary, then we have

$$\|A\|_2 = \|UA\|_2 = \|AV\|_2 = \|UAV\|_2$$

for every $A \in \mathbb{C}^{m,n}$. Norms and singular values are closely related. The following theorem is proven, e.g., in Lecture 5 of [TB97].

Theorem A.2 *Let $A \in \mathbb{C}^{m,n}$ have singular values $\sigma_1 \ge \ldots \ge \sigma_{\min\{m,n\}} \ge 0$. Then*

$$\|A\|_2 = \sigma_1.$$

In case $m = n$, A is invertible if and only if $\sigma_n > 0$. In this case

$$\|A^{-1}\|_2 = \frac{1}{\sigma_n}.$$

Let $\mathbb{M}_k$ be the set of matrices from $\mathbb{C}^{m,n}$ having rank lower than k ($\mathbb{M}_1$ contains only the nullmatrix). Then, for $k = 1, \ldots, \min\{m, n\}$

$$\min \left\{ \|A - X\|_2; \ X \in \mathbb{M}_k \right\} = \sigma_k. \tag{A.4}$$

From Eq. (A.4) we see that $\sigma_{\min\{m,n\}} \le \varepsilon$ is a warning that within the ε-vicinity of A there are matrices with deficient rank. Being a discontinuous function of the matrix components, the rank of a matrix is almost impossible to compute numerically, at least when less than $\min\{m, n\}$. Singular values, in contrast, are stable (see Theorem A.3 below) and can be computed reliably. Thus, computing the smallest singular values of a matrix answers best the question for its rank.

Theorem A.3 (Sensitivity of Singular Values) *Let $A, \delta A \in \mathbb{C}^{m,n}$. Let $\sigma_1 \geq \ldots \geq \sigma_{\min\{m,n\}} \geq 0$ be the singular values of A and let $\tilde{\sigma}_1 \geq \ldots \geq \tilde{\sigma}_{\min\{m,n\}} \geq 0$ be the singular values of $A + \delta A$. Then*

$$|\sigma_i - \tilde{\sigma}_i| \leq \|\delta A\|_2, \quad i = 1, \ldots, \min\{m, n\}.$$

The upper bound is sharp.

Proof See, e.g., [Dem97, p. 198]. $\qquad\qquad\square$

Appendix B
Function Spaces

In all inverse problems treated in this book one is asking for an unknown *function*. For an abstract formulation of inverse problems as equations in vector spaces—as in Sect. 1.2—we interpret functions as vectors. This does not help much with the practical solution of inverse problems. However, it is an adequate language to formulate inverse problems and to characterize many of the difficulties one encounters with their solution. It can also help to develop intuition, like in Example B.14 below, where approximation by Fourier sums is described as a projection into a function space. Below we will use the notation $\mathbb{K}$ to mean either $\mathbb{R}$ or $\mathbb{C}$, when it is not necessary nor desirable to be more specific.

B.1 Linear Spaces

A **vector space** or **linear space** X over $\mathbb{K}$ is a nonempty set of elements, which are called **vectors**, for which two kinds of operations are defined, a **vector addition**, and a **scalar multiplication**. Vector addition takes two vectors, $\mathbf{x} \in X$ and $\mathbf{y} \in X$, say, and defines a new vector $\mathbf{x} + \mathbf{y} \in X$. Scalar multiplication takes a vector $\mathbf{x} \in X$ and an element $\alpha \in \mathbb{K}$ and defines a new vector $\alpha \cdot \mathbf{x} \in X$. The numbers $\alpha \in \mathbb{K}$ are called **scalars**. Vector addition has to obey the commutative law $\mathbf{x} + \mathbf{y} = \mathbf{y} + \mathbf{x}$ and the associative law $\mathbf{x} + (\mathbf{y} + \mathbf{z}) = (\mathbf{x} + \mathbf{y}) + \mathbf{z}$ for all $\mathbf{x}, \mathbf{y}, \mathbf{z} \in X$. The set X must include a zero vector $\mathbf{0}$, the neutral element of addition, for which $\mathbf{x} + \mathbf{0} = \mathbf{x}$ holds for all $\mathbf{x} \in X$. Also, every vector must have an additive inverse, denoted $-\mathbf{x}$, such that $\mathbf{x} - \mathbf{x} := \mathbf{x} + (-\mathbf{x}) = \mathbf{0}$. Scalar multiplication must obey the associative law $\alpha \cdot (\beta \cdot \mathbf{x}) = (\alpha\beta) \cdot \mathbf{x}$. It must also obey the distributive laws $\alpha \cdot (\mathbf{x} + \mathbf{y}) = \alpha \cdot \mathbf{x} + \alpha \cdot \mathbf{y}$ and $(\alpha + \beta) \cdot \mathbf{x} = \alpha \cdot \mathbf{x} + \beta \cdot \mathbf{x}$. Finally, it must follow the rules $1 \cdot \mathbf{x} = \mathbf{x}$ and $0 \cdot \mathbf{x} = \mathbf{0}$. If $\mathbb{R} = \mathbb{K}$, then X is called a **real vector space**. It is called a **complex vector space**, if $\mathbb{K} = \mathbb{C}$. The most important example of a linear space is the **Euclidean space** $\mathbb{R}^s$, $s \in \mathbb{N}$. The following example introduces a more abstract space.

© The Author(s), under exclusive license to Springer Nature Switzerland AG 2020
M. Richter, *Inverse Problems*, Lecture Notes in Geosystems
Mathematics and Computing, https://doi.org/10.1007/978-3-030-59317-9

Example B.1 Let $\emptyset \neq \Omega \subset \mathbb{R}^s$ and let

$$\mathscr{F}(\Omega, \mathbb{K}) := \{f : \Omega \to \mathbb{K}\}$$

be the set of all $\mathbb{K}$-valued functions defined on Ω. We can define the addition ("superposition") $f + g$ of two functions $f, g \in \mathscr{F}(\Omega, \mathbb{K})$ by $(f + g)(t) := f(t) + g(t)$ for all $t \in \Omega$. Note that $f + g$ is a function, whereas $f(t) + g(t)$ is the sum of two numbers $f(t), g(t) \in \mathbb{K}$. Likewise we can define a scalar multiplication $\lambda \cdot f$ for $f \in \mathscr{F}(\Omega, \mathbb{K})$ and $\lambda \in \mathbb{K}$ by $(\lambda \cdot f)(t) := \lambda f(t)$. The zero function

$$0 : \Omega \to \mathbb{K}, \quad t \mapsto 0(t) := 0,$$

will not be distinguished notationally from the number $0 \in \mathbb{K}$. The additive inverse of $f \in \mathscr{F}(\Omega, \mathbb{K})$ is given by $-f$, defined by $(-f)(t) := -f(t)$ for all $t \in \Omega$. We have $f + g = g + f$ for $f, g \in \mathscr{F}(\Omega, \mathbb{K})$, since $f(t) + g(t) = g(t) + f(t) \in \mathbb{K}$ for all $t \in \Omega$. In the same way one verifies that all associative, commutative, and distributive laws hold in $\mathscr{F}(\Omega, \mathbb{K})$. Thus, $\mathscr{F}(\Omega, \mathbb{K})$ is a linear space, the vectors being functions, vector addition being function superposition, and scalar multiplication being defined as scaling of function values. The sine function $\sin : \mathbb{R} \to \mathbb{R}$ is an element or "point" in the space $\mathscr{F}(\mathbb{R}, \mathbb{R})$ and we write $\sin \in \mathscr{F}(\mathbb{R}, \mathbb{R})$ in the same way as we write $(1, 1, 1)^\top \in \mathbb{R}^3$ for a point in the Euclidean space $\mathbb{R}^3$.
$\Diamond$

A **function space** is a linear space whose elements are functions. $\mathscr{F}(\Omega, \mathbb{K})$ is a function space. If X is a linear space and $U \subseteq X$ is a subset with the property $\mathbf{x} + \mathbf{y} \in U$ for all $\mathbf{x}, \mathbf{y} \in U$ and the property $\alpha \cdot \mathbf{x} \in U$ for all $\mathbf{x} \in U$ and all $\alpha \in \mathbb{K}$, then U is called a **subspace** of X. It is clear that a subspace of a linear space must be a linear space by itself, since all the laws concerning addition and scalar multiplication will automatically hold.

For $s \in \mathbb{N}$, an s-dimensional vector

$$\alpha = (\alpha_1, \ldots, \alpha_s) \in \mathbb{N}_0^s$$

will be called a **multi-index**. We set $|\alpha| = \sum_{i=1}^s \alpha_i$. Multi-indices are useful to express partial derivatives. Given a function $v : \mathbb{R}^s \to \mathbb{R}$, its partial derivatives of order $|\alpha|$ at $x \in \mathbb{R}^s$ may be written as

$$D^\alpha v(x) = \frac{\partial^{|\alpha|} v}{\partial x_1^{\alpha_1} \cdots \partial x_s^{\alpha_s}}(x).$$

Example B.2 (Spaces of Continuously Differentiable Functions) Let $\emptyset \neq \Omega \subseteq \mathbb{R}^s$ be some set. We define

$$C(\Omega) := \{v : \Omega \to \mathbb{R}; \ v \text{ continuous}\}.$$

The set $C(\Omega)$ is a subspace of the linear space $\mathscr{F}(\Omega, \mathbb{R})$, since the sum of two continuous functions is a continuous function and since the scalar multiple of a continuous function is a continuous function as well. We write $v \in C(\Omega)$ to say that v is a scalar-valued continuous function defined on Ω. Now let $\Omega \subseteq \mathbb{R}^s$ be an *open* set. For any $k \in \mathbb{N}_0$ one defines

$$C^k(\Omega) := \left\{ v : \Omega \to \mathbb{R}; \ D^\alpha v \in C(\Omega) \text{ for } \alpha \in \mathbb{N}_0^s, \ |\alpha| \le k \right\} \tag{B.1}$$

with $C^0(\Omega) := C(\Omega)$. The set $C^k(\Omega)$ is called the space of k times continuously differentiable functions (defined on Ω). It is another example of a subspace of $\mathscr{F}(\Omega, \mathbb{R})$. The set Ω was chosen to be open, so it does not contain its boundary (a formal definition of the boundary of a set will follow shortly) and we do not run into difficulties when considering differential quotients $D^\alpha v(x)$. In the one-dimensional case, however, we can define derivatives at the boundaries of an interval as one-sided differential quotients. In this case, it makes sense to speak of differentiable functions defined on closed intervals and we will use the abbreviated notations

$$C^k(a, b) := C^k(]a, b[) \quad \text{and} \quad C^k[a, b] := C^k([a, b]), \qquad k \in \mathbb{N}_0,$$

for the spaces of k times continuously differentiable functions $u :]a, b[\to \mathbb{R}$ and $u : [a, b] \to \mathbb{R}$, respectively. In the latter case, one-sided derivatives have to be taken for $x = a$ and for $x = b$. It will always be assumed that $-\infty < a < b < \infty$ in this case. $\diamond$

The set of all complex valued, continuous functions defined on a nonempty set $\Omega \subseteq \mathbb{R}^s$ is denoted by $C(\Omega, \mathbb{C})$. Moreover, for open sets Ω we use the notation

$$C^k(\Omega, \mathbb{C}) := \left\{ v : \Omega \to \mathbb{R}; \ D^\alpha v \in C(\Omega, \mathbb{C}) \text{ for } \alpha \in \mathbb{N}_0^s, \ |\alpha| \le k \right\}.$$

B.2 Operators

Let X and Y be linear spaces and let $D \subseteq X$. A mapping $T : D \to Y$ is called an **operator**. An operator $T : D \to Y$ is called **linear**, if D is a linear space and if the equalities

$$T(\mathbf{x} + \mathbf{y}) = T(\mathbf{x}) + T(\mathbf{y}) \quad \text{and} \quad T(\alpha \cdot \mathbf{x}) = \alpha \cdot T(\mathbf{x})$$

hold for all $\mathbf{x}, \mathbf{y} \in D$ and all $\alpha \in \mathbb{K}$.

Example B.3 (Integration as an Operator) The mapping

$$I : C[a, b] \to C^1[a, b], \quad x \mapsto y, \quad y(s) = \int_a^s x(t)\, dt, \quad a \le s \le b,$$

defining an antiderivative for every $x \in C[a, b]$, is a linear operator. $\qquad \Diamond$

Example B.4 (Differential Operators) Let $\emptyset \neq \Omega \subseteq \mathbb{R}^s$ be open and let $\alpha \in \mathbb{N}_0^s$ be a multi-index of length $|\alpha| \le k \in \mathbb{N}$. Then,

$$T : C^k(\Omega) \to C(\Omega), \quad v \mapsto D^\alpha v,$$

is a linear operator, a so-called differential operator. More generally,

$$T : C^k(\Omega) \to C(\Omega), \quad u \mapsto \sum_{|\alpha| \le k} c_\alpha D^\alpha u,$$

where $c_\alpha \in C(\Omega)$, is a linear differential operator. Using this operator, a partial differential equation of the form

$$\sum_{|\alpha| \le k} c_\alpha(x) D^\alpha u(x) = w(x), \quad x \in \Omega,$$

can be written concisely as an equation in function space, namely $T(u) = w$. $\qquad \Diamond$

For an operator $T : D \to Y$ we will use the notation

$$T(D) := \{\mathbf{y} = T(\mathbf{x}); \ \mathbf{x} \in D\}$$

to designate the so-called **range** of T. The operator T is called surjective, if $T(D) = Y$, i.e., if every element $\mathbf{y} \in Y$ is the image of some $\mathbf{x} \in D$—such an element $\mathbf{x} \in D$ is called a **pre-image** of $\mathbf{y} \in Y$. An operator $T : D \to Y$ is called **injective**, if the implication

$$T(\mathbf{x}_1) = T(\mathbf{x}_2) \quad \Longrightarrow \quad \mathbf{x}_1 = \mathbf{x}_2$$

holds. This means that every image $\mathbf{y} \in T(D)$ has a unique pre-image $\mathbf{x}$. The operator T is called **bijective** or **invertible**, if it is surjective and injective at the same time. In this case, there exists an **inverse operator**

$$T^{-1} : Y \to D, \quad \mathbf{y} \mapsto T^{-1}(\mathbf{y}),$$

such that

$$T^{-1}(T(\mathbf{x})) = \mathbf{x} \quad \text{and} \quad T(T^{-1}(\mathbf{y})) = \mathbf{y}$$

hold for all $\mathbf{x} \in D$ and all $\mathbf{y} \in Y$. We will also write

$$T^{-1}(\mathbf{y}) := \{\mathbf{x} \in D; \ T(\mathbf{x}) = \mathbf{y}\}$$

for the set of all pre-images of $\mathbf{y} \in Y$. Using the latter notation means no statement about the invertibility of T.

Example B.5 (Inversion of Operators) The differential operator $T : C^1[a, b] \to C[a, b]$, $v \mapsto v'$, is surjective, since for every $u \in C[a, b]$, we have $u = v' = T(v)$ for the function $v \in C^1[a, b]$ defined by

$$v(x) := \int_a^x u(t)\, dt,$$

It is not injective, though, since $T(v) = T(v + c)$ for every $v \in C^1[a, b]$ and any constant $c \in \mathbb{R}$. Therefore, T cannot be inverted. However, we may choose the subset

$$X_0 := \{u \in C^1[a, b]; \ u(a) = 0\} \subset C^1[a, b],$$

and consider the restriction

$$T|_{X_0} : X_0 \to C[a, b], \quad u \mapsto u'.$$

$T|_{X_0}$ is bijective and thus invertible. Next we consider the initial value problem

$$w'(x) = u(x) \cdot w(x), \ a \le x \le b, \quad w(a) = w_0 > 0,$$

compare Example 1.1. As is well known, this problem has a unique solution $w \in C^1[a, b]$ for any $u \in C[a, b]$. Therefore, there exists an operator $T : C[a, b] \to C^1[a, b]$, which maps a (parameter) function $u \in C[a, b]$ to the solution $w \in C^1[a, b]$ of the corresponding initial value problem. This operator is not linear, since for $u_1, u_2 \in C[a, b]$, we do not even have $T(u_1) + T(u_2) \in T(C[a, b])$ (initial value!). The operator is injective, and we can even state an explicit formula for pre-images, namely

$$T^{-1}(w) = \frac{w'}{w} = \frac{d}{dx}\,(\ln(w)).$$

The operator is not surjective, though, since $T(u)$ is a positive function for any $u \in C[a, b]$, and therefore $T(C[a, b]) \ne C^1[a, b]$. However, we may restrict the range and consider T as an operator from $C[a, b]$ to $Y := T(C[a, b])$. *This* operator can be inverted.   ◊

B.3   Normed Spaces

Let X be a linear space over the field $\mathbb{K}$. A mapping

$$\| \bullet \| : X \to [0, \infty[, \quad x \mapsto \|x\|,$$

is called a **norm** on X, if it has following properties:

(1) $\|x\| = 0 \iff x = 0$,
(2) $\|\lambda x\| = |\lambda| \|x\|$ for all $\lambda \in \mathbb{K}$ and $x \in X$, and
(3) $\|x + y\| \le \|x\| + \|y\|$ for all $x, y \in X$ (**triangle inequality**).

If $\| \bullet \|$ is a norm on X, then $(X, \| \bullet \|)$ is called a **normed space**. If property (1) is replaced by the weaker condition

$$\|x\| = 0 \Longleftarrow x = 0,$$

then $\| \bullet \|$ is called **semi-norm**. In this case, $\|x\| = 0$ can happen even if $x \neq 0$, see Example B.10 below.

Example B.6 In $\mathbb{R}^s$, a norm is given by

$$\|x\|_2 := \sqrt{\sum_{i=1}^{s} |x_i|^2}.$$

This is the so-called **Euclidean norm**. If $\varnothing \neq \Omega \subset \mathbb{R}^s$ is compact (closed and bounded), then any continuous function $u \in C(\Omega)$ takes its maximal value on Ω. In this case, we can introduce a norm on $C(\Omega)$ by setting

$$\|v\|_{C(\Omega)} := \sup\{|v(x)|; \ x \in \Omega\}.$$

This is the so-called **maximum-norm**.   ◇

If $\Omega \subseteq \mathbb{R}^s$ is open, there is no guarantee that $v \in C(\Omega)$ takes a maximal value on Ω, it may not even be bounded. So we can use the maximum-norm only in favorable situations. To name these, we need some preparation. Let $\Omega \subseteq \mathbb{R}^s$ be any set. With $B(x, \varepsilon) := \{z \in \mathbb{R}^s; \ \|x - z\|_2 < \varepsilon\}$ for $x \in \mathbb{R}^s$ and $\varepsilon > 0$, we define the **boundary** $\partial\Omega$ of Ω as the set

$$\partial\Omega := \{x \in \mathbb{R}^s; \ B(x, \varepsilon) \cap \Omega \neq \varnothing \text{ and } B(x, \varepsilon) \cap \Omega^C \neq \varnothing \text{ for all } \varepsilon > 0\},$$

where $\Omega^C := \mathbb{R}^s \setminus \Omega$ is the complement of Ω. We call $\overline{\Omega} := \Omega \cup \partial\Omega$ the **closure** of Ω. It can be shown that $\overline{\Omega}$ always is a closed set. If $\Omega \subset \mathbb{R}^s$ is bounded, then $\overline{\Omega}$ is compact. For an open subset $\Omega \in \mathbb{R}^s$, we say that a function $u \in C(\Omega)$ has a **continuous extension on** $\overline{\Omega}$, if there exists a continuous function $v \in C(\overline{\Omega})$ such that $u(x) = v(x)$ for all $x \in \Omega$. If such a function does exist, then it is unique, and we will call it u again, although this leads to an ambiguity: $u \in C(\overline{\Omega})$ can mean that u is a continuous function defined on $\overline{\Omega}$, but at the same time it can also mean that $u \in C(\Omega)$ has a continuous extension on $\overline{\Omega}$. For example, the function

$$u : \Omega = \mathbb{R} \setminus \{0\} \to \mathbb{R}, \quad x \mapsto \frac{\sin(x)}{x},$$

has a continuous extension defined on $\overline{\Omega} = \mathbb{R}$. This extension will also be called u and takes the value $u(0) = 1$. The function

$$u : \Omega = \mathbb{R} \setminus \{0\} \to \mathbb{R}, \quad x \mapsto \frac{1}{x},$$

has no continuous extension on $\overline{\Omega} = \mathbb{R}$.

Example B.7 Let $\Omega \subseteq \mathbb{R}^s$ be open. We already defined the set $C^k(\Omega)$ of all k times continuously differentiable functions in (B.1). If $\Omega \subset \mathbb{R}^s$ is open and bounded, then its closure $\overline{\Omega}$ is compact and we can equip the linear space

$$C^k(\overline{\Omega}) := \left\{ v \in C^k(\Omega); \ D^\alpha v \in C(\overline{\Omega}) \text{ for all } \alpha \in \mathbb{N}_0^s, \ |\alpha| \le k \right\} \tag{B.2}$$

with a norm defined by

$$\|v\|_{C^k(\overline{\Omega})} := \sum_{|\alpha| \le k} \|D^\alpha v\|_{C(\overline{\Omega})}, \quad v \in C^k(\overline{\Omega}). \tag{B.3}$$

Since $C^k(\overline{\Omega})$ is a subspace of $C(\overline{\Omega})$, the maximum-norm $\| \bullet \|_{C(\overline{\Omega})}$ can be used on $C^k(\overline{\Omega})$ as well. $\diamond$

Let $\varnothing \ne \Omega \subseteq \mathbb{R}^s$. We define the **support** of $u : \Omega \to \mathbb{K}$ to be the set

$$\mathrm{supp}(u) := \overline{\{x \in \Omega; \ u(x) \ne 0\}}.$$

The support of a function u is always closed. If it is also bounded, then u is said to have **compact support**. For an open, nonempty set $\Omega \subseteq \mathbb{R}^s$ we introduce the notation

$$C_0^k(\Omega) := \{v \in C^k(\Omega); \ \mathrm{supp}(v) \text{ is compact}\}. \tag{B.4}$$

If $v \in C^k(\Omega)$ has compact support $S \subset \Omega$, then the support of all derivatives $D^\alpha v$ necessarily also is compact and contained in S. Then, (B.3) can be used as a norm on $C_0^k(\Omega)$. Sometimes we need

$$C^\infty(\Omega) = \bigcap_{k \in \mathbb{N}_0} C^k(\Omega) \quad \text{and} \quad C_0^\infty(\Omega) = \{v \in C^\infty(\Omega); \ \text{supp}(v) \text{ is compact}\},$$

the elements of these sets being called **infinitely often differentiable functions (with compact support)**. We also write

$$C^k(\Omega, \mathbb{K}^m) := \{f = (f_1, \ldots, f_m); \ f_j \in C^k(\Omega) \text{ for } j = 1, \ldots, m\},$$

for the linear space of vector valued functions on Ω with k times continuously differentiable, $\mathbb{K}$-valued component functions.

A sequence $(x_n)_{n \in \mathbb{N}} \subset X$ in a normed space $(X, \| \bullet \|)$ is said to **converge** to $x \in X$, if $\lim_{n \to \infty} \|x_n - x\| = 0$. Often one uses the shorthand notation $x_n \to x$ but has to be aware that convergence does not only depend on the sequence $(x_n)_{n \in \mathbb{N}}$ itself, but also on the norm used. A sequence $(x_n)_{n \in \mathbb{N}_0} \subseteq X$ is called a **Cauchy sequence**, if for every $\varepsilon > 0$ there is a $N \in \mathbb{N}$ such that

$$\|x_n - x_m\| < \varepsilon \quad \text{for all} \quad n, m \geq N.$$

Every sequence convergent in X is a Cauchy sequence. If the converse is also true, the normed space $(X, \| \bullet \|)$ is called **complete**. A complete normed space is called a **Banach space**. The spaces $C^k(\overline{\Omega})$ (for bounded Ω) and $C_0^k(\Omega)$ introduced above are complete with respect to the norm (B.3).

Let $(X, \| \bullet \|_X)$ and $(Y, \| \bullet \|_Y)$ be normed spaces over $\mathbb{K}$. An operator $T : D \subseteq X \to Y$ is called **continuous** at $x_0 \in D$, if the following implication holds for any sequence $(x_n)_{n \in \mathbb{N}} \subseteq D$:

$$\lim_{n \to \infty} \|x_n - x_0\|_X = 0 \quad \Longrightarrow \quad \lim_{n \to \infty} \|T(x_n) - T(x_0)\|_Y = 0. \tag{B.5}$$

T is called **continuous on** D, if it is continuous at every $x_0 \in D$.

A linear operator $T : X \to Y$ is called **bounded**, if there is a constant C such that $\|Tx\|_Y \leq C\|x\|_X$ holds for all $x \in X$. For every linear bounded operator $T : X \to Y$ we can define its **operator norm**:

$$\|T\| := \sup_{x \in X \setminus \{0\}} \frac{\|Tx\|_Y}{\|x\|_X} < \infty.$$

$\|T\|$ depends on both, $\| \bullet \|_X$ and $\| \bullet \|_Y$, but this is not made explicit notationally. For a linear bounded operator we evidently have $\|Tx\|_Y \leq \|T\| \cdot \|x\|_X$ for all $x \in X$. Thus, every linear bounded operator is continuous (everywhere). We even have

$$T \text{ is continuous} \quad \Longleftrightarrow \quad T \text{ is bounded}$$

when T is linear. A proof can be found in any textbook on functional analysis.

Example B.8 Consider $(X = C[a, b], \| \bullet \|_{C[a,b]})$ and $(Y = C^1[a, b], \| \bullet \|_{C[a,b]}))$. The integral operator $I : X \to Y$ from Example B.3 is bounded:

$$\|Ix\|_{C[a,b]} = \max_{a \le s \le b} \left\{ \left| \left| \int_a^s x(t)\, dt \right| \right| \right\} \le (b - a) \max_{a \le s \le b} \{|x(t)|\} = (b - a)\|x\|_{C[a,b]},$$

and thus $\|I\|_{C[a,b]} \le (b - a)$ (one could even show $\|I\|_{C[a,b]} = (b - a)$). Consequently, I is continuous. $\Diamond$

Continuity of an operator essentially depends on the chosen norms !

Proposition B.9 (Continuity and Discontinuity of Differentiation) *With respect to the norms $\| \bullet \|_X = \| \bullet \|_{C[a,b]}$ on $X = C^1[a, b]$ and $\| \bullet \|_Y = \| \bullet \|_{C[a,b]}$ on $Y = C[a, b]$, differentiation*

$$D : C^1[a, b] \to C[a, b], \quad x \mapsto Dx := x',$$

is a discontinuous operator. With respect to $\| \bullet \|_X = \| \bullet \|_{C^1[a,b]}$ on X and $\| \bullet \|_Y$ as above, the same operator is continuous.

Proof Take the sequence $(x_n)_{n \in \mathbb{N}} \subset C^1[a, b]$ of functions

$$x_n : [a, b] \to \mathbb{R}, \quad t \mapsto x_n(t) = \frac{1}{\sqrt{n}} \sin(nt)$$

with derivatives $Dx_n(t) = (x_n)'(t) = \sqrt{n} \cos(nt)$. This sequence converges to the zero function $x = 0$ with respect to the norm $\| \bullet \|_{C[a,b]}$, since $\|x_n - 0\|_{C[a,b]} \le 1/\sqrt{n} \to 0$. However, $\|Dx_n - Dx\|_{C[a,b]} = \sqrt{n} \to \infty$ for $n \to \infty$, showing that D is not continuous with respect to this norm on X. On the other hand, for any sequence $(x_n)_{n \in \mathbb{N}} \subset C^1[a, b]$ and $x_0 \in C^1[a, b]$:

$$\|x_n - x_0\|_{C^1[a,b]} = \|x_n - x_0\|_{C[a,b]} + \|Dx_n - Dx_0\|_{C[a,b]} \overset{n \to \infty}{\longrightarrow} 0$$

$$\implies \|Dx_n - Dx_0\|_{C[a,b]} \overset{n \to \infty}{\longrightarrow} 0,$$

which proves continuity with respect to the other norm on X. $\square$

A (linear) operator $T : X \to \mathbb{R}$ is called a (linear) **functional**. A norm on $\mathbb{R}$ is given by the absolute value $| \bullet |$. $\mathbb{R}$ will always be equipped with this norm.

B.4   Inner Product Spaces

A **scalar product** or **inner product** on a linear space X is a mapping

$$\langle\bullet|\bullet\rangle : X \times X \to \mathbb{K}, \quad (x, y) \mapsto \langle x|y\rangle,$$

when the following holds for all $x, y, z \in X$ and $\lambda \in \mathbb{K}$:

(1) $\langle x + y|z\rangle = \langle x|z\rangle + \langle y|z\rangle$,
(2) $\langle \lambda x|y\rangle = \lambda\langle x|y\rangle$,
(3) $\langle x|y\rangle = \overline{\langle y|x\rangle}$, and
(4) $\langle x|x\rangle > 0$ for $x \neq 0$.

In (3), $\overline{\langle y|x\rangle}$ means the complex conjugate of the number $\langle y|x\rangle \in \mathbb{K}$. In real spaces ($\mathbb{K} = \mathbb{R}$) this has no effect since $\bar{z} = z$ for $z \in \mathbb{R}$. In this case, the inner product is linear in both arguments. If $\mathbb{K} = \mathbb{C}$, linearity in the second argument does no longer hold, since $\langle x|\lambda y\rangle = \bar{\lambda}\langle x|y\rangle$. When $\langle\bullet|\bullet\rangle$ is a scalar product on X, then $(X, \langle\bullet|\bullet\rangle)$ is called **pre-Hilbert space** or **inner product space**.

In every inner product space $(X, \langle\bullet|\bullet\rangle)$ the scalar product **induces** a norm:

$$\|x\| := \sqrt{\langle x|x\rangle} \quad \text{for all} \quad x \in X,$$

such that the **Cauchy–Schwarz inequality**

$$|\langle x|y\rangle| \leq \|x\|\|y\| \quad \text{for all} \quad x, y \in X$$

holds. An inner product space which is complete with respect to the induced norm is called a **Hilbert space**.

Example B.10 (The Space L_2) Let $\Omega \in \mathbb{R}^s$ be a **domain**, i.e., an open and connected set.[1] If Ω is a "simple" bounded geometric object like a polytope or a ball and if $v \in C(\overline{\Omega})$, then it is clear what is meant by the integral

$$I_\Omega(v) = \int_\Omega v(x)\, dx.$$

But this integral can even be given a meaning as a so-called **Lebesgue integral** for any open set Ω and any nonnegative, "measurable" function $v : \Omega \to \mathbb{R}$. We will not give a definition of what is meant by "measurability." Non-measurable functions are so exotic that we do not have to consider them—we simply claim that all functions needed for our purposes are measurable. Also, we do not go into Lebesgue integration theory, since whenever we actually want to carry out an integration for a specific function v and on a specific domain Ω, both will be regular enough to make the Lebesgue integral coincide with the Riemann integral. In general, the integral value $I_\Omega(v)$ may be finite or infinite; if it is finite, we call v integrable. A function $v : \Omega \to \mathbb{R}$ (which may also take negative values) is called **integrable**, if $I_\Omega(|v|) < \infty$. In this case one sets

---

[1] "Connected" means that Ω cannot be written as the union of two disjoint open sets.

$$I_\Omega(v) = \int_\Omega v(x)\,dx = \int_\Omega v^+(x)\,dx - \int_\Omega v^-(x)\,dx,$$

where

$$v = v^+ - v^-, \quad v^+(x) = \begin{cases} v(x), & \text{if } v(x) \geq 0 \\ 0, & \text{else} \end{cases}, \quad v^-(x) = \begin{cases} 0, & \text{if } v(x) \geq 0 \\ -v(x), & \text{else} \end{cases}.$$

A complex valued function $w = u + iv : \Omega \to \mathbb{C}$ with real and imaginary part $u, v : \Omega \to \mathbb{R}$ will also be called integrable if $I_\Omega(|v|) < \infty$. Its integral is a complex value defined by $I_\Omega(w) = I_\Omega(u) + iI_\Omega(v)$ with i being the imaginary unit. A subset $N \subset \Omega$ will be called a **nullset**, if it has volume 0.[2] Two functions $v_1, v_2 : \Omega \to \mathbb{R}$ which are equal except on a nullset are called **equal almost everywhere (a.e.)** For example, the function $v : \mathbb{R} \to \mathbb{R}$ with $v(x) = 1$ for $x = 0$ and $v(x) = 0$ for $x \neq 0$ is equal a.e. to the null function. Two integrable functions v_1 and v_2 being equal a.e. will produce the same integral value $I_\Omega(v_1) = I_\Omega(v_2)$. Let us now define, for $p \in \mathbb{N}$,

$$\|v\|_{L_p(\Omega)} := \left(\int_\Omega |v(x)|^p \, dx \right)^{1/p} \tag{B.6}$$

(which may become infinite without further restrictions on v) and, provisionally,

$$L_p(\Omega) := \{v : \Omega \to \mathbb{K}; \ \|v\|_{L_p(\Omega)} < \infty\}, \quad p \in \mathbb{N}. \tag{B.7}$$

To the latter definition, we add a supplementary agreement: two functions $v_1, v_2 \in L_p(\Omega)$ shall be *identified*, whenever they are equal almost everywhere.[3] Only with this supplementary agreement the implication $(\|v\|_{L_p(\Omega)} = 0) \Rightarrow (v = 0)$ becomes true and $(L_p(\Omega), \| \bullet \|_{L_p(\Omega)})$ becomes a normed space, even a Banach space. As a drawback of the above supplementary agreement it is meaningless to speak of the value $v(x_0)$, $x_0 \in \Omega$, of a function $v \in L_p(\Omega)$, since v has to be

---

[2]This is a rather careless statement, since one cannot reasonably define a volume for *any* set $N \subseteq \mathbb{R}^s$. Defining a volume $|I| := (b_1 - a_1) \cdot \ldots \cdot (b_s - a_s)$ for an s-dimensional interval $I =]a_1, b_1[\times \ldots \times]a_s, b_s[$ with $a_j \leq b_j$ for all j, a precise statement would be: $N \subset \Omega$ is a nullset, if for every $\varepsilon > 0$ there exists a sequence $(I_j)_{j=1,2,\ldots}$ of s-dimensional intervals such that

$$N \subset \bigcup_{j=1}^\infty I_j \quad \text{and} \quad \sum_{j=1}^\infty |I_j| \leq \varepsilon.$$

[3]In a mathematically clean way, the elements of $L_p(\Omega)$ therefore have to be defined not as functions, but as equivalence classes of functions being equal a.e.

identified with any other function agreeing with it except at x_0. The elements of $L_1(\Omega)$ are called **integrable functions (on Ω)**, whereas the elements of $L_2(\Omega)$ are called **square integrable functions (on Ω)**. The space $L_2(\Omega)$ is special, as it even is a Hilbert space with associated inner product

$$\langle u|v \rangle_{L_2(\Omega)} := \int_\Omega u(x)\overline{v(x)}\, dx. \tag{B.8}$$

Here, $\overline{v(x)}$ means again the complex conjugate value of $v(x)$. For a real valued function $v : \Omega \to \mathbb{R}$, we have $\overline{v(x)} = v(x)$. The triangle inequality in $(L_2(\Omega), \langle \bullet|\bullet \rangle_{L_2(\Omega)})$ is called **Minkowski's inequality** and the Cauchy–Schwarz inequality is called **Hölder's inequality**. It is stated as follows:

$$u, v \in L_2(\Omega) \implies u \cdot v \in L_1(\Omega), \quad \|uv\|_{L_1(\Omega)} \le \|u\|^2_{L_2(\Omega)} \|v\|^2_{L_2(\Omega)}. \tag{B.9}$$

In the one-dimensional case, we will write $L_2(a, b)$ instead of $L_2(]a, b[)$.   ◊

Example B.11 (The Sobolev Spaces H^k) Let $\Omega \subset \mathbb{R}^s$ be a domain with boundary $\partial\Omega$ smooth enough to allow an application of the divergence theorem needed below and let $k \in \mathbb{N}_0$. For $v \in C^1(\overline{\Omega})$ integration by parts (the divergence theorem) shows that

$$\int_\Omega \frac{\partial v}{\partial x_j}(x)\varphi(x)\, dx = -\int_\Omega v(x)\frac{\partial \varphi}{\partial x_j}(x)\, dx \quad \text{for all} \quad \varphi \in C_0^\infty(\Omega).$$

More generally, if $v \in C^k(\overline{\Omega})$, then repeated integration by parts shows that

$$\int_\Omega D^\alpha v(x)\varphi(x)\, dx = (-1)^{|\alpha|}\int_\Omega v(x)D^\alpha \varphi(x)\, dx \quad \text{for all} \quad \varphi \in C_0^\infty(\Omega)$$

for any multi-index $\alpha \in \mathbb{N}_0^s$ with $|\alpha| \le k$. For $v \in L_2(\Omega)$ derivatives $D^\alpha v$ cannot be defined in the usual way as differential quotients. However, motivated by the above formula, whenever a function $u \in L_2(\Omega)$ exists such that

$$\int_\Omega u(x)\varphi(x)\, dx = (-1)^{|\alpha|}\int_\Omega v(x)D^\alpha \varphi(x)\, dx \quad \text{for all} \quad \varphi \in C_0^\infty(\Omega), \tag{B.10}$$

then u will be called **weak derivative** or **generalized derivative** of v and we will write $u = D^\alpha v \in L_2(\Omega)$. If a weak derivative of v exists, then it is uniquely defined by (B.10). Also, if $u \in C^k(\overline{\Omega})$, then its weak derivative coincides with its derivative in the usual sense.[4] For example, the function $v :]-1, 1[\to \mathbb{R}, x \mapsto |x|$, which

---

[4]"Coincides" is to be understood as equality of $L_2(\Omega)$-functions, not as pointwise equality.

belongs to $C[-1, 1] \subset L_2(-1, 1)$, is not differentiable. It is weakly differentiable, though, the weak derivative being (identifiable with) the Heaviside function

$$u(x) = \begin{cases} 1, & \text{for } x \in]0, 1[, \\ 0, & \text{for } x = 0, \\ -1, & \text{for } x \in]-1, 0[\end{cases}.$$

The Heaviside function in turn is *not even weakly differentiable*, since its weak derivative would have to coincide a.e. with the zero function. Then the left hand side of (B.10) would be zero for every $\varphi \in C_0^\infty(\Omega)$, but not so the right hand side.

We now define the **Sobolev space**

$$H^k(\Omega) := \{v \in L_2(\Omega); \ D^\alpha v \in L_2(\Omega) \text{ for } |\alpha| \le k\}, \quad k \in \mathbb{N}. \tag{B.11}$$

It can be equipped with the scalar product

$$\langle u|v \rangle_{H^k(\Omega)} := \sum_{|\alpha| \le k} \langle D^\alpha u | D^\alpha v \rangle_{L_2(\Omega)}, \quad u, v \in H^k(\Omega), \tag{B.12}$$

which induces the **Sobolev norm**

$$\|u\|_{H^k(\Omega)} = \sqrt{\sum_{|\alpha| \le k} \|D^\alpha u\|_{L_2(\Omega)}^2}, \quad u \in H^k(\Omega).$$

It can be shown that $(H^k(\Omega), \langle \bullet|\bullet \rangle_{H^k(\Omega)})$ is a Hilbert space. In the one-dimensional case, we will write $H^k(a, b)$ instead of $H^k(]a, b[)$.   ◊

B.5 Convexity, Best Approximation

Let $(X, \langle \bullet|\bullet \rangle)$ be an inner product space over the scalar field $\mathbb{K}$. Two vectors $x, y \in X$ are called **orthogonal**, if $\langle x|y \rangle = 0$. For real inner product spaces we define an angle $\alpha \in [0, \pi]$ between two vectors $x, y \in X \setminus \{0\}$ by requiring

$$\cos \alpha = \frac{\langle x|y \rangle}{\|x\| \|y\|},$$

where $\| \bullet \|$ is the norm induced by the scalar product. Orthogonality then means $\alpha = \pi/2$, as it should. A subset $C \subset X$ of a vector space is called **convex**, if the following implication holds:

$$x, y \in C \text{ and } \lambda \in \mathbb{R}, \ 0 \le \lambda \le 1 \implies (1 - \lambda)x + \lambda y \in C.$$

Fig. B.1 Geometric
interpretation of the
projection theorem

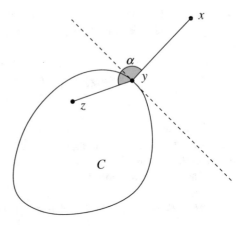

Convexity is essential for the existence of a unique solution of an important
optimization problem, stated in the following **projection theorem**. A proof can be
found in (all) text books on functional analysis.

Theorem B.12 (Projection Theorem) *Let $(X, \langle \bullet | \bullet \rangle)$ be a Hilbert space over the
scalar field $\mathbb{K}$ with induced norm $\| \bullet \|$ and let $C \subset X$ be a nonempty, closed, and
convex subset. Then for every $x \in X$ there exists a unique element $y \in C$ such that*

$$\| x - y \| \leq \| x - z \| \quad \text{for all} \quad z \in C.$$

The vector y is uniquely characterized by

$$Re\langle x - y | z - y \rangle \leq 0 \quad \text{for all} \quad z \in C. \tag{B.13}$$

In the above theorem, $\mathrm{Re}(z)$ is the real part of a complex number $z \in \mathbb{C}$. In case
$z \in \mathbb{R}$ one has $\mathrm{Re}(z) = z$. For real Hilbert spaces, (B.13) means that the angle α
between $x - y$ and all vectors $z - y$ is obtuse or equal to $\pi/2$ which immediately leads
to a geometric interpretation of the projection theorem, as illustrated in Fig. B.1. A
finite-dimensional subspace

$$X_n = \mathrm{span}\{\hat{x}_1, \ldots, \hat{x}_n\} \subset X, \quad \hat{x}_1, \ldots, \hat{x}_n \text{ linearly independent,}$$

is a nonempty, closed, and convex set. In this case, the projection theorem reads:

Theorem B.13 (Projection   Theorem,   Finite-Dimensional   Subspaces) *Let
$(X, \langle \bullet | \bullet \rangle)$ be a Hilbert space with induced norm $\| \bullet \|$ and let $X_n \subset X$ be an
n-dimensional subspace with basis $\{\hat{x}_1, \ldots, \hat{x}_n\}$. Then for every $x \in X$ there is a
unique $x_n \in X_n$ such that*

$$\| x - x_n \| \leq \| x - v \| \quad \text{for all} \quad v \in X_n.$$

The vector x_n is uniquely characterized by the equations

$$\langle x - x_n | \hat{x}_i \rangle = 0 \quad for \quad i = 1, \ldots, n. \tag{B.14}$$

By Eq. (B.14), the residual $x - x_n$ is orthogonal to all basis vectors of X_n and thus to all vectors in X_n. Thus we get x_n by an orthogonal projection of x onto X_n—exactly the same is done when solving the linear least squares problem, see Fig. 3.1. Actually, completeness of X is not needed in the projection theorem, it suffices that C is a complete set. This is always the case in the situation of Theorem B.13 with $C = X_n$. The following example makes use of this generalization.

Example B.14 Let $\mathbb{K} = \mathbb{C}$ and let $X = C[0, 1]$ be the space of continuous, complex valued functions on $[0, 1]$. We define a scalar product:

$$\langle f | g \rangle := \int_0^1 f(t)\overline{g(t)} \, dt, \quad f, g \in X,$$

which makes $(X, \langle \bullet | \bullet \rangle)$ a pre-Hilbert space. The functions

$$e_k : [0, 1] \to \mathbb{C}, \quad t \mapsto e^{2\pi i k t}, \quad k \in \mathbb{Z},$$

(with i the imaginary unit) are pairwise orthogonal and even **orthonormal**, since $\|e_k\| = \sqrt{\langle e_k | e_k \rangle} = 1$. By Theorem B.13, for every $f \in X$ there is a unique

$$f_n \in \mathbb{T}_n := \left\{ p = \sum_{k=-n}^{n} c_k e_k; \ c_k \in \mathbb{C} \right\}$$

satisfying $\|f - f_n\| \leq \|f - p\|$ for all $p \in \mathbb{T}_n$. To compute f_n, we make the ansatz

$$f_n(t) = \sum_{k=-n}^{n} c_k(f) e^{2\pi i k t}$$

with unknown coefficients $c_k(f)$ depending on f. From (B.14) and by orthonormality, we get

$$\langle f_n | e_k \rangle = \sum_{j=-n}^{n} c_j(f) \langle e_j | e_k \rangle = c_k(f) = \langle f | e_k \rangle = \int_0^1 f(t) e^{-2\pi i k t} \, dt.$$

The coefficients $c_k(f)$ are called **Fourier coefficients** of f and f_n is called n-th Fourier polynomial of f. $\diamond$

Appendix C
The Fourier Transform

We define the space of s-variate Lebesgue integrable functions

$$L_1(\mathbb{R}^s) := \{v : \mathbb{R}^s \to \mathbb{C}; \int_{\mathbb{R}^s} |v(x)|\, dx < \infty\}, \tag{C.1}$$

adding two supplementary agreements, just as in Appendix B. First, it is always assumed that functions v are "measurable," non-measurable functions being so exotic that we do not have to consider them. Second, two functions $v_1, v_2 : \mathbb{R}^s \to \mathbb{C}$ being equal almost everywhere and thus having the same integral value, are to be identified. The set $L_1(\mathbb{R}^s)$ does of course also include real valued functions. For every function $f \in L_1(\mathbb{R}^s)$ we define its **Fourier transform** as the function $\mathscr{F}f$ given by

$$(\mathscr{F}f)(y) := \hat{f}(y) := \int_{\mathbb{R}^s} f(x)e^{-2\pi i x \cdot y}\, dx, \quad y \in \mathbb{R}^s. \tag{C.2}$$

Here, $x \cdot y = \sum_{j=1}^{s} x_j y_j$ and i is the imaginary unit. The Fourier transform in general is a complex valued function, even if $f \in L_1(\mathbb{R}^s)$ is real valued. The **inverse Fourier transform** is defined by

$$(\overline{\mathscr{F}}f)(x) := \int_{\mathbb{R}^s} f(y)e^{+2\pi i x \cdot y}\, dy, \quad x \in \mathbb{R}^s. \tag{C.3}$$

In case $f \in L_1(\mathbb{R}^s)$ *and* $\hat{f} \in L_1(\mathbb{R}^s)$, the function f is related to its Fourier transform by the **Fourier inversion formula** $f \stackrel{a.e.}{=} \overline{\mathscr{F}}\mathscr{F}f$, i.e.,

M. Richter, *Inverse Problems*, Lecture Notes in Geosystems
Mathematics and Computing, https://doi.org/10.1007/978-3-030-59317-9

$$f(x) \overset{\text{a.e.}}{=} \int_{\mathbb{R}^s} \hat{f}(y) e^{2\pi i x \cdot y} \, dy, \quad x \in \mathbb{R}^s. \tag{C.4}$$

If $f \in L_1(\mathbb{R}^s) \cap L_2(\mathbb{R}^s)$, then it can be shown that

$$\hat{f} \in L_2(\mathbb{R}^s) \quad \text{and} \quad \|f\|_{L_2(\mathbb{R}^s)} = \|\hat{f}\|_{L_2(\mathbb{R}^s)} \tag{C.5}$$

($L_2(\mathbb{R}^s)$ and $\| \bullet \|_{L_2(\mathbb{R}^s)}$ are introduced in Appendix B). This can be used to define the Fourier transform on the space $L_2(\mathbb{R}^s)$: for every function $f \in L_2(\mathbb{R}^s)$ there exists a sequence $f_n \in C_0^\infty(\mathbb{R}^s) \subset L_1(\mathbb{R}^s) \cap L_2(\mathbb{R}^s)$ such that $\|f_n - f\|_{L_2(\mathbb{R}^s)} \to 0$ (which is not trivial to see, but well known). By (C.5), the sequence $(\hat{f}_n)_{n \in \mathbb{N}}$ of Fourier transforms is a Cauchy sequence in the complete space $L_2(\mathbb{R}^s)$ and thus converges to a function $g \in L_2(\mathbb{R}^s)$. We continue to write $\hat{f} := \mathscr{F}f := g$ and call $\hat{f} = g$ the Fourier transform of f. But strictly speaking, we now have *two* maps $\mathscr{F}$—one defined on $L_1(\mathbb{R}^s)$ by (C.2) and one defined on $L_2(\mathbb{R}^s)$ by taking the L_2-limit of a sequence of functions. We use the notation $\mathscr{F}$ for both maps, although they only coincide on $L_1(\mathbb{R}^s) \cap L_2(\mathbb{R}^s)$. By construction, $\hat{f} \in L_2(\mathbb{R}^s)$ for $f \in L_2(\mathbb{R}^s)$. Then, the map $\overline{\mathscr{F}}$ from (C.3) can be applied to $\hat{f}$ in the same sense as $\mathscr{F}$ was applied to f. Moreover, for the sequence of functions $f_n \in C_0^\infty(\mathbb{R}^s)$ as above, one can show that $\hat{f}_n$ is integrable and that $f_n = \overline{\mathscr{F}} \hat{f}_n$ holds for all $n \in \mathbb{N}$. Taking the L_2-limits, one sees that

$$\overline{\mathscr{F}} \mathscr{F} f = f$$

holds for $f \in L_2(\mathbb{R}^s)$. This will also be written in the form

$$f(x) \multimap \hat{f}(y),$$

although L_2-functions are not defined pointwise. To summarize, the Fourier transform is a linear, continuous operator on $L_2(\mathbb{R}^s)$ and we have

$$\|\hat{f}\|_{L_2(\mathbb{R}^s)} = \|f\|_{L_2(\mathbb{R}^s)} \quad \text{for} \quad f(x) \multimap \hat{f}(y), \tag{C.6}$$

which is known as **Plancherel's identity**.

The **convolution** of $f, g \in L_2(\mathbb{R}^s)$ is the function $f * g$ defined by

$$(f * g)(x) := \int_{\mathbb{R}^s} f(x - y) g(y) \, dy. \tag{C.7}$$

In general, we cannot assume that $f * g \in L_2(\mathbb{R}^s)$. But in case that $f, g \in L_2(\mathbb{R}^s)$ and $f * g \in L_2(\mathbb{R}^s)$, the following relation holds, which is the basis of efficient methods to solve Fredholm equations of the convolutional type:

$$(f * g)(x) \multimap \hat{f}(y)\hat{g}(y) \quad \text{for} \quad f(x) \multimap \hat{f}(y), \; g(x) \multimap \hat{g}(y). \tag{C.8}$$

The following example shows how convolution can be used to model multipath propagation in signal transmission.

Example C.1 If an analog signal (i.e., a function of time) $u \in C_0(\mathbb{R})$ is transmitted, e.g., in mobile communication, multipath propagation leads to interferences. Figure C.1 shows an example. The mathematical model for multipath propagation is the following:

$$w(s) = \int_0^\ell g(t)u(s - t)\, dt. \tag{C.9}$$

Here,

- w is the received signal.
- $u(. - t)$ is the transmitted signal, delayed by t units (seconds). The delay corresponds to a propagation time from transmitter (TX) to receiver (RX).
- $g(t)$ is a damping factor corresponding to a loss of signal power. The function $g : [0, \ell] \to \mathbb{R}$ models the communication channel.
- ℓ is the channel length. Signals delayed by more than ℓ seconds are so weak that they are neglected.

Setting $g(t) := 0$ for $t \notin [0, \ell]$, function g is made a member of $L_2(\mathbb{R})$, as is u. Then, (C.9) can be replaced by

$$w(s) = \int_{-\infty}^{\infty} g(t)u(s - t)\, dt,$$

which is a convolution of g and u. The task of an **equalizer** is to find out u, when w and g are known. This can be done by taking the Fourier transform on both sides of the convolution equation, which, according to (C.8), leads to

$$\hat{w}(y) = \hat{g}(y)\hat{u}(y), \quad y \in \mathbb{R}.$$

If $\hat{g}(y) \neq 0$ for all $y \in \mathbb{R}$, one can get u by applying the inverse Fourier transform to $\hat{u} = \hat{w}/\hat{g}$. In practice, some regularization has to be added to make this solution method work, see Sect. 3.8. $\Diamond$

Fig. C.1 Interferences for
multipath propagation

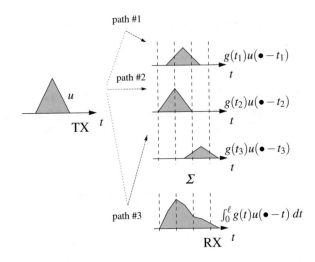

C.1 One-Dimensional Discrete Fourier Transform

Practical computation of Fourier transforms can only be done approximately by
means of discretization. We consider the Fourier transform of a function $u \in C_0(\mathbb{R})$
with $\text{supp}(u) \subset [-a, a]$ for some $a > 0$, assuming moreover that for some even
$N \in \mathbb{N}$, equidistant sample values

$$u_j := u(t_j), \quad t_j := jh, \quad h := \frac{2a}{N}, \quad j \in W := \left\{ -\frac{N}{2}, \dots, \frac{N}{2} - 1 \right\} \quad (C.10)$$

of u are given. Using the linear B-Spline ("hat function")

$$B_2(t) := \begin{cases} t + 1, & -1 \le t \le 0 \\ 1 - t, & 0 \le t \le 1 \\ 0, & \text{else} \end{cases}, \quad (C.11)$$

(compare to (2.3), where the different notation $N_{j,2}$ was used) an approximation of
u is given by

$$u_N(t) := \sum_{j=-N/2}^{N/2-1} u_j B_2(t/h - j). \quad (C.12)$$

This polygonal line, which interpolates u at the sample points, can be Fourier
transformed *exactly*:

$$\widehat{u_N}(y) = \underbrace{\left(\frac{\sin(\pi h y)}{\pi h y}\right)^2}_{=:\,\sigma(y)} \cdot \underbrace{\left(h \sum_{j=-N/2}^{N/2-1} u_j e^{-2\pi i j h y}\right)}_{=:\,U(y)}. \tag{C.13}$$

The so-called **attenuation factors** $\sigma(y)$ are data independent and determine the decay rate of $\widehat{u_N}(y)$ to zero for $|y| \to \infty$, while $U(y)$ is periodic with period $N/2a$. If the exact continuity class of u is known, by the appropriate choice of an interpolation scheme $u \mapsto u_I$ (not necessarily piecewise linear interpolation) one can achieve that $\hat{u}$ and $\widehat{u_I}$ have the same rate of decay at infinity, see [Gau72]. If one only knows M *non*-equidistant samples

$$u(t_j), \quad -a \le t_0 < \ldots < t_{M-1} \le a,$$

it is still possible to approximate $\hat{u}$ by taking the Fourier transform of some spline-approximant, but one can no longer split off attenuation factors like in (C.13). Alternatively, one can directly approximate the integral defining the Fourier transform, e.g., using the trapezoidal rule. This leads to

$$\hat{u}(y) \approx \sum_{j=0}^{M-1} u_j e^{-2\pi i y t_j}, \tag{C.14}$$

with

$$u_j = \begin{cases} \frac{1}{2} u(t_0)(t_1 - t_0), & j = 0 \\ \frac{1}{2} u(t_j)(t_{j+1} - t_{j-1}), & j = 1, \ldots, M-2 \\ \frac{1}{2} u(t_{M-1})(t_{M-1} - t_{M-2}), & j = M-1 \end{cases}.$$

An efficient evaluation of formula (C.13) is best possible for the special choice $y = k/(2a)$, $k \in \mathbb{Z}$, in which case both, samples of the function u *and* of its Fourier transform $\widehat{u_N}$ are equidistant. We get

$$\widehat{u_N}\left(\frac{k}{2a}\right) = 2a \cdot \underbrace{\left(\frac{\sin(\pi k/N)}{\pi k/N}\right)^2}_{=:\,\sigma_k} \cdot \underbrace{\left(\frac{1}{N} \sum_{j=-N/2}^{N/2-1} u_j e^{-2\pi i j k/N}\right)}_{=:\,U_k}, \quad k \in \mathbb{Z}.$$

$$\tag{C.15}$$

Note that because of their periodicity, only U_k, $k = -N/2, \ldots, N/2 - 1$, need to be computed to get all values $\widehat{u_N}(k/2a)$, $k \in \mathbb{Z}$. By direct calculation one can show that

$$U_k = \frac{1}{N} \sum_{j=-N/2}^{N/2-1} u_j e^{-2\pi ijk/N}, \quad k = -\frac{N}{2}, \ldots, \frac{N}{2} - 1,$$

$$\Longleftrightarrow \quad u_j = \sum_{k=-N/2}^{N/2-1} U_k e^{2\pi ijk/N}, \quad j = -\frac{N}{2}, \ldots, \frac{N}{2} - 1. \quad (C.16)$$

Computing the above values U_k from u_j is called **discrete Fourier transform (DFT)** and computing the values u_j from U_k is called **inverse discrete Fourier transform (IDFT)**. The famous **FFT algorithm**—of which many variants do exist—is an efficient implementation of DFT und IDFT. In case N is a power of 2, it requires only $\mathcal{O}(N \log(N))$ instead of N^2 arithmetical operations. For a description of this algorithm see [PTVF92, p. 504 ff].

C.2   Discrete Fourier Transform for Non-equidistant Samples

We come back to the evaluation of (C.13) at arbitrary frequencies and to the evaluation of (C.14), cases in which the FFT algorithm cannot be directly used. One way to make it come into play nevertheless was proposed in [Fou99] and is based on the following

Lemma C.2 *Let $p > 1$ be such that $pN \in \mathbb{N}$ and let $1 \leq \alpha < 2 - 1/p$. Let $\Phi : \mathbb{R} \to \mathbb{R}$ be an even, piecewise continuously differentiable function being continuously differentiable on $[-N/2, N/2]$ and having the properties*

$$supp(\Phi) \subset [-\alpha pN/2, \alpha pN/2] \quad and \quad \Phi(x) > 0 \text{ for } |x| \leq N/2.$$

Then

$$e^{-2\pi ij\xi/N} = \frac{1}{pN} \frac{1}{\Phi(j)} \sum_{\ell \in \mathbb{Z}} \widehat{\Phi}\left(\frac{\xi - \ell/p}{N}\right) e^{-2\pi ij\ell/(pN)} \quad (C.17)$$

for all $j \in \mathbb{Z}$ with $|j| \leq N/2$.

Proof Consider

$$g : \mathbb{R} \to \mathbb{C}, \quad x \mapsto g(x) := \sum_{\ell \in \mathbb{Z}} \Phi(x + pN\ell) e^{-2\pi i(x+pN\ell)\xi/N}.$$

By the assumptions on Φ and α, we have

$$g(j) = \Phi(j) e^{-2\pi ij\xi/N} \quad \text{for} \quad |j| \leq N/2.$$

Since g is periodic with period pN, it can be expressed as a Fourier series

$$g(x) = \sum_{\ell \in \mathbb{Z}} c_\ell e^{-2\pi i x \ell/(pN)}$$

(equality valid for $|x| \leq N/2$, convergence of the series everywhere) with Fourier coefficients

$$c_\ell = \frac{1}{pN} \int\limits_{-pN/2}^{pN/2} g(s) e^{2\pi i \ell s/(pN)} \, ds$$

$$= \frac{1}{pN} \int\limits_{-pN/2}^{pN/2} \left(\sum_{k \in \mathbb{Z}} \Phi(s + pNk) e^{-2\pi i (s+pNk)\xi/N} \right) e^{2\pi i \ell s/(pN)} \, ds$$

$$= \frac{1}{pN} \int\limits_{\mathbb{R}} \Phi(s) e^{-2\pi i s \xi/N} e^{2\pi i \ell s/(pN)} \, ds$$

$$= \frac{1}{pN} \int\limits_{\mathbb{R}} \Phi(s) e^{-2\pi i s(\xi - \ell/p)/N} \, ds = \frac{1}{pN} \hat{\Phi} \left(\frac{\xi - \ell/p}{N} \right), \quad \ell \in \mathbb{Z},$$

from which the result follows.   □

Lemma C.2 can be used in (C.13) to compute approximations of the Fourier transform of an equidistantly sampled function at arbitrary frequencies, i.e., to compute

$$\hat{u}_k := U\left(\frac{\xi_k}{2a} \right) = \frac{2a}{N} \sum_{j=-N/2}^{N/2-1} u_j e^{-2\pi i j \xi_k/N}, \quad k = 0, \ldots, M-1. \tag{C.18}$$

Note that $N = M$ is *not* required. Making use of (C.17) and rearranging leads to

$$\hat{u}_k = \frac{2a}{N} \sum_{\ell \in \mathbb{Z}} \hat{\Phi} \left(\frac{\xi_k - \ell/p}{N} \right) \cdot \frac{1}{pN} \sum_{j=-N/2}^{N/2} \frac{u_j}{\Phi(j)} e^{-2\pi i j \ell/(pN)}. \tag{C.19}$$

This double sum can be computed approximately in three steps:

(1) Define values

$$\tilde{u}_j := \begin{cases} u_j/\Phi(j), & |j| \leq N/2 \\ 0, & N/2 < |j| \leq pN/2 \end{cases}.$$

(2) Compute values

$$\tilde{U}_\ell = \frac{1}{pN} \sum_{j=-pN/2}^{pN/2-1} \tilde{u}_j e^{-2\pi i j \ell/(pN)}, \quad \ell \in \mathbb{Z}.$$

Since $(\tilde{U}_\ell)_{\ell \in \mathbb{Z}}$ is periodic, only pN values are to be computed. This can be done efficiently using an FFT of length pN.

(3) For $k = 0, \ldots, M-1$, approximately compute $\hat{u}_k$ as

$$\hat{u}_k \approx \frac{2a}{N} \sum_{\ell=\ell_0-K+1}^{\ell_0+K} \widehat{\Phi}\left(\frac{\xi_k - \ell/p}{N}\right) \tilde{U}_\ell,$$

where ℓ_0 is the largest integer smaller than or equal to $p\xi_k$ and where $K \in \mathbb{N}$ is a constant.

For reasons of efficiency, K must be a small number. But for small values of K we can only expect the approximation to be good if Φ is chosen such that $|\widehat{\Phi}(v)|$ decays rapidly for v away from zero. In Lemma C.2, the requirement $\mathrm{supp}(\Phi) \subset [-\alpha pN/2, \alpha pN/2]$ instead of $\mathrm{supp}(\Phi) \subset [-N/2, N/2]$ was chosen having in mind that the "broader" Φ, the more "peaked" its Fourier transform $\widehat{\Phi}$ can be. In [Fou99] the following choice is proposed: $p = 2$, $\alpha = 1.49$, K between 3 and 6 and

$$\Phi(x) := \begin{cases} I_0\left(\frac{\pi K}{\gamma}\sqrt{\gamma^2 - x^2}\right), & |x| \le \gamma := \alpha pN/2 \\ 0, & |x| > \gamma \end{cases},$$

where I_0 is the modified Bessel function of order 0:

$$I_0(x) = \sum_{k=0}^{\infty} \frac{(x^2/4)^k}{k!(k+1)!}.$$

The Fourier transform of Φ is

$$\widehat{\Phi}(y) = \frac{2\gamma}{\pi} \frac{\sinh(\pi\sqrt{K^2 - (2\gamma y)^2})}{\sqrt{K^2 - (2\gamma y)^2}}.$$

Lemma C.2 can also be used in (C.14) to compute

$$\hat{u}\left(\frac{k}{2a}\right) \approx \hat{u}_k := \sum_{j=0}^{M-1} u_j e^{-2\pi i k t_j/(2a)}, \quad k = -N/2, \ldots, N/2-1. \tag{C.20}$$

These values approximate equidistant samples of the Fourier transform of a function which itself is sampled non-equidistantly. We define

$$\tau_j := \frac{N}{2a} t_j \quad \Longrightarrow \quad \tau_j \in [-N/2, N/2], \quad j = 0, \ldots, M-1,$$

and rewrite

$$\hat{u}_k = \sum_{j=0}^{M-1} u_j e^{-2\pi i k \tau_j / N}.$$

Inserting (C.17) with $\xi = \tau_j$ and j replaced by k leads to

$$\hat{u}_k = \frac{1}{\Phi(k)} \frac{1}{pN} \sum_{\ell \in \mathbb{Z}} e^{-2\pi i k \ell / (pN)} \sum_{j=0}^{M-1} u_j \hat{\Phi} \left(\frac{\tau_j - \ell/p}{N} \right). \tag{C.21}$$

The ℓ-sum is computed approximately by summing over $\ell \in \{-pN/2, \ldots, pN/2 - 1\}$ only. For each such ℓ, the j-sum is computed approximately by summing over those j with $|p\tau_j - \ell| \leq K$. Then the ℓ-sum is computed by an FFT of length pN, the results of which are scaled by factors $1/\Phi(k)$.

The ideas presented above can be generalized to the computation of two- and higher dimensional Fourier transforms for non-equidistant samples, see [Fou99]. Fourmont also gives error estimates.

C.3   Error Estimates for Fourier Inversion in Sect. 2.5

In this paragraph we include two lemmas containing technical details showing that Fourier inversion of convolution equations is a convergent discretization.

Lemma C.3 (Fourier Domain Error of Spline Interpolant) *Let $s \in \{1, 2\}$, let $a > 0$, let $Q =\,]-a, a[^s$ and let N, h, and W be as in (2.85) for $s = 2$ and as in (C.10) for $s = 1$. Let $w \in H^1(\mathbb{R}^s) \cap C(\mathbb{R}^s)$ with Fourier transform $\hat{w}$. Assume the decay condition*

$$|w(x)| + |\hat{w}(x)| \leq C(1 + \|x\|_2)^{-s-\varepsilon} \quad \text{for all} \quad x \in \mathbb{R}^s$$

holds for some constants $C, \varepsilon > 0$. Let $w_N \in C_0(\mathbb{R}^s)$ be defined as in (2.100). Then there is a constant $C > 0$ depending on w such that

$$\left| \hat{w}\left(\frac{\beta}{2a}\right) - \widehat{w_N}\left(\frac{\beta}{2a}\right) \right| \leq Ch^2 + h^2 \sum_{\alpha \notin W} |w(\alpha h)|. \tag{C.22}$$

Remarks The first term on the right hand side of (C.22) tends to zero for $h \to 0$. The second term is a Riemann sum converging to

$$\int_{\mathbb{R}^s \setminus Q} |w(x)| \, dx$$

and *does not vanish* when the discretization level is increased, unless the support of w is contained in Q (in this case $\widehat{w}$ would be called a "band limited" function in signal processing). Otherwise, the total error (C.22) can be made arbitrarily small only by choosing N *and* Q large enough.

Proof In the following proof some formulae are explicitly written only for $s = 2$ in order to keep the notation simple. Consider the function g defined by

$$g(y) := \sum_{\alpha \in \mathbb{Z}^s} \widehat{w} \left(y - \frac{\alpha}{h} \right). \tag{C.23}$$

Our assumptions about w are sufficient for w to be continuous, have a continuous Fourier transform, and make the sum (C.23) converge absolutely everywhere by the decay condition. Function g is periodic (in all coordinate directions) with period $1/h$ and can be developed into a Fourier series:

$$g(y) = \sum_{\beta \in \mathbb{Z}^s} g_\beta e^{-2\pi i h \beta \cdot y}. \tag{C.24}$$

A short calculation shows that

$$g_\beta := h^s \int_{[-1/(2h), 1/(2h)]^s} g(y) e^{+2\pi i h \beta \cdot y} \, dy = h^s w(\beta h), \quad \beta \in \mathbb{Z}^s. \tag{C.25}$$

From (C.25) one sees that because of the assumed decay condition for w the sum $\sum_{\beta \in \mathbb{Z}^s} |g_\beta|$ converges. Therefore, the sum in (C.24) converges uniformly and pointwise equality holds. Equating (C.23) with (C.24) we get the so-called **generalized Poisson summation formula**

$$\sum_{\alpha \in \mathbb{Z}^s} \widehat{w} \left(y - \frac{\alpha}{h} \right) = h^s \sum_{\alpha \in \mathbb{Z}^s} w(\alpha h) e^{-2\pi i h \alpha \cdot y}. \tag{C.26}$$

Next we introduce the approximant

$$w_\infty(x) := \sum_{\alpha \in \mathbb{Z}^s} w_\alpha \Phi(x/h - \alpha), \quad w_\alpha := w(\alpha h), \quad \alpha \in \mathbb{Z}^s,$$

of w, which is Fourier transformed to

$$\widehat{w_\infty}(y) = \sigma(y)W(y),$$

where (for $s = 2$)

$$\sigma(y) = \left(\frac{\sin(\pi h y_1)}{\pi h y_1}\right)^2 \cdot \left(\frac{\sin(\pi h y_2)}{\pi h y_2}\right)^2 \quad \text{and} \quad W(y) = h^2 \sum_{\alpha \in \mathbb{Z}^2} w_\alpha e^{-2\pi i h \alpha \cdot y}.$$

Comparing with (C.26), one sees that $\widehat{w_\infty}(y) = \sigma(y)g(y)$. Decomposing $g = \widehat{w} + \widehat{\rho}$ with

$$\widehat{\rho}(y) = \sum_{\alpha \neq 0} \widehat{w}\left(y - \frac{\alpha}{h}\right),$$

we find—using (2.89)—that

$$\widehat{w}(y) - \widehat{w_N}(y) = \widehat{w}(y) - \widehat{w_\infty}(y) + \widehat{w_\infty}(y) - \widehat{w_N}(y)$$

$$= (1 - \sigma(y))\widehat{w}(y) - \sigma(y)\widehat{\rho}(y) + \sigma(y)h^2 \sum_{\alpha \notin W} w_\alpha e^{-2\pi i h \alpha \cdot y} \quad \text{(C.27)}$$

We will estimate the size of this difference for $y = \beta/2a$, $\beta \in W$. To do so, we need two auxiliary results. The first one is the estimate

$$|\widehat{w}(y)| \leq \frac{C}{1 + \|y\|_2^2}, \qquad \text{(C.28)}$$

which immediately follows from the decay condition. The second one is the expansion

$$\frac{\sin(\pi x)}{\pi x} = \sum_{n=0}^{\infty} \frac{(-\pi^2 x^2)^n}{(2n + 1)!} = 1 - \frac{\pi^2 x^2}{3!} + R_x, \qquad \text{(C.29)}$$

which is valid for all $x \in \mathbb{R}$ and where $|R_x| \leq \pi^4 |x|^4/5!$ for $|x| \leq 1$ according to Leibniz' criterion. From (C.29) one derives for $s = 2$ the estimate

$$|1 - \sigma(\beta/2a)| \leq A_1 \frac{\|\beta\|_2^2}{N^2} + A_2 \frac{\|\beta\|_2^4}{N^4} + A_3 \frac{\|\beta\|_2^6}{N^6} + \ldots + A_8 \frac{\|\beta\|_2^{16}}{N^{16}} \qquad \text{(C.30)}$$

for some constants $A_1, \ldots, A_8$ (similar estimate in case $s = 1$). Also, from (C.28) we get $|\widehat{w}(\beta/2a)| \leq A/(4a^2 + \|\beta\|_2^2)$ with some constant A. Since $\|\beta\|_2/N \leq 1$ for $\beta \in W$, this means that the first summand on the right hand side of (C.27) can be estimated for $\beta \in W$:

$$\left\| \left[1 - \sigma \left(\frac{\beta}{2a} \right) \right] \widehat{w} \left(\frac{\beta}{2a} \right) \right\| \leq D_1 h^2 \tag{C.31}$$

for some constant D_1. Also, one deduces from (C.28) that

$$\left| \widehat{w} \left(\frac{\beta}{2a} - \frac{\alpha}{h} \right) \right| \leq h^2 \frac{C}{\|\beta/N - \alpha\|_2^2} \quad \text{for} \quad \beta \in W, \alpha \neq 0.$$

Together with $|\sigma(y)| \leq 1$ this shows that the second term on the right hand side of (C.27) can be estimated by

$$\left| \sigma \left(\frac{\beta}{2a} \right) \widehat{\rho} \left(\frac{\beta}{2a} \right) \right| \leq D_2 h^2$$

for some constant D_2. The last term on the right hand side of (C.27) evidently can be estimated by

$$h^2 \sum_{\alpha \notin W} |w(\alpha h)|$$

and this ends the proof.   □

In Fig. C.2 we show Q together with $Q^+ =]-a-h, a[^s$, which contains the support of u_N defined by (2.99). Function u_N can only be expected to be a good approximant of u on the domain $Q^- =]-a, a-h[^s$ contained in Q.

Lemma C.4 (Error from Fourier Interpolation) *Let $s \in \{1, 2\}$, let $a > 0$, let $Q =]-a, a[^s$, let $Q^- :=]-a, a-h[^s$, and let $u \in H^s(Q)$ be extended by zero values to become a function $u \in L_2(\mathbb{R}^s)$ with integrable Fourier transform $\widehat{u}$. Let $h = 2a/N$, $N \in \mathbb{N}$ an even number, and determine $u_N \in C_0(\mathbb{R}^s)$ via (2.93) and (2.94) for $s = 2$ or via (C.12), (C.15) and (C.16) for $s = 1$. Then there is a constant $C > 0$ such that*

$$\|u - u_n\|_{L_2(Q^-)} \leq Ch\|u\|_{H^1(Q)}. \tag{C.32}$$

Proof For $u \in L_2(\mathbb{R}^s) \cap H^s(Q)$ define the function $g : \mathbb{R}^s \to \mathbb{R}$,

$$g(x) = \sum_{\alpha \in \mathbb{Z}^s} u(x - 2a\alpha), \tag{C.33}$$

which is periodic with period $2a$ and can be developed into a Fourier series, namely

$$g(x) = \sum_{\beta \in \mathbb{Z}^s} g_\beta e^{+2\pi i \beta \cdot x/(2a)}, \quad x \in Q. \tag{C.34}$$

A short calculation shows that

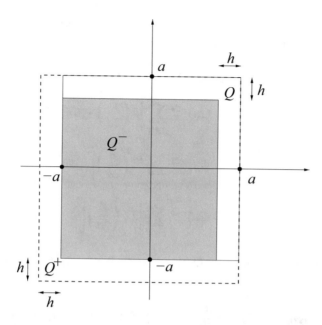

Fig. C.2 Rectangle Q, support Q^+ of u_N, and "domain of approximation" Q^-

$$g_\beta := \left(\frac{1}{2a}\right)^s \int_Q g(x)e^{-2\pi i\beta\cdot x/(2a)}\,dx = \left(\frac{1}{2a}\right)^s \widehat{u}\left(\frac{\beta}{2a}\right), \quad \beta \in \mathbb{Z}^s. \qquad (C.35)$$

Equating (C.33) with (C.34) we get a(nother) generalized Poisson summation formula

$$\sum_{\alpha\in\mathbb{Z}^s} u\,(x - 2a\alpha) = \left(\frac{1}{2a}\right)^s \sum_{\beta\in\mathbb{Z}^s} \widehat{u}\left(\frac{\beta}{2a}\right) e^{+2\pi i\beta\cdot x/(2a)}. \qquad (C.36)$$

Identity (C.36) will be used for $u - u_N$ instead of u. For arguments $x \in Q^-$ all summands on the left hand side then vanish except for $\alpha = 0$ (see Fig. C.2). We therefore get from (C.36)

$$\|u - u_N\|_{L_2(Q^-)} \leq \left(\frac{1}{2a}\right)^s \left\|\sum_{\beta\in\mathbb{Z}^s} \left(\widehat{u}\left(\frac{\beta}{2a}\right) - \widehat{u}_N\left(\frac{\beta}{2a}\right)\right) e^{2\pi i\beta\cdot\bullet/(2a)}\right\|_{L_2(Q^-)}$$

$$\leq \left(\frac{1}{2a}\right)^s \left\|\sum_{\beta\in\mathbb{Z}^s} \left(\widehat{u}\left(\frac{\beta}{2a}\right) - \widehat{u}_N\left(\frac{\beta}{2a}\right)\right) e^{2\pi i\beta\cdot\bullet/(2a)}\right\|_{L_2(Q)} \qquad (C.37)$$

Because of the orthonormality of the functions $x \mapsto e^{2\pi i \beta \cdot x/(2a)}$ with respect to the scalar product $\langle \bullet | \bullet \rangle_{L_2(Q)}$, we get from (C.37)

$$\|u - u_N\|^2_{L_2(Q^-)} \leq S_1 + S_2 + S_3, \quad \text{where} \tag{C.38}$$

$$S_1 := (2a)^{-2s} \sum_{\beta \in W} \left| \widehat{u} \left(\frac{\beta}{2a} \right) - \widehat{u_N} \left(\frac{\beta}{2a} \right) \right|^2,$$

$$S_2 := (2a)^{-2s} \sum_{\beta \notin W} \left| \widehat{u} \left(\frac{\beta}{2a} \right) \right|^2, \quad \text{and}$$

$$S_3 := (2a)^{-2s} \sum_{\beta \notin W} \left| \widehat{u_N} \left(\frac{\beta}{2a} \right) \right|^2.$$

From the interpolation conditions (2.95) we conclude $S_1 = 0$. Differentiating (C.36) we get

$$\frac{\partial}{\partial x_j} u(x) = (2a)^{-s} \sum_{\beta \in \mathbb{Z}^s} 2\pi i \cdot \frac{\beta_j}{2a} \cdot \widehat{u} \left(\frac{\beta}{2a} \right) e^{2\pi i \beta \cdot x/(2a)}, \quad x \in Q, \ j = 1, \ldots, s.$$

From this, we get

$$\|u\|^2_{H^1(Q)} = \sum_{|\alpha| \leq 1} \|D^\alpha u\|^2_{L_2(Q)} = (2a)^{-2s} \sum_{\beta \in \mathbb{Z}^s} \left(1 + \left[\frac{\pi}{a} \right]^2 \|\beta\|^2_2 \right) \left| \widehat{u} \left(\frac{\beta}{2a} \right) \right|^2.$$
$$\tag{C.39}$$

Now observe that $\|\beta\|^2_2 \geq (N/2)^2 = (a/h)^2$ for $\beta \notin W$. Consequently,

$$1 + \left[\frac{\pi}{a} \right]^2 \|\beta\|^2_2 \geq 1 + \left[\frac{\pi}{a} \right]^2 \left(\frac{a}{h} \right)^2 \geq \frac{1}{h^2} \quad \text{for} \quad \beta \notin W. \tag{C.40}$$

Therefore we can estimate

$$S_2 \leq h^2 (2a)^{-2s} \sum_{\beta \notin W} \left(1 + \left[\frac{\pi}{a} \right]^2 \|\beta\|^2_2 \right) \left| \widehat{u} \left(\frac{\beta}{2a} \right) \right|^2 \overset{(\text{C}.39)}{\leq} h^2 \|u\|^2_{H^1(Q)}.$$

It remains to estimate S_3. For notational simplicity this will only be done for $s = 2$. The case $s = 1$ can be treated analogously. Reusing (C.40) we at first get

$$S_3 \leq h^2 (2a)^{-4} \sum_{\beta \notin W} \left(1 + \left[\frac{\pi}{a} \right]^2 \|\beta\|^2_2 \right) \left| \widehat{u_N} \left(\frac{\beta}{2a} \right) \right|^2. \tag{C.41}$$

Every index $\beta \notin W$ can uniquely be written in the form $\beta = \alpha + N\gamma$, where $\alpha \in W$ and $\gamma \in \mathbb{Z}^2$, $\gamma \neq 0$. From (2.90) and (2.91) we know that[1]

$$\widehat{u_N}(\beta/(2a)) = \sigma_\beta U_\beta = \sigma_{\alpha+N\gamma} U_\alpha$$

because of the discrete Fourier coefficients' periodicity. Rewriting the right hand side of (C.41), we get

$$S_3 \leq h^2 (2a)^{-4} \sum_{\alpha \in W} \sum_{\gamma \neq 0} \left(1 + \left[\frac{\pi}{a}\right]^2 \|\alpha + N\gamma\|_2^2\right) |\sigma_{\alpha+N\gamma} U_\alpha|^2. \qquad (C.42)$$

It is easy to see that $\sigma_{\alpha+N\gamma} = 0$ for $(\alpha_1 = 0$ and $\gamma_1 \neq 0)$ or $(\alpha_2 = 0$ and $\gamma_2 \neq 0)$. The corresponding terms on the right hand side of (C.42) vanish. For the non-vanishing terms we get

$$\sigma_{\alpha+N\gamma} = \sigma_\alpha \cdot \frac{1}{(1 + N\gamma_1/\alpha_1)^2 (1 + N\gamma_2/\alpha_2)^2},$$

thereby setting $0/0 := 0$ in case $\alpha_1 = \gamma_1 = 0$ or $\alpha_2 = \gamma_2 = 0$. Thus we find for all pairs (α, γ) such that $\sigma_{\alpha+N\gamma}$ does not vanish and also using $0/0 := 0$:

$$\left(1 + \left[\frac{\pi}{a}\right]^2 \|\alpha + N\gamma\|_2^2\right) |\sigma_{\alpha+N\gamma}|^2$$

$$= |\sigma_\alpha|^2 \left(\frac{1 + \left[\frac{\pi}{a}\right]^2 \left(\alpha_1^2(1 + \gamma_1 N/\alpha_1)^2 + \alpha_2^2(1 + \gamma_2 N/\alpha_2)^2\right)}{(1 + N\gamma_1/\alpha_1)^4 (1 + N\gamma_2/\alpha_2)^4}\right)$$

$$\leq |\sigma_\alpha|^2 \left[1 + \left(\frac{\pi}{a}\right)^2 \|\alpha\|_2^2\right] \frac{1}{(1 + N\gamma_1/\alpha_1)^2 (1 + N\gamma_2/\alpha_2)^2}$$

$$\leq |\sigma_\alpha|^2 \left[1 + \left(\frac{\pi}{a}\right)^2\right] \|\alpha\|_2^2 \frac{1}{(2|\gamma_1| - 1)^2 (2|\gamma_2| - 1)^2},$$

the final estimate being true without any restriction for $\alpha \in W$ and $\gamma \in \mathbb{Z}^2 \setminus \{0\}$. Inserting into (C.42) we arrive at

$$S_3 \leq h^2 (2a)^{-4} \sum_{\alpha \in W} \left(1 + \left[\frac{\pi}{a}\right]^2 \|\alpha\|_2^2\right) |\sigma_\alpha U_\alpha|^2 \left(\sum_{\gamma \in \mathbb{Z}^2} \frac{1}{(2|\gamma_1| - 1)^2 (2|\gamma_2| - 1)^2}\right).$$

Since the sum over γ converges and since $\sigma_\alpha U_\alpha = \widehat{u_N}(\alpha/(2a)) = \widehat{u}(\alpha/(2a))$ for all $\alpha \in W$, we finally get

---

[1] In case $s = 1$ one rather uses (C.15).

$$S_3 \le Ch^2(2a)^{-4} \sum_{\alpha \in W} \left(1 + \left[\frac{\pi}{a}\right]^2 \|\alpha\|_2^2\right) \left|\widehat{u}\left(\frac{\alpha}{2a}\right)\right|^2 \le Ch^2 \|u\|_{H^1(Q)}^2$$

and this ends the proof.   □

Lemma C.4 hypothetically assumes knowledge of exact Fourier transform values $\widehat{u}(\beta/(2a))$, $\beta \in W$. In reality we only know approximate values

$$\frac{\widehat{w_N}\,(\beta/(2a))}{\widehat{k}\,(\beta/(2a))} \approx \widehat{u}\left(\frac{\beta}{2a}\right) = \frac{\widehat{w}\,(\beta/(2a))}{\widehat{k}\,(\beta/(2a))}, \quad \beta \in W.$$

Then the sum S_1 in (C.38) does no longer vanish. Rather, approximation errors $\widehat{w}(\beta/(2a)) - \widehat{w_N}(\beta/(2a))$, magnified by the division through $\widehat{k}(\beta/(2a))$, contribute to the overall error.

Appendix D
Regularization Property of CGNE

Theorem 3.41 showed that CGNE formally (i.e., in the sense of Definition 3.14) is a regularization method for the least squares problem. It did not show, however, that stopping CGNE after a number of iterations can produce a result closer to $\hat{x} = A^+ b$ than is the solution $\bar{x} = A^+ b^\delta$ of the least squares problem for perturbed data—compare Theorem 3.6. Theorem 3.41 therefore cannot justify using CGNE for practical regularization of (finite-dimensional) least squares problems. The following proof of the regularizing property of CGNE is modelled after Sect. 4.3.3 of [Lou89].

We have already seen in the proof of Theorem 3.41 that (3.177) guarantees a stop of CGNE according to the criterion (3.178) after k iterations, with $0 < k \le n$. Let now k be fixed accordingly. If $x_k = \bar{x}$ (i.e., CGNE has minimized $\|b^\delta - Ax\|_2$ exactly after k steps), then $\hat{x} - x_k = (A^\top A)^{-1} A^\top (b - b^\delta)$ and $\|\hat{x} - x_k\|_2 \le \|(A^\top A)^{-1} A^\top\|_2 \|b - b^\delta\|_2 \le \delta/\sigma_n$ with σ_n being the smallest singular value of A. This confirms estimate (3.179) from Theorem 3.41, but no regularization beyond this can be found. In the following, we rather assume that CGNE will be stopped by (3.178) for some index $k < n$, but $x_k \ne \bar{x}$ and that therefore

$$\rho_0, \ldots, \rho_k \ne 0. \tag{D.1}$$

Every $x \in \mathcal{K}_k$ (see (3.175)) can be written in the form $x = p(A^\top A) A^\top b^\delta$, where $p \in P_{k-1}$ is a polynomial of degree $k - 1$ or less, uniquely defined by x. Using an SVD $A = U \Sigma V^\top$ (the columns of U are $u_1, \ldots, u_m \in \mathbb{R}^m$, the columns of V are $v_1, \ldots, v_n \in \mathbb{R}^n$, and $\sigma_1 \ge \cdots \ge \sigma_n > 0$ are the singular values of A) and a reduced SVD $A = \hat{U} \hat{\Sigma} V^\top$ of A (omitting the last $m - n$ columns of U and rows of Σ), a short calculation shows

$$\|b^\delta - Ax\|_2^2 = \|b^\delta - Ap(A^\top A)A^\top b^\delta\|_2^2$$

$$= \sum_{i=1}^{n} \left(1 - \sigma_i^2 p(\sigma_i^2)\right)^2 (u_i^\top b^\delta)^2 + \sum_{i=n+1}^{m} (u_i^\top b^\delta)^2 \qquad \text{(D.2)}$$

for every $x \in \mathscr{K}_k$, where $p \in P_{k-1}$ is defined by x. CGNE produces the vector $x = x_k \in \mathscr{K}_k$ minimizing $\|b^\delta - Ax\|_2$. The corresponding polynomial from P_{k-1}, which minimizes (D.2) and *which depends on b^δ*, will be designated by p_{k-1}. Thus,

$$x_k = p_{k-1}(A^\top A)A^\top b^\delta = \sum_{i=1}^{n} \sigma_i p_{k-1}(\sigma_i^2)(u_i^\top b^\delta)v_i, \qquad \text{(D.3)}$$

as compared to the unregularized solution

$$\bar{x} = A^+ b^\delta = \sum_{i=1}^{n} \frac{1}{\sigma_i}(u_i^\top b^\delta)v_i. \qquad \text{(D.4)}$$

Now let

$$P_k^0 := \{q \in P_k; \; q(0) = 1\}.$$

Every polynomial $q \in P_k^0$ can uniquely be written in the form $q(t) = 1 - tp(t)$ for some $p \in P_{k-1}$. Especially set $q_k(t) := 1 - tp_{k-1}(t)$, with $p_{k-1} \in P_{k-1}$ as above. With $q_k \in P_k^0$ defined this way, the residual corresponding to x_k can be expressed in the form

$$\rho_k = A^\top b^\delta - A^\top Ax_k = A^\top b^\delta - A^\top Ap_{k-1}(A^\top A)A^\top b^\delta = q_k(A^\top A)A^\top b^\delta.$$

From this one computes the Euclidean scalar products

$$\langle \rho_j | \rho_\ell \rangle = \left[q_j(A^\top A)A^\top b^\delta\right]^\top \left[q_\ell(A^\top A)A^\top b^\delta\right]$$

$$= \sum_{i=1}^{n} \sigma_i^2 q_j(\sigma_i^2)q_\ell(\sigma_i^2)(u_i^\top b^\delta)^2 \quad \text{for } 1 \le j, \ell \le k, \qquad \text{(D.5)}$$

where we have used the SVD of A. From Lemma 3.39 one knows that $\langle \rho_j | \rho_\ell \rangle = 0$ for $j \ne \ell$ and this shows that the polynomials $q_1, \ldots, q_k \in P_k^0$ are orthogonal with respect to the scalar product defined on P_k by

$$\langle q | \tilde{q} \rangle_0 := \sum_{i=1}^{n} \sigma_i^2 q(\sigma_i^2)\tilde{q}(\sigma_i^2)(u_i^\top b^\delta)^2, \quad q, \tilde{q} \in P_k, \qquad \text{(D.6)}$$

which depends on b^δ.[1] From $\hat{x} = \sum(1/\sigma_i)(u_i^\top b)v_i$ and from (D.3) one derives

$$\|\hat{x} - x_k\|_2^2 = \sum_{i=1}^{n} \frac{1}{\sigma_i^2}\left[(u_i^\top b) - \sigma_i^2 p_{k-1}(\sigma_i^2)(u_i^\top b^\delta)\right]^2 \tag{D.10}$$

$$= \sum_{\sigma_i \geq \tau} \cdots + \sum_{\sigma_i < \tau} \cdots ,$$

where τ is a freely chosen positive number, used to split the sum (D.10) into two parts, which will be bounded independently of each other. Concerning the first sum, begin with

---

[1] It is evident that $\langle \bullet | \bullet \rangle_0$ is a symmetric bilinear form on P_k. Positive definiteness is established if

$$\langle q | q \rangle_0 = \sum_{i=1}^{n} \sigma_i^2 \left[q(\sigma_i^2)\right]^2 (u_i^\top b^\delta)^2 > 0 \quad \text{for} \quad q \in P_k \setminus \{0\} \tag{D.7}$$

can be shown. Since $\sigma_i > 0$ for $i = 1, \ldots, n$,

$$\langle q | q \rangle_0 = 0 \iff q(\sigma_i^2)(u_i^\top b^\delta) = 0, \quad i = 1, \ldots, n. \tag{D.8}$$

Assume now that σ_{k_j}, $j = 1, \ldots s$, are s *pairwise different* singular values of A with multiplicities $\mu_1, \ldots, \mu_s$, respectively. Numbering multiple singular values multiple times, we denote the above by $\sigma_{i_1}, \ldots \sigma_{i_\ell}$, $\ell = \mu_1 + \ldots + \mu_s$. Let u_{i_j}, $j = 1, \ldots, \ell$ be the corresponding columns from $U \in \mathbb{R}^{m,m}$. Assume further that

$$b^\delta \in \langle u_{i_1}, \ldots, u_{i_\ell}, u_{n+1}, \ldots, u_m \rangle.$$

In this case, repeating the argument that led to (D.2), one gets for $x = p(A^\top A)A^\top b^\delta \in \mathscr{K}_k$

$$b^\delta - Ax = \sum_{j=1}^{\ell} \underbrace{(1 - \sigma_{i_j}^2 p(\sigma_{i_j}^2))}_{q(\sigma_{i_j}^2)}(u_{i_j}^\top b^\delta)u_{i_j} + \sum_{i=n+1}^{m} (u_i^\top b^\delta)u_i. \tag{D.9}$$

Now if $s \leq k$, then one could choose

$$q_s(t) := \prod_{j=1}^{s}\left(1 - \frac{t}{\sigma_{k_j}^2}\right) \in P_s^0 \subseteq P_k^0$$

to achieve $b^\delta - Ax_s \in \langle u_{n+1}, \ldots, u_m \rangle = \mathscr{R}_A^\perp$ for the vector $x_s \in \mathscr{K}_s$ corresponding to q_s. But this would mean $A^\top(b^\delta - Ax_s) = 0$ and we would have $x_s = \bar{x}$. On the other hand, from (D.9) it can immediately be seen that the polynomial q_s just defined minimizes $\|b^\delta - Ax\|_2$ over $\mathscr{K}_s \subset \mathscr{K}_k$ and this means that x_s is exactly the vector produced by CGNE after s steps: CGNE would stop after $s \leq k$ steps in contradiction to (D.1). If CGNE does *not* stop after k iterations this means that at least $k + 1$ terms from $(u_i^\top b^\delta)$, $i = 1, \ldots, n$, belonging to *different* singular values σ_i must not vanish. But then in fact (D.8) cannot happen for $q \neq 0$, since there is no polynomial $q \in P_k \setminus \{0\}$ having $k + 1$ different zeros.

$$\sum_{\sigma_i \geq \tau} \cdots \leq \tau^{-2} \sum_{\sigma_i \geq \tau} \left[u_i^\top (b - b^\delta) + (1 - \sigma_i^2 p_{k-1}(\sigma_i^2)) u_i^\top b^\delta \right]^2.$$

Making use of $(a + b)^2 \leq 2a^2 + 2b^2$, of (3.177), and of

$$\sum_{\sigma_i \geq \tau} (1 - \sigma_i^2 p_{k-1}(\sigma_i^2))^2 (u_i^\top b^\delta)^2 \quad \leq \quad \sum_{i=1}^{n} (1 - \sigma_i^2 p_{k-1}(\sigma_i^2))^2 (u_i^\top b^\delta)^2$$

$$\stackrel{(D.2)}{\leq} \| b^\delta - A x_k \|_2^2 \stackrel{(3.178)}{\leq} \delta^2,$$

this leads to

$$\sum_{\sigma_i \geq \tau} \cdots \leq 2\tau^{-2} \| b - b^\delta \|_2^2 + 2\tau^{-2}\delta^2 \leq 4\tau^{-2}\delta^2. \tag{D.11}$$

The second sum on the right hand side of (D.10) equals zero for $\tau \leq \sigma_n$. In this case one ends up with the same (qualitative) result as in Theorem 3.41. But one may as well choose $\tau > \sigma_n$, in which case the second sum does not vanish. An upper bound for this sum can be split into two summands:

$$\sum_{\sigma_i < \tau} \cdots \leq 2 \underbrace{\sum_{\sigma_i < \tau} \frac{(u_i^\top b)^2}{\sigma_i^2}}_{=: S_1} + 2 \underbrace{\sum_{\sigma_i < \tau} \sigma_i^2 \left[p_{k-1}(\sigma_i^2) \right]^2 (u_i^\top b^\delta)^2}_{=: S_2} \tag{D.12}$$

From $\hat{x} = A^+ b = \sum (1/\sigma_i)(u_i^\top b) v_i$ one gets

$$S_1 \leq \frac{\tau^2}{\sigma_n^2} \sum_{\sigma_i < \tau} \frac{(u_i^\top b)^2}{\sigma_i^2} \leq \frac{\tau^2}{\sigma_n^2} \| \hat{x} \|_2^2. \tag{D.13}$$

We also get

$$S_2 = \sum_{\sigma_i < \tau} \sigma_i^2 \left[p_{k-1}(\sigma_i^2) \right]^2 (u_i^\top b + u_i^\top (b^\delta - b))^2$$

$$\leq 2 \sum_{\sigma_i < \tau} \left[\sigma_i p_{k-1}(\sigma_i^2) \right]^2 (u_i^\top b)^2 + 2 \sum_{\sigma_i < \tau} \left[\sigma_i p_{k-1}(\sigma_i^2) \right]^2 (u_i^\top b^\delta - u_i^\top b)^2$$

$$\leq 2\tau^2 [Q_k(\tau)]^2 \sum_{\sigma_i < \tau} (u_i^\top b)^2 + 2\tau^2 [Q_k(\tau)]^2 \sum_{\sigma_i < \tau} (u_i^\top (b^\delta - b))^2,$$

where the following abbreviation was used

$$Q_k(\tau) := \max_{0 \le t \le \tau} |p_{k-1}(t^2)|. \tag{D.14}$$

Making use of the orthogonality of the polynomials $q_1, \ldots, q_k$, from which it follows that q_k has k distinct real zeros

$$0 < t_{k,1} < t_{k,2} < \cdots < t_{k,k} < \|A\|_2^2, \tag{D.15}$$

it is proven in Lemma 4.3.15 of [Lou89] that

$$Q_k(\tau) = \max_{0 \le t \le \tau} |p_{k-1}(t^2)| = p_{k-1}(0) = \sum_{j=1}^{k} \frac{1}{t_{k,j}} \ge \frac{1}{t_{k,1}}.$$

Consequently

$$\tau^2 [Q_k(\tau)]^2 \le 1 \quad \text{if} \quad \tau \le [p_{k-1}(0)]^{-1} \le t_{k,1}, \tag{D.16}$$

and the above estimate can be continued as follows:

$$S_2 \le 2 \sum_{\sigma_i < \tau} \sigma_i^4 \frac{(u_i^\top b)^2}{\sigma_i^4} + 2\|b^\delta - b\|_2^2 \le 2\tau^4 \frac{\|\hat{x}\|_2^2}{\sigma_n^2} + 2\delta^2. \tag{D.17}$$

Putting (D.11), (D.12), (D.13), and (D.16) together, one arrives at

$$\|\hat{x} - x_k\|_2^2 \le 4\delta^2(1 + \tau^{-2}) + 2\frac{\|\hat{x}\|_2^2}{\sigma_n^2}(\tau^2 + 2\tau^4).$$

For $\tau \le \min\{1, [p_{k-1}(0)]^{-1}\}$ one further estimates

$$\|\hat{x} - x_k\|_2^2 \le 8 \left(\tau^{-2}\delta^2 + \tau^2 \frac{\|\hat{x}\|_2^2}{\sigma_n^2} \right). \tag{D.18}$$

This estimate shows that the choice of a (not too) small value τ may counterbalance the negative effect of a small singular value σ_n. Setting, for example,

$$\tau = \theta \cdot \sqrt{\frac{\sigma_n \cdot \delta}{\|\hat{x}\|_2}}, \quad \theta \text{ such that } \tau \le \min\{1, [p_{k-1}(0)]^{-1}\},$$

one derives the estimate

$$\|\hat{x} - x_k\|_2^2 \le \left(8\theta^2 + 8\theta^{-2} \right) \frac{\|\hat{x}\|_2}{\sigma_n} \delta$$

or, equivalently,

$$\|\hat{x} - x_k\|_2 \leq C \sqrt{\frac{\|\hat{x}\|_2}{\sigma_n}} \cdot \sqrt{\delta}, \quad C = \sqrt{8\theta^2 + 8\theta^{-2}}. \tag{D.19}$$

As to the regularizing effect of CGNE, the polynomial $q_k(t) = 1 - t p_{k-1}(t)$ has k distinct zeroes $0 < t_{k,1} < \cdots < t_{k,k} < \|A\|_2^2 = \sigma_1^2$, see (D.15). This means that p_{k-1} interpolates the function $t \mapsto 1/t$ at these zeroes. Consequently, the polynomial $t p_{k-1}(t^2)$ also interpolates $t \mapsto 1/t$ at $t_{k,1}, \ldots, t_{k,k}$, but has a zero at $t = 0$, where $t \mapsto 1/t$ has a pole. In view of the two representations (D.3) and (D.4) this explains how CGNE achieves a regularization of least squares problems.

Appendix E
Existence and Uniqueness Theorems for Waveform Inversion

We collect some mathematical results related to waveform inversion.

E.1 Wave Equation with Constant Coefficient

The results presented in this paragraph can be found, e.g., in [Eva98, Sec. 2.4]. To begin with, consider the one-dimensional, homogeneous initial-value problem

$$\begin{cases} u_{tt} - c^2 u_{xx} = 0, & (x, t) \in \mathbb{R} \times]0, \infty[, \\ u(x, 0) = \varphi(x), \quad u_t(x, 0) = \psi(x), & x \in \mathbb{R}, \end{cases} \tag{E.1}$$

where $c > 0$ is some constant and where $\varphi \in C^2(\mathbb{R})$ and $\psi \in C^1(\mathbb{R})$. The unique solution of (E.1) reads

$$u(x, t) = \frac{1}{2} \left(\varphi(x - ct) + \varphi(x + ct) \right) + \frac{1}{2c} \int_{x-ct}^{x+ct} \psi(\zeta) \, d\zeta. \tag{E.2}$$

This is **d'Alembert's formula**. From (E.2), we learn that the solution of (E.1) has the general form

$$u(x, t) = F(x - ct) + G(x + ct)$$

for appropriate functions F and G. This is interpreted as the superposition of two waves, one traveling at speed c to the left, and the other traveling at speed c to the right. We can also see that the smoothness of F and G—which in turn is determined by φ and ψ—directly translates into the smoothness of the solution u. In case φ is twice continuously differentiable and ψ once, the function u is twice continuously differentiable. Assuming smoother initial values, the solution u will

© The Author(s), under exclusive license to Springer Nature Switzerland AG 2020
M. Richter, *Inverse Problems*, Lecture Notes in Geosystems
Mathematics and Computing, https://doi.org/10.1007/978-3-030-59317-9

also be smoother. On the other hand, we could define some "weak" solution of (E.1), requiring restricted smoothness properties of F and G. This will be pursued below.

Next we move the start time from $t = 0$ to $t = s \geq 0$. Thus, we consider the initial-value problem

$$\begin{cases} u_{tt} - c^2 u_{xx} = 0, & (x, t) \in \mathbb{R} \times]s, \infty[, \\ u(x, s) = \varphi(x), \quad u_t(x, s) = \psi(x), & x \in \mathbb{R}. \end{cases} \tag{E.3}$$

According to (E.1), we obtain the unique solution

$$u(x, t; s) = \frac{1}{2} [\varphi(x - c(t - s)) + \varphi(x + c(t - s))] + \frac{1}{2c} \int_{x-c(t-s)}^{x+c(t-s)} \psi(\zeta) \, d\zeta \tag{E.4}$$

of (E.3), which depends on the parameter s. With $\varphi \in C^2(\mathbb{R})$ and $\psi \in C^1(\mathbb{R})$, we have $u(\cdot, \cdot; s) \in C^2(\mathbb{R} \times [s, \infty[)$ for every $s \geq 0$. We will now use formula (E.4) to solve the *inhomogeneous* initial-value problem

$$\begin{cases} u_{tt} - c^2 u_{xx} = h(x, t), & (x, t) \in \mathbb{R} \times]0, \infty[, \\ u(x, 0) = 0, \quad u_t(x, 0) = 0, & x \in \mathbb{R}. \end{cases} \tag{E.5}$$

For $h \in C^1(\mathbb{R} \times [0, \infty[)$, the unique solution u of (E.5) reads

$$u(x, t) = \int_0^t u(x, t; s) \, ds = \frac{1}{2c} \int_0^t \int_{x-c(t-s)}^{x+c(t-s)} h(\zeta, s) \, d\zeta \, ds. \tag{E.6}$$

The technique of computing the solution of an inhomogeneous problem as the superposition of appropriately chosen homogeneous problems is known as **Duhamel's principle**.

Proof Choose $\varphi(x) = 0$ and $\psi(x) = h(x, s)$ in (E.3). The corresponding solution (E.4) reads

$$u(x, t; s) = \frac{1}{2c} \int_{x-c(t-s)}^{x+c(t-s)} h(\zeta, s) \, d\zeta. \tag{E.7}$$

Since $h \in C^1(\mathbb{R} \times [0, \infty[)$, we have $u(\cdot, \cdot; s) \in C^2(\mathbb{R} \times [s, \infty[)$ for every $s \geq 0$. Consequently, $u \in C^2(\mathbb{R} \times [0, \infty[)$ for u as defined by (E.6). From (E.6), we immediately get $u(x, 0) = 0$, as required by (E.5). Differentiating (E.7) according to Leibniz's formula gets us

$$u_t(x, t) = u(x, t; t) + \int_0^t u_t(x, t; s) \, ds = \int_0^t u_t(x, t; s) \, ds,$$

$$u_{tt}(x, t) = u_t(x, t; t) + \int_0^t u_{tt}(x, t; s)\, ds$$

$$= h(x, t) + c^2 \int_0^t u_{xx}(x, t; s)\, ds,$$

$$u_{xx}(x, t) = \int_0^t u_{xx}(x, t; s)\, ds.$$

These equations show that (E.5) holds.   □

Next we look at the intial/boundary-value problem

$$\begin{cases} u_{tt} - c^2 u_{xx} = 0, & x > 0,\ t > 0, \\ u(x, 0) = 0, \quad u_t(x, 0) = \psi(x), & x \geq 0, \\ u_x(0, t) = 0, & t \geq 0, \end{cases} \tag{E.8}$$

which will be solved using a technique called **reflection method**. To this end, we use the even continuation of ψ to a function

$$\tilde{\psi}(x) := \begin{cases} \psi(x), & x \geq 0 \\ \psi(-x), & x \leq 0 \end{cases}$$

defined on the real line. We have $\tilde{\psi} \in C^1(\mathbb{R})$, iff $\psi \in C^1[0, \infty[$ and $\psi'(0) = 0$. Let $\tilde{u}$ be the solution of (E.1) for $\varphi = 0$ and for $\tilde{\psi}$ instead of ψ. According to d'Alembert's formula, we get

$$\tilde{u}(x, t) = \frac{1}{2c} \int_{x-ct}^{x+ct} \tilde{\psi}(\zeta)\, d\zeta. \tag{E.9}$$

Straightforward calculation shows that the restriction of $\tilde{u}$ to the half-space $x \geq 0$ solves (E.8). Conversely, a solution u of (E.8) can be extended to a function

$$\tilde{u}(x, t) := \begin{cases} u(x, t), & x \geq 0,\ t \geq 0 \\ u(-x, t), & x \leq 0,\ t \geq 0 \end{cases}$$

which solves (E.1) for $\tilde{\psi}$ instead of ψ. This shows that (E.8) has a unique solution. Writing (E.9) in terms of ψ only, we explicitly get

$$u(x, t) = \begin{cases} \frac{1}{2c} \int_{x-ct}^{x+ct} \psi(\zeta)\, d\zeta, & x \geq ct \geq 0 \\ \frac{1}{2c} \int_0^{x+ct} \psi(\zeta)\, d\zeta + \frac{1}{2c} \int_0^{ct-x} \psi(\zeta)\, d\zeta, & 0 \leq x \leq ct \end{cases} \tag{E.10}$$

as the solution of (E.8). Note that this solution does not belong to C^2, unless $\psi'(0) = 0$.

Finally, we consider the inhomogeneous, initial/boundary-value problem

$$\begin{cases} u_{tt} - c^2 u_{xx} = h(x, t), & x > 0, \ t > 0, \\ u(x, 0) = 0, \quad u_t(x, 0) = 0, & x \geq 0, \\ u_x(0, t) = 0, & t \geq 0. \end{cases} \qquad \text{(E.11)}$$

According to Duhamel's principle, the solution will be obtained in the form

$$u(x, t) = \int_0^t u(x, t; s) \, ds,$$

where $u(x, t; s)$ solves the parameterized problem

$$\begin{cases} u_{tt} - c^2 u_{xx} = 0, & x > 0, \ t > s \geq 0, \\ u(x, s) = 0, \quad u_t(x, s) = h(x, s) & x \geq 0, \\ u_x(0, t) = 0, & t \geq s. \end{cases}$$

According to (E.10) the solution of the latter reads

$$u(x, t; s) = \begin{cases} \frac{1}{2c} \int_{x-c(t-s)}^{x+c(t-s)} h(y, s) \, dy, & x \geq c(t - s) \geq 0 \\ \frac{1}{2c} \int_0^{x+c(t-s)} h(y, s) \, dy + \frac{1}{2c} \int_0^{c(t-s)-x} h(y, s) \, dy, & 0 \leq x \leq c(t - s). \end{cases}$$

If $h(\cdot, s)$ is continuously differentiable and $h_x(0, s) = 0$ holds, then $u(\cdot, \cdot; s) \in C^2(]0, \infty[\times]s, \infty[)$. Within the region defined by $x \geq ct$, where $x \geq c(t - s)$ holds a fortiori for all $s \in [0, t]$, we thus derive

$$u(x, t) = \frac{1}{2c} \int_0^t \int_{x-c(t-s)}^{x+c(t-s)} h(y, s) \, dy \, ds. \qquad \text{(E.12)}$$

Within the region $x \leq ct$, we have to distinguish, whether $x \geq c(t - s)$ or $x \leq c(t - s)$. We thus get

$$u(x, t) = \int_0^{t-x/c} u(x, t; s) \, ds + \int_{t-x/c}^t u(x, t; s) \, ds,$$

which leads to the explicit formula

$$u(x, t) = \frac{1}{2c} \left(\int_0^{t-x/c} \left[\int_0^{x+c(t-s)} h(y, s)\, dy + \int_0^{c(t-s)-x} h(y, s)\, dy \right] ds \right.$$

$$\left. + \int_{t-x/c}^{t} \int_{x-c(t-s)}^{x+c(t-s)} h(y, s)\, dy\, ds \right).$$

$$(E.13)$$

The solution u defined by (E.12) and (E.13) is twice continuously differentiable if h is continuously differentiable with $h_x(0, s) = 0$. For less smooth functions h like the one from (1.69), the integrals in (E.12) and (E.13) may still make sense and define a solution with reduced differentiability.

E.2 Identifiability of Acoustic Impedance

In this paragraph, we will consider the initial/boundary-value problem for a partial differential equation of second order:

$$\sigma(x) \frac{\partial^2 y(x, t)}{\partial t^2} - \frac{\partial}{\partial x} \left(\sigma(x) \frac{\partial y(x, t)}{\partial x} \right) = 0, \quad x \in]0, X_0[, \ t \in]0, T_0[, \quad (1.44)$$

$$y(x, 0) = 0, \quad \frac{\partial y(x, 0)}{\partial t} = 0, \qquad\qquad\qquad x \in]0, X_0[, \quad (1.45)$$

$$-\sigma(0) \frac{\partial y(0, t)}{\partial x} = g(t), \quad \frac{\partial y(X_0, t)}{\partial x} = 0 \qquad\qquad t \in]0, T_0[. \quad (1.46)$$

We will formulate results from sections 4 and 5 of [BCL77] about

- The existence and uniqueness of a solution y for given functions g and σ.
- The identifiability of σ from knowledge of y and g.

To start with, the above system will be transformed into an initial/boundary-value problem for two differential equations of first order. To do so, we have to assume that σ is continuously differentiable and bounded from below and above:

$$\sigma \in C^1[0, X_0] \quad \text{and} \quad 0 < \sigma_- \leq \sigma(x) \leq \sigma_+.$$

Moreover, we will assume g to be continuous. Let $y : [0, X_0] \times [0, T_0] \to \mathbb{R}$ be a solution of (1.44)–(1.46)—this means that y is continuously differentiable on $[0, X_0] \times [0, T_0]$ (with appropriate one-sided differential quotients at the boundary) and two times continuously differentiable on $]0, X_0[\times]0, T_0[$. Then we can define C^0-functions $u, v : [0, X_0] \times [0, T_0] \to \mathbb{R}$ by

$$u(x, t) = \left(\frac{\sigma(x)}{2}\right)^{1/2} \left(\frac{\partial y}{\partial t}(x, t) + \frac{\partial y}{\partial x}(x, t)\right) \tag{E.14}$$

and

$$v(x, t) = \left(\frac{\sigma(x)}{2}\right)^{1/2} \left(\frac{\partial y}{\partial t}(x, t) - \frac{\partial y}{\partial x}(x, t)\right), \tag{E.15}$$

which are continuously differentiable on $]0, X_0[\times]0, T_0[$. It can readily be verified that

$$\frac{\partial y}{\partial t} = \frac{1}{\sqrt{2\sigma}}(u + v), \qquad \frac{\partial y}{\partial x} = \frac{1}{\sqrt{2\sigma}}(u - v), \tag{E.16}$$

and that u and v solve the following equations:

$$\frac{\partial u}{\partial t} - \frac{\partial u}{\partial x} + \frac{\sigma'(x)}{2\sigma(x)} v = 0, \qquad\qquad x \in]0, X_0[, \; t \in]0, T_0[, \tag{E.17}$$

$$\frac{\partial v}{\partial t} + \frac{\partial v}{\partial x} - \frac{\sigma'(x)}{2\sigma(x)} u = 0, \qquad\qquad x \in]0, X_0[, \; t \in]0, T_0[, \tag{E.18}$$

$$u(x, 0) = 0, \quad v(x, 0) = 0, \qquad\qquad x \in]0, X_0[, \tag{E.19}$$

$$-\left(\frac{\sigma(0)}{2}\right)^{1/2} (u(0, t) - v(0, t)) = g(t), \qquad\qquad t \in]0, T_0[, \tag{E.20}$$

$$u(X_0, t) - v(X_0, t) = 0, \qquad\qquad t \in]0, T_0[. \tag{E.21}$$

Conversely, assume u and v are continuously differentiable solutions of (E.17)–(E.21). By (E.17) and (E.18), the vector field

$$F = \begin{pmatrix} F_1 \\ F_2 \end{pmatrix} = \frac{1}{\sqrt{2\sigma}} \begin{pmatrix} u - v \\ u + v \end{pmatrix}$$

meets the integrability condition $\partial F_1/\partial t = \partial F_2/\partial x$. Thus there exists a function y on $]0, X_0[\times]0, T_0[$, having F as its gradient—as stated by (E.16). It is straightforward then to show that y is a solution of (1.44), (1.45), and (1.46).

It can be shown that for continuous g and under the above condition on σ a unique solution of (E.17)–(E.21) exists. But such a result also holds under weaker assumptions, as will be stated next. Setting

$$R =]0, X_0[\times]0, T_0[$$

and using partial integration, for $u \in C^1(R)$ and $\varphi \in C_0^\infty(R)$ one obtains

$$\int_R \left(\frac{\partial u}{\partial t} - \frac{\partial u}{\partial x} \right) \varphi \, dx \, dt = \int_R u \left(\frac{\partial \varphi}{\partial x} - \frac{\partial \varphi}{\partial t} \right) dx \, dt.$$

For the integral on the right hand side to exist, we only have to assume u to be integrable or square integrable. In analogy to (B.10), we will say that $u \in L_2(R)$ has a weak directional derivative $\frac{\partial u}{\partial t} - \frac{\partial u}{\partial x}$, if there exists some function $w \in L_1(R)$ such that

$$\int_R w\varphi \, dx \, dt = \int_R u \left(\frac{\partial \varphi}{\partial x} - \frac{\partial \varphi}{\partial t} \right) dx \, dt$$

holds for all $\varphi \in C_0^\infty(R)$. We will then write $w = \partial u/\partial t - \partial u/\partial x$. Let us define

$$U := \left\{ u \in L_2(R); \ \frac{\partial u}{\partial t} - \frac{\partial u}{\partial x} \in L_1(R) \right\}, \tag{E.22}$$

where the directional derivative is to be understood in the weak sense. Analogously, set

$$V := \left\{ v \in L_2(R); \ \frac{\partial v}{\partial t} + \frac{\partial v}{\partial x} \in L_1(R) \right\}. \tag{E.23}$$

Since R is bounded, using (B.9) one easily shows $u \in L_1(R)$ for any $u \in U$. Alike, $v \in L_1(R)$ if $v \in V$. It can be shown [BCL77, section 4.2] that $u(\bullet, t) \in L_1(0, X_0)$ for $u \in U$ and for every fixed value $t \in [0, T_0]$. Moreover, the mapping

$$[0, T_0] \to L_1(0, X_0), \quad t \mapsto u(\bullet, t),$$

is continuous. These facts are expressed by saying that $u \in C([0, T_0], L_1(0, X_0))$—the space of continuous, L_1-valued mappings defined on $[0, T_0]$. Analogously, $u(x, \bullet) \in L_1(0, T_0)$ for every $x \in [0, X_0]$ and the mapping

$$[0, X_0] \to L_1(0, T_0), \quad x \mapsto u(x, \bullet),$$

is continuous, which will be expressed by saying $u \in C([0, X_0], L_1(0, T_0))$. Analogous results hold for V, so that

$$U, V \quad \subset \quad C([0, X_0], L_1(0, T_0)) \cap C([0, T_0], L_1(0, X_0)). \tag{E.24}$$

Acoustic impedances will be assumed to lie in the space

$$\mathscr{S} := \left\{ \sigma \in H^1(0, X_0); \ 0 < \sigma_- \overset{a.e.}{\leq} \sigma(x) \overset{a.e.}{\leq} \sigma_+, \ \|\sigma'\|_{L_2(0, X_0)} \leq M \right\}. \tag{E.25}$$

Since all functions from the Sobolev space $H^1(0, X_0)$ are (almost everywhere) continuous on $[0, X_0]$, the above condition on uniform boundedness makes sense. If $\sigma \in \mathscr{S}$, then $\sigma'/(2\sigma) \in L_2(0, X_0)$.

We now consider (E.17)–(E.21) for $g \in L_2(0, T_0)$ and $\sigma \in \mathscr{S}$. Under these assumptions, we call (u, v) a **weak solution** of (E.17)–(E.21), if

- $u \in U$, $v \in V$,
- (E.17) holds as an equality of the $L_1(R)$-functions $\partial u/\partial t - \partial u/\partial x$ and $-[\sigma'/(2\sigma)]v$ (note (B.9)); Eq. (E.18) is to be interpreted in the same sense,
- (E.19)–(E.21) hold as equalities of L_2-functions; these equations make sense because of (E.24).

From [BCL77], we take

Theorem E.1 *For $g \in L_2(0, T_0)$ and $\sigma \in \mathscr{S}$, the system (E.17)–(E.21) admits a unique weak solution $(u, v) \in U \times V$.*

According to (1.36), because of $\partial u(0, t)/\partial t = \partial y(0, t)/\partial t$, and because of (E.16), the function

$$d : [0, T_0] \to \mathbb{R}, \quad t \mapsto \frac{1}{\sqrt{2\sigma(0)}} (u(0, t) + v(0, t)) \qquad (E.26)$$

is to be interpreted as the observed seismogram. Theorem E.1 assures that seismograms d are determined by acoustic impedances $\sigma \in \mathscr{S}$, when the function g is assumed to be fixed. In other words,

$$T : \mathscr{S} \to L_2(0, T_0), \quad \sigma \mapsto T(\sigma) = d, \qquad (E.27)$$

where d is defined by (E.26) and where u and v are defined by Theorem E.1, indeed is a function. Also from [BCL77], we take

Theorem E.2 *If $0 \in supp(g)$, the operator T defined by (E.27) is injective.*

In case $g(t) = 0$ for $0 \le t \le \tau$ for some $\tau > 0$, the solution y of (1.44), (1.45), and (1.46) would read $y(x, t) = 0$ for $0 \le t \le \tau$. Therefore, we can always assume that $0 \in supp(g)$ is fulfilled. Theorem E.2 means that in principle $\sigma \in \mathscr{S}$ can be derived from knowledge of a corresponding (exact) seismogram d. This result is in accordance with (1.44), (1.45), and (1.46): For a known, smooth function y, (1.44) means an ordinary differential equation for σ, which is uniquely determined by the additional initial value defined by (1.46).

References

[BACO02] H. Bertete-Aguirre, E. Cherkaev, M. Oristaglio, Non-smooth gravity problem with total variation penalization functional. Geophys. J. Int. **149**, 499–507 (2002)

[BCF19] H. Barucq, G. Chavent, F. Faucher, A priori estimates of attraction basins for nonlinear least squares, with application to Helmholtz seismic inverse problem. Inverse Problems **35**, 115004 (2019)

[BCL77] A. Bamberger, G. Chavent, P. Lailly, Etude mathématique et numérique d'un problème inverse pour l'équation des ondes à une dimension. Rapport LABORIA nr. 226, IRIA (1977)

[BCL79] A. Bamberger, G. Chavent, P. Lailly, About the stability of the inverse problem in 1-D wave equations—application to the interpretation of seismic profiles. Appl. Math. Optim. **5**, 1–47 (1979)

[BCL99] M.A. Branch, T.F. Coleman, Y. Li, A subspace, interior, and conjugate gradient method for large-scale bound-constrained minimization problems. SIAM J. Sci. Comput. **21**(1), 1–23 (1999)

[BG09] C. Burstedde, O. Ghattas, Algorithmic strategies for full waveform inversion: 1D experiments. Geophysics **74**, WCC37–WCC46 (2009)

[Bjö96] Å. Björck, *Numerical Methods for Least Squares Problems* (SIAM, Philadelphia, 1996)

[BO05] M. Burger, S. Osher, A survey on level set methods for inverse problems and optimal design. Eur. J. Appl. Math. **16**, 263–301 (2005)

[Bra07] D. Braess, *Finite Elements*, 3rd edn. (Cambridge University Press, Cambridge, 2007)

[BSZC95] C. Bunks, F.M. Salek, S. Zaleski, G. Chavent, Multiscale seismic waveform inversion. Geophysics **60**, 1457–1473 (1995)

[CCG01] F. Cément, G. Chavent, S. Gómez, Migration-based traveltimes waveform inversion of 2-D simple structures: A synthetic example. Geophysics **66**, 845–860 (2001)

[Cha74] G. Chavent, Identification of functional parameters in partial differential equations, in *Identification of Parameters in Distributed Systems*, ed. by R.E. Goodson, M. Polis (The American Society of Mechanical Engineering, New York, 1974), pp. 31–48

[Cha09] G. Chavent, *Nonlinear Least Squares for Inverse Problems* (Springer, Dordrecht, 2009)

[DB74] G. Dahlquist, Å. Björck, *Numerical Methods* (Prentice Hall, Englewood Cliffs, 1974)

[dB90] C. de Boor, *Splinefunktionen* (Birkhäuser, Basel, 1990)

[Dem97] J.W. Demmel, *Applied Numerical Linear Algebra* (SIAM, Philadelphia, 1997)

[EHN96] H.W. Engl, M. Hanke, A. Neubauer, *Regularization of Inverse Problems* (Kluwer Academic Publishers, Dordrecht, 1996)

© The Author(s), under exclusive license to Springer Nature Switzerland AG 2020
M. Richter, *Inverse Problems*, Lecture Notes in Geosystems Mathematics and Computing, https://doi.org/10.1007/978-3-030-59317-9

[EKN89] H.W. Engl, K. Kunisch, A. Neubauer, Convergence rates for Tikhonov regularisation of non-linear ill-posed problems. Inverse Prob. **5**, 523–540 (1989)

[Eld77] L. Eldén, Algorithms for the regularisation of ill-conditioned least squares problems. BIT **17**, 134–145 (1977)

[Eld82] L. Eldén, A weighted pseudoinverse, generalized singular values, and constrained least squares problems. BIT **22**, 487–502 (1982)

[Eng97] H.W. Engl, *Integralgleichungen* (Springer, Wien, 1997)

[Eva98] L.C. Evans, *Partial Differential Equations* (AMS, Philadelphia, 1998)

[FCBC20] F. Faucher, G. Chavent, H. Barucq, H. Calandra, A priori estimates of attraction basins for velocity model reconstruction by time-harmonic Full Waveform Inversion and Data-Space Reflectivity formulation. Geophysics **85**, 1–126 (2020)

[Fic11] A. Fichtner, *Full Seismic Waveform Modelling and Inversion* (Springer, Berlin, 2011)

[FM12] D. Fischer, V. Michel, Sparse regularization of inverse gravimetry – case study: spatial and temporal mass variations in South America. Inverse Prob. **28** (2012). https://doi.org/10.1088/0266-5611/28/6/065012

[FN18a] W. Freeden, M.Z. Nashed, Inverse Gravimetry: background material and multiscale mollifier approaches. Int. J. Geomath. **9**, 199–264 (2018)

[FN18b] W. Freeden, M.Z. Nashed, Operator theoretic and regularization approaches to ill-posed problems. Int. J. Geomath. **9**, 9–115 (2018)

[FNS10] W. Freeden, M.Z. Nashed, T. Sonar, *Handbook of Geomathematics* (Springer, Heidelberg, 2010)

[Fou99] K. Fourmont, Schnelle Fourier-Transformation bei nichtäquidistanten Gittern und tomographische Anwendungen. Ph. D. Thesis, Universität Münster, Germany, 1999

[Gau72] W. Gautschi, Attenuation factors in practical Fourier analysis. Numer. Math. **18**, 373–400 (1972)

[GHW79] G.H. Golub, M. Heath, G. Wahba, Generalized cross validation as a method for choosing a good ridge parameter. Technometrics **21**, 215–224 (1979)

[Han92] P.C. Hansen, Analysis of discrete ill-posed problems by means of the L-curve. SIAM Rev. **34**, 561–580 (1992)

[Han97a] M. Hanke, A regularizing Levenberg-Marquardt scheme, with applications to inverse groundwater filtration problems. Inverse Prob. **13**, 79–95 (1997)

[Han97b] M. Hanke, Regularizing properties of a truncated Newton-CG algorithm for nonlinear inverse problems. Numer. Funct. Anal. Optim. **18**, 971–993 (1997)

[Hof99] B. Hofmann, *Mathematik inverser Probleme* (Teubner, Stuttgart, 1999)

[ILQ11] V. Isakov, S. Leung, J. Qian, A fast local level set method for inverse gravimetry. Commun. Comput. Physics **10**, 1044–1070 (2011)

[Isa90] V. Isakov, *Inverse Source Problems* (AMS, Philadelphia, 1990)

[Isa06] V. Isakov, *Inverse Problems for Partial Differential Equations*, 2nd edn. (Springer, New York, 2006)

[JBC+89] M. Jannane, W. Beydoun, E. Crase, D. Cao, Z. Koren, E. Landa, M. Mendes, A. Pica, M. Noble, G. Roeth, S. Singh, R. Snieder, A. Tarantola, D. Trezeguet, M. Xie, Wavelengths of Earth structures that can be resolved from seismic reflection data. Geophysics **54**, 906–910 (1989)

[Kir96] A. Kirsch, *An Introduction to the Mathematical Theory of Inverse Problems* (Springer, 1996)

[Kir11] A. Kirsch, *An Introduction to the Mathematical Theory of Inverse Problems*, 2nd edn. (Springer, New York, 2011)

[KNS08] B. Kaltenbacher, A. Neubauer, O. Scherzer, *Iterative Regularization Methods for Nonlinear Ill-Posed Problems*. (de Gruyter, Berlin, 2008)

[KR14] A. Kirsch, A. Rieder, Seismic tomography is locally ill-posed. Inverse Prob. **30**, 125001 (2014)

[Lou89] A.K. Louis, *Inverse und schlecht gestellte Probleme* (Teubner, Stuttgart, 1989)

[Lou96] A.K. Louis, Approximate inverse for linear and some nonlinear problems. Inverse Prob. **12**, 175–190 (1996)

[LT03] S. Larsson, V. Thomée, *Partial Differential Equations with Numerical Methods* (Springer, Berlin, 2003)

[LY08] D. Luenberger, Y. Ye, *Linear and Nonlinear Programming* (Springer, New York, 2008)

[Mat14] Matlab, *Release 2014b* (The MathWorks Inc., Natick, 2014)

[MBM$^+$16] L. Métivier, R. Brossier, Q. Mérigot, E. Oudet, J. Virieux, Measuring the misfit between seismograms using an optimal transport distance: application to full waveform inversion. Geophys. J. Int. **205**, 345–337 (2016)

[MF08] V. Michel, A.S. Fokas, A unified approach to various techniques for the non-uniqueness of the inverse gravimetric problem and wavelet-based methods. Inverse Prob. **24**, 1–23 (2008)

[Mic05] V. Michel, Regularized wavelet-based multiresolution recovery of the harmonic mass density distribution from data of the Eearth's gravitational field at satellite height. Inverse Prob. **21**, 997–1025 (2005)

[Mor78] J.J. Moré, The Levenberg-Marquardt algorithm: implementation and theory, in *Numerical Analysis. Proceedings Biennial Conference Dundee 1977, Lecture Notes in Mathematics*, ed. by G.A. Watson, vol. 630 (Springer, Berlin, 1978), pp. 105–116

[MS20] V. Michel, N. Schneider, A first approach to learning a best basis for gravitational field modelling. GEM Int. J. Geomath. **11**, 1 (2020)

[Nat77] F. Natterer, Regularisierung schlecht gestellter Probleme durch Projektionsverfahren. Numer. Math. **28**, 329–341 (1977)

[Nie86] Y. Nievergelt, Elementary inversion of Radon's transform. SIAM Rev. **28**, 79–84 (1986)

[Pra99] R.G. Pratt, Seismic waveform inversion in the frequency domain, Part I: theory and verification in a physical scale model. Geophysics **64**, 888–901 (1999)

[PTVF92] W.H. Press, S.A. Teukolsky, W.T. Vetterling, B.B. Flannery, *Numerical Recipes in C*, 2nd edn. (Cambridge University Press, Cambridge, 1992)

[Ram02] R. Ramlau, Morozov's discrepancy principle for Tikhonov regularization of nonlinear operators. Numer. Funct. Anal. Optim. **23**, 147–172 (2002)

[Rei67] C.H. Reinsch, Smoothing by spline functions. Numer. Math. **10**, 177–183 (1967)

[ROF92] L.I. Rudin, S. Osher, E. Fatemi, Nonlinear total variation based noise removal algorithms. Physica D **60**, 259–268 (1992)

[Sam11] D. Sampietro, GOCE exploitation for Moho modeling and applications, in *Proceedings of 4th International GOCE User Workshop, Munich*, ed. by L. Ouwehand (ESA Communications, Oakville, 2011)

[SC91] W.W. Symes, J.J. Carazzone, Velocity inversion by differential semblance optimization. Geophysics **56**, 654–663 (1991)

[Sch07] G.T. Schuster, Basics of seismic wave theory. http://utam.gg.utah.edu/tomo06/06_seg/basicseisbook.pdf (2007)

[Sch12] S. Schäffler, *Global Optimization. A Stochastic Approach* (Springer, New York, 2012)

[SP04] L. Sirgue, R.G. Pratt, Efficient waveform inversion and imaging: a strategy for selecting temporal frequencies. Geophysics **69**, 231–248 (2004)

[Sym09] W.W. Symes, The seismic reflection inverse problem. Inverse Prob. **25** (2009). https://doi.org/10.1088/0266-5611/25/12/123008

[TB97] L.N. Trefethen, D. Bau, *Numerical Linear Algebra* (SIAM, Philadelphia, 1997)

[Tre96] L.N. Trefethen, Finite difference and spectral methods for ordinary and partial differential equations. Unpublished text (1996). Available at http://people.maths.ox.ac.uk/trefethen/pdetext.html

[Wal98] W. Walter, *Ordinary Differential Equations* (Springer, New York, 1998)

[WR09] Y. Wang, Y. Rao, Reflection seismic waveform tomography. J. Geophys. Res. **114** (2009). https://doi.org/10.1029/2008JB005916

Index

© The Author(s), under exclusive license to Springer Nature Switzerland AG 2020
M. Richter, *Inverse Problems*, Lecture Notes in Geosystems
Mathematics and Computing, https://doi.org/10.1007/978-3-030-59317-9

Printed in the United States
by Baker & Taylor Publisher Services